ArcGIS Blueprints

Explore the robust features of Python to create
real-world ArcGIS applications through exciting,
hands-on projects

Eric Pimpler

[PACKT] open source *
PUBLISHING   community experience distilled

BIRMINGHAM - MUMBAI

ArcGIS Blueprints

First published: December 2015

Production reference: 1151215

Published by Packt Publishing Ltd.
Livery Place
35 Livery Street
Birmingham B3 2PB, UK.

ISBN 978-1-78528-622-3

www.packtpub.com

Credits

Author
Eric Pimpler

Reviewers
Chad Cooper
Eleza Boban Kollannur
Prasad Lingam
Shaik Shavali

Commissioning Editor
Akram Hussain

Acquisition Editor
Vinay Argekar

Content Development Editor
Siddhesh Salvi

Technical Editor
Vishal Mewada

Copy Editor
Stuti Srivastava

Project Coordinator
Nidhi Joshi

Proofreader
Safis Editing

Indexer
Priya Sane

Graphics
Kirk D'Penha

Production Coordinator
Shantanu N. Zagade

Cover Work
Shantanu N. Zagade

About the Author

Eric Pimpler is the founder and owner of GeoSpatial Training Services (http://geospatialtraining.com/) and has over 20 years of experience implementing and teaching GIS solutions using Esri, Google Earth, Google Maps, and open source technologies. Currently, he focuses on ArcGIS application development with Python and the development of custom ArcGIS Server web and mobile applications using JavaScript.

Eric is the author of *Programming ArcGIS 10.1 with Python Cookbook*, *Programming ArcGIS with Python Cookbook - Second Edition*, *Building Web and Mobile ArcGIS Server Applications with JavaScript*, and the soon to be published *ArcGIS Blueprints*.

He has a bachelor's degree in geography from Texas A&M University and a master of applied geography degree with a concentration in GIS from Texas State University.

About the Reviewers

Chad Cooper has worked in the geographic information systems realm for 13 years in technician, analyst, and developer roles. He is currently a solutions engineer with Geographic Information Services, Inc. (http://www.gisinc.com/), where he works on a variety of projects for the state and local government teams. Chad has been published in Esri's ArcUser magazine and the Python magazine. He lives in northwest Arkansas with his beautiful wife and three children.

Eleza Boban Kollannur is an architect and environmental planner working as a GIS analyst in the water and waste water sector for more than 6 years. She is passionate about the development of automation and programming solutions through model building and Python scripting. She has been involved in master planning and coastal and marine projects for the preparation of a water balance model, watershed analysis, network model builds, and spatial analysis. She has worked with MWH Global and DHI India for various projects in the Middle East, India, and UK. Eleza is interested in building custom user interface (UI) geoprocessing tools for ArcGIS desktop and server with Python.

Prasad Lingam has been passionately exploring geoinformation technologies for almost 10 years. He has gained knowledge in the application of geoinformatics to areas such as urban planning, transportation, utilities, environment, and construction management, thus leveraging his civil engineering background. He is currently working at MWH (for more information, visit http://www.mwhglobal.com/) in the water and waste water domain, implementing geospatial analysis in Desktop GIS and promoting web- and mobile-based GIS Applications to solve operational and planning issues. His work experience spans project locations such as Perth, Middle East, New Zealand, Fiji, and India. He is keen about studying the confluence of geospatial technologies with technologies such as big data, BIM, geovisualization, and so on.

Shaik Shavali is a senior GIS developer at Dar Al-Riyadh. He has 7 years of experience in the field of geospatial technologies and projects. His areas of expertises are developing custom GIS web and mobile applications using the latest ESRI technologies. He was one of the lead developers for Emergency Response Management Systems for the largest Islamic pilgrimage (Hajj), gathering nearly 2 million people. Currently, he is actively taking part in designing and developing GIS projects for the government sector in Saudi Arabia.

He has received his bachelor's in engineering degree, and presently, he is pursuing his master's in GIS through UNIGIS.

Firstly, I would like to thank Allah for his countless blessings. I would also like to thank my parents (Akbar Saheb and Fathima), in-laws (Ehasanulla and Shahnaz), and family for their emotional support, guidance, and prayers. Finally, I would like to thank my better-half, Farheen Ehasanulla, for her love and support, which always pushed me to do better. She is my backbone, best critic, and most importantly, my best friend.

www.PacktPub.com

Support files, eBooks, discount offers, and more

For support files and downloads related to your book, please visit www.PacktPub.com.

Did you know that Packt offers eBook versions of every book published, with PDF and ePub files available? You can upgrade to the eBook version at www.PacktPub.com and as a print book customer, you are entitled to a discount on the eBook copy. Get in touch with us at service@packtpub.com for more details.

At www.PacktPub.com, you can also read a collection of free technical articles, sign up for a range of free newsletters and receive exclusive discounts and offers on Packt books and eBooks.

https://www2.packtpub.com/books/subscription/packtlib

Do you need instant solutions to your IT questions? PacktLib is Packt's online digital book library. Here, you can search, access, and read Packt's entire library of books.

Why subscribe?

- Fully searchable across every book published by Packt
- Copy and paste, print, and bookmark content
- On demand and accessible via a web browser

Free access for Packt account holders

If you have an account with Packt at www.PacktPub.com, you can use this to access PacktLib today and view 9 entirely free books. Simply use your login credentials for immediate access.

Table of Contents

Preface

ArcGIS Desktop 10.3 is the leading desktop solution for GIS analysis and mapping. The ArcPy site package, which contains the ArcPy mapping and Data Access modules, enables Python programmers to access all the GIS functionalities provided through ArcGIS Desktop. ArcPy can be integrated with other open source Python libraries to enhance GUI development; create stunning reports, charts, and graphs; access REST web services; perform statistics analysis; and more. This book will teach you how to take your ArcGIS Desktop application development skills to the next level by integrating the functionality provided by ArcPy with open source Python libraries to create advanced ArcGIS Desktop applications.

In addition to working with the ArcPy, ArcPy Mapping, and ArcPy Data Access modules, the book also covers the ArcGIS REST API and a wide variety of open source Python modules, including requests, csv, plotly, tweepy, simplekml, wxPython, and others.

What this book covers

Chapter 1, *Extracting Real-Time Wildfire Data from ArcGIS Server with the ArcGIS REST API*, describes how to use the ArcGIS REST API with Python to extract real-time wildfire information from an ArcGIS Server map service and write the data to a local geodatabase. A Python Toolbox will be created in ArcGIS Desktop to hold the tools.

Chapter 2, *Tracking Elk Migration Patterns with GPS and ArcPy*, teaches you how to read a CSV file containing the GPS coordinates of elk migration patterns and write the data to a local geodatabase. The ArcPy mapping module will then be used to visualize time-enabled data and create a series of maps that show the migration patterns over time.

Chapter 3, Automating the Production of Map Books with Data Driven Pages and ArcPy, shows you how to use the Data Driven Pages functionality in ArcGIS Desktop along with the ArcPy mapping module to automate the production of a map book. The use of Python add-ins for ArcGIS Desktop for user interface development will also be introduced.

Chapter 4, Analyzing Crime Patterns with ArcGIS Desktop, ArcPy, and Plotly - Part 1, is the first of two chapters that covers the creation of ArcPy scripts for crime analysis. In this first chapter, you'll learn how to use the Python requests module to extract crime data from the city of Seattle's open database and write to a local geodatabase. You'll then write custom script tools to aggregate the crime data at various geographic levels and create maps that can be printed or exported to a PDF format.

Chapter 5, Analyzing Crime Patterns with ArcGIS Desktop, ArcPy, and Plotly - Part 2, is the second of two chapters that covers the creation of ArcPy scripts for crime analysis. In this chapter, you'll learn how to use the Plotly platform and the Python module to create compelling graphs and charts of crime data that can be integrated into the ArcGIS Desktop layout view for printing and export.

Chapter 6, Viewing and Querying Parcel Data, teaches you how to use the wxPython module to create advanced graphical user interface (GUI) applications for ArcGIS Desktop using Python within the context of an application that queries and views parcel data. Python add-ins for ArcGIS Desktop will also be used in the creation of the application.

Chapter 7, Using Python with the ArcGIS REST API and GeoEnrichment Service for Retail Site Selection, teaches you how to use the ArcGIS Online GeoEnrichment Service with Python to retrieve demographic and lifestyle information to support the site selection process of a new store location. You'll also build tools to interactively select the potential geographic location of stores based on demographic factors.

Chapter 8, Supporting Search and Rescue Operations with ArcPy, Python Add-Ins, and simplekml, teaches you how to build a Search and Rescue (SAR) application that identifies the last known location of the subject, creates search sectors in the support of operations, and exports the data to Google Earth for visualization purposes.

Chapter 9, Real Time Twitter Mapping with Tweepy, ArcPy, and the Twitter API, covers the mining of a live stream of tweets containing specific terms and hash tags. Tweets that contain geographic coordinates will be written to a local geodatabase for further analysis. In addition, several tools will be created to enable the analysis of this social media data. Finally, the results will be shared with the public through the ArcGIS Online service.

Chapter 10, *Integrating Smart Phone Photos with ArcGIS Desktop and ArcGIS Online*, covers the creation of a real estate application that reads photo metadata, extracts the coordinate information, retrieves the nearest address to the photo, and writes this information to a local feature class. In addition, the photos will be copied to a Dropbox account using the Python Dropbox module so that the photos can be accessed through a web application. Finally, the property feature class will be uploaded to ArcGIS Online, integrated with the Dropbox photos, and shared as a web-based map.

What you need for this book

ArcGIS Blueprints is written for ArcGIS Desktop 10.3. However, ArcGIS Desktop 10.2 can be used for most of the chapters as well. Python 2.7, along with the IDLE development environment, is installed along with ArcGIS Desktop, so no additional installations of Python should be performed. If desired, you can use your preferred Python development environment. I recommend PyScripter if you don't have a preference.

Who this book is for

ArcGIS Blueprints is written for intermediate-level ArcGIS Desktop programmers who wish to take their development skills to the next level. This book will cover intermediate to advanced level ArcGIS Desktop development topics with ArcPy and a variety of open source Python libraries to create applications for a wide array of topics.

Conventions

In this book, you will find a number of text styles that distinguish between different kinds of information. Here are some examples of these styles and an explanation of their meaning.

Code words in text, database table names, folder names, filenames, file extensions, pathnames, dummy URLs, user input, and Twitter handles are shown as follows: "The ArcPy data access module that is arcpy.da."

A block of code is set as follows:

```
def getParameterInfo(self):
    """Define parameter definitions"""
    param0 = arcpy.Parameter(displayName = "ArcGIS Server Wildfire
URL", \
                        name="url", \
                        datatype="GPString", \
                        parameterType="Required",\
                        direction="Input")
```

When we wish to draw your attention to a particular part of a code block, the relevant lines or items are set in bold:

```
                        parameterType="Required",\
                        direction="Input")
    params = [param0, param1]
    return params
```

Any command-line input or output is written as follows:

```
# cp /usr/src/asterisk-addons/configs/cdr_mysql.conf.sample
    /etc/asterisk/cdr_mysql.conf
```

New terms and **important words** are shown in bold. Words that you see on the screen, for example, in menus or dialog boxes, appear in the text like this: "You can create a Python Toolbox in a folder by right-clicking on the **Folder** and navigating to **New | Python Toolbox**."

> Warnings or important notes appear in a box like this.

> Tips and tricks appear like this.

Reader feedback

Feedback from our readers is always welcome. Let us know what you think about this book—what you liked or disliked. Reader feedback is important for us as it helps us develop titles that you will really get the most out of.

To send us general feedback, simply e-mail `feedback@packtpub.com`, and mention the book's title in the subject of your message.

If there is a topic that you have expertise in and you are interested in either writing or contributing to a book, see our author guide at `www.packtpub.com/authors`.

Customer support

Now that you are the proud owner of a Packt book, we have a number of things to help you to get the most from your purchase.

Downloading the example code

You can download the example code files from your account at `http://www.packtpub.com` for all the Packt Publishing books you have purchased. If you purchased this book elsewhere, you can visit `http://www.packtpub.com/support` and register to have the files e-mailed directly to you.

Downloading the color images of this book

We also provide you with a PDF file that has color images of the screenshots/diagrams used in this book. The color images will help you better understand the changes in the output. You can download this file from `https://www.packtpub.com/sites/default/files/downloads/ArcGISBlueprints_ColoredImages.pdf`.

Errata

Although we have taken every care to ensure the accuracy of our content, mistakes do happen. If you find a mistake in one of our books—maybe a mistake in the text or the code—we would be grateful if you could report this to us. By doing so, you can save other readers from frustration and help us improve subsequent versions of this book. If you find any errata, please report them by visiting `http://www.packtpub.com/submit-errata`, selecting your book, clicking on the **Errata Submission Form** link, and entering the details of your errata. Once your errata are verified, your submission will be accepted and the errata will be uploaded to our website or added to any list of existing errata under the Errata section of that title.

To view the previously submitted errata, go to `https://www.packtpub.com/books/content/support` and enter the name of the book in the search field. The required information will appear under the **Errata** section.

Piracy

Piracy of copyrighted material on the Internet is an ongoing problem across all media. At Packt, we take the protection of our copyright and licenses very seriously. If you come across any illegal copies of our works in any form on the Internet, please provide us with the location address or website name immediately so that we can pursue a remedy.

Please contact us at copyright@packtpub.com with a link to the suspected pirated material.

We appreciate your help in protecting our authors and our ability to bring you valuable content.

Questions

If you have a problem with any aspect of this book, you can contact us at questions@packtpub.com, and we will do our best to address the problem.

1

Extracting Real-Time Wildfire Data from ArcGIS Server with the ArcGIS REST API

The ArcGIS platform, which contains a number of different products, including ArcGIS Desktop, ArcGIS Pro, ArcGIS for Server, and ArcGIS Online, provides a robust environment to perform geographic analysis and mapping. The content produced by this platform can be integrated using the ArcGIS REST API and a programming language such as Python. Many of the applications we'll build in this book use the ArcGIS REST API as the *bridge* to exchange information between software products.

We're going to start by developing a simple ArcGIS Desktop custom script tool in `ArcToolbox` that connects to an ArcGIS Server map service to retrieve real-time wildfire information. The wildfire information will be retrieved from a USGS map service that provides real-time wildfire data. For this chapter and all other chapters in this book, the reader is expected to have intermediate-level experience of Python and `ArcPy`. Ideally, you should be running version 10.3 or 10.2 of ArcGIS Desktop. Previous versions of ArcGIS Desktop have some significant differences that may cause problems in the development of some applications in the book.

We'll use the ArcGIS REST API and the Python requests module to connect to the map service and request the data. The response from the map service will contain data that will be written to a feature class stored in a local geodatabase using the `ArcPy` data access module.

This will all be accomplished with a custom script tool attached to an ArcGIS Python Toolbox. ArcGIS Python toolboxes are relatively new; they were first introduced in version 10.1 of ArcGIS Desktop. They provide a Python-centric method to create custom toolboxes and tools. The older method to create toolboxes in ArcGIS Desktop, while still relevant, requires a combination of Python and a wizard-based approach to create tools.

In this chapter, we will cover the following topics:

- ArcGIS Desktop Python's toolboxes
- The ArcGIS Server map and feature services
- The Python requests module
- The Python JSON module
- The ArcGIS REST API
- The ArcPy data access module that is `arcpy.da`

A general overview of the Python libraries for ArcGIS is provided in the appendix of this book. It is recommended that you read this chapter before continuing with the appendix and other chapters.

Design

Before we start building the application, we'll spend some time planning what we'll build. This is a fairly simple application, but it serves to illustrate how ArcGIS Desktop and ArcGIS Server can be easily integrated using the ArcGIS REST API. In this application, we'll build an ArcGIS Python Toolbox that serves as a container for a single tool called `USGSDownload`. The `USGSDownload` tool will use the Python requests, **JavaScript Object Notation (JSON)**, and `ArcPy` da modules to request real-time wildfire data from a USGS map service. The response from the map service will contain information including the location of the fire, the name of the fire, and some additional information that will then be written to a local geodatabase.

The communication between the ArcGIS Desktop Python Toolbox and the ArcGIS Server map service will be accomplished through the ArcGIS REST API and the Python language.

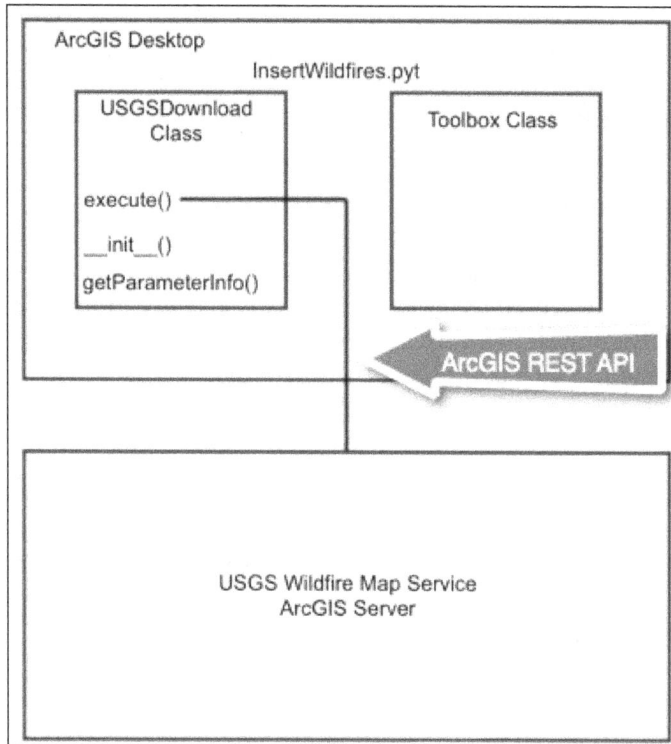

Let's get started and build the application.

Creating the ArcGIS Desktop Python Toolbox

There are two ways to create toolboxes in ArcGIS: script tools in custom toolboxes and script tools in Python toolboxes. Python toolboxes encapsulate everything in one place: parameters, validation code, and source code. This is not the case with custom toolboxes, which are created using a wizard and a separate script that processes the business logic.

A Python Toolbox functions like any other toolbox in ArcToolbox, but it is created entirely in Python and has a file extension of .pyt. It is created programmatically as a class named Toolbox. In this section, you will learn how to create a Python Toolbox and add a tool. You'll only create the basic structure of the toolbox and tool that will ultimately connect to an ArcGIS Server map service containing the wildfire data. In a later section, you'll complete the functionality of the tool by adding code that connects to the map service, downloads the current data, and inserts it into a feature class. Take a look at the following steps:

1. **Open ArcCatalog**: You can create a Python Toolbox in a folder by right-clicking on the **Folder** and navigating to **New | Python Toolbox**. In ArcCatalog, there is a folder called **Toolboxes**, and inside it, there is a **My Toolboxes** folder, as shown in the following screenshot. Right-click on this folder and navigate to **New | Python Toolbox**.

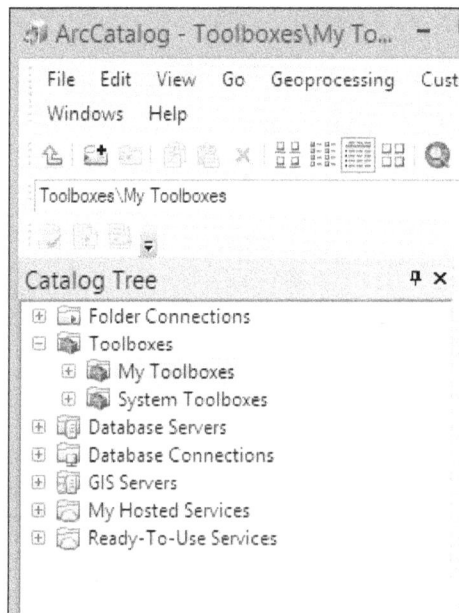

2. The name of the toolbox is controlled by the filename. Name the toolbox `InsertWildfires.pyt`.

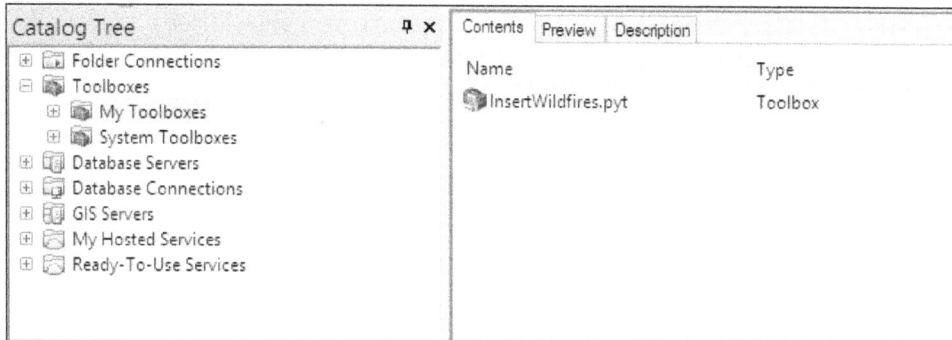

3. The Python Toolbox file (`.pyt`) can be edited in any text or code editor. By default, the code will open in Notepad. However, you will want to use a more advanced Python development environment, such as PyScripter, IDLE, and so on. You can change this by setting the default editor for your script by navigating to **Geoprocessing | Geoprocessing Options** and going to the **Editor** section. In the following screenshot, you'll notice that I have set my editor to PyScripter, which is my preferred environment. You may want to change this to **IDLE** or whichever development environment you are currently using.

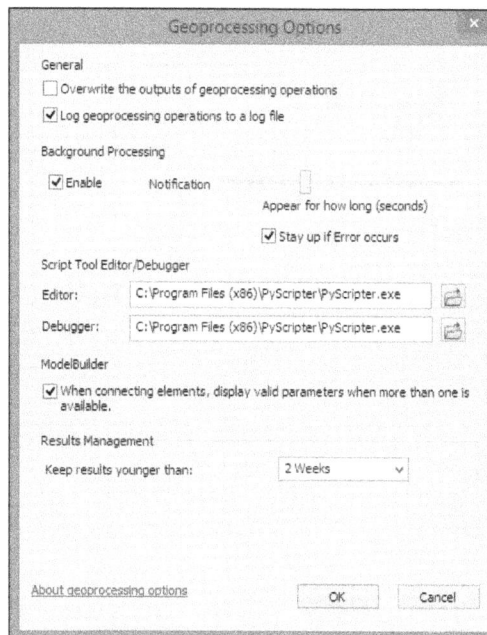

4. For example, to find the path to the executable for the **IDLE** development environment, you can navigate to **Start | All Programs | ArcGIS | Python 2.7 | IDLE**. Right-click on **IDLE** and select **Properties** to display the properties window. Inside the **Target** textbox, you should see a path to the executable, as shown in the following screenshot. You will want to copy and paste only the actual path starting with C:\Python27 and not the quotes that surround the path.

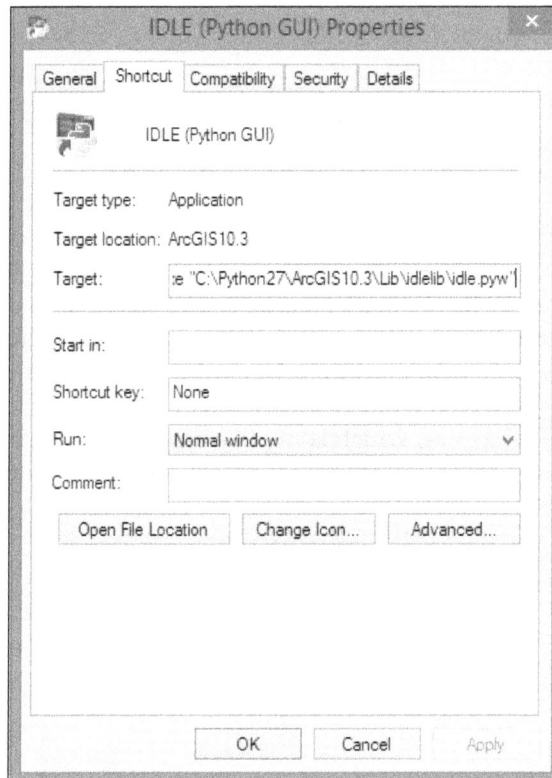

5. Copy and paste the path into the **Editor** and **Debugger** sections inside the **Geoprocessing Options** dialog box.

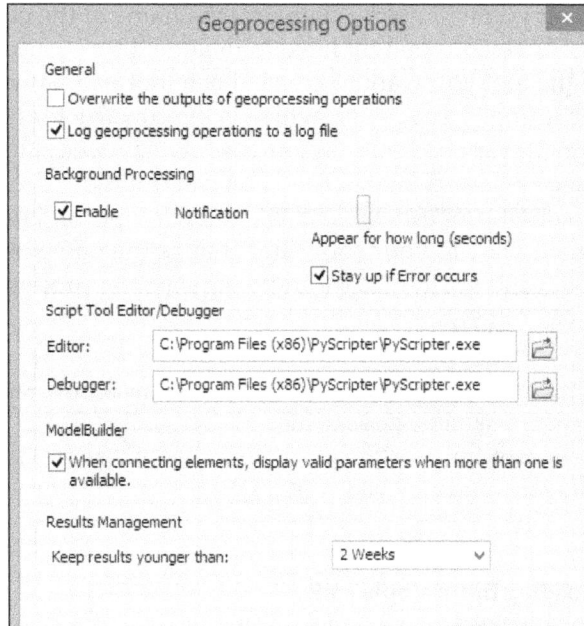

6. Right-click on `InsertWildfires.pyt` and select **Edit**. This will open the development environment you defined earlier, as shown in the following screenshot. Your environment will vary depending upon the editor that you have defined:

```
import arcpy


class Toolbox(object):
    def __init__(self):
        """Define the toolbox (the name of the toolbox is the name of the
        .pyt file)."""
        self.label = "Toolbox"
        self.alias = ""

        # List of tool classes associated with this toolbox
        self.tools = [Tool]


class Tool(object):
    def __init__(self):
        """Define the tool (tool name is the name of the class)."""
        self.label = "Tool"
        self.description = ""
        self.canRunInBackground = False

    def getParameterInfo(self):
        """Define parameter definitions"""
        params = None
        return params
```

7. Remember that you will not be changing the name of the class, which is Toolbox. However, you will have to rename the Tool class to reflect the name of the tool you want to create. Each tool will have various methods, including __init__(), which is the constructor for the tool along with getParameterInfo(), isLicensed(), updateParameters(), updateMessages(), and execute(). You can use the __init__() method to set initialization properties, such as the tool's label and description. Find the class named Tool in your code, and change the name of this tool to USGSDownload, and set the label and description properties:

```python
class USGSDownload(object):
  def __init__(self):
    """Define the tool (tool name is the name of the class)."""
    self.label = "USGS Download"
    self.description = "Download from USGS ArcGIS Server instance"
    self.canRunInBackground = False
```

> **Downloading the example code.**
> You can download the example code files from your account at http://www.packtpub.com for all the Packt Publishing books you have purchased. If you purchased this book elsewhere, you can visit http://www.packtpub.com/support and register to have the files e-mailed directly to you

You can use the Tool class as a template for other tools that you'd like to add to the toolbox by copying and pasting the class and its methods. We're not going to do that in this chapter, but I wanted you to be aware of this.

8. You will need to add each tool to the tools property (the Python list) in the Toolbox class. Add the USGSDownload tool, as shown in the following code snippet:

```python
  def __init__(self):
    """Define the toolbox (the name of the toolbox is the name of
the .pyt file."""
    self.label = "Toolbox"
    self.alias = ""

    #List of tool classes associated with this toolbox
    self.tools = [USGSDownload]
```

9. Save your code.

10. When you close the code editor, your toolbox should be automatically refreshed. You can also manually refresh a toolbox by right-clicking on the **InsertWildfires.pyt** and selecting **Refresh**. If a syntax error occurs in your code, the toolbox icon will change, as shown in the following screenshot. Note the red **X** next to the toolbox. Your tool may not visible inside the toolbox either. If you've coded everything correctly, you will not see this icon, as shown in the following screenshot:

11. To see the error, right-click on **InsertWildfires.pyt** and select **Check Syntax**.

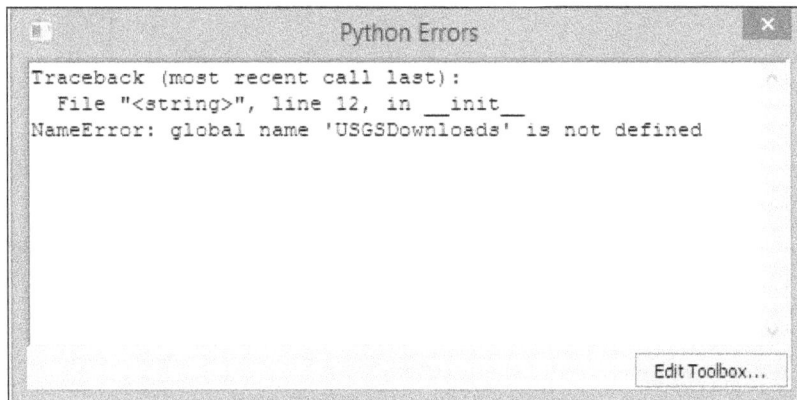

Assuming that you don't have any syntax errors, you should see the following screenshot of the Toolbox/Tool structure:

Working with tool parameters

Almost all tools have parameters. The parameter values will be set by the end user with the **Tool** dialog box, or they will be hardcoded within the script. When the tool is executed, the parameter values are sent to your tool's source code. Your tool reads these values and proceeds with its work. You use the getParameterInfo() method to define the parameters for your tool. Individual parameter objects are created as part of this process. If necessary, open InsertWildfires.pyt in your code editor and find the getParameterInfo() function in the USGSDownload class. Add the following parameters, and then we'll discuss what the code is doing:

```
def getParameterInfo(self):

        """"Define parameter definitions"""
        # First parameter
        param0 = arcpy.Parameter(displayName="ArcGIS Server Wildfire
URL",
                        name="url",
                        datatype="GPString",
                        parameterType="Required",
                        direction="Input")
        param0.value = "http://wildfire.cr.usgs.gov/arcgis/rest/
services/geomac_dyn/MapServer/0/query"

        # Second parameter
        param1 = arcpy.Parameter(displayName="Output Feature Class",
                        name="out_fc",
                        datatype="DEFeatureClass",
                        parameterType="Required",
                        direction="Input")
```

Each parameter is created using `arcpy.Parameter` and is passed a number of arguments that define the object. For the first Parameter object (`param0`), we are going to capture a URL to an ArcGIS Server map service containing real-time wildfire data. We give it a display name **ArcGIS Server Wildfire URL**, which will be displayed on the dialog box for the tool. A name for the parameter, a datatype, a parameter type, and a direction. In the case of the first parameter (`param0`), we also assign an initial value, which is the URL to an existing map service containing the wildfire data. For the second parameter, we're going to define an output feature class where the wildfire data that is read from the map service will be written. An empty feature class to store the data has already been created for you.

Next we'll add both the parameters to a Python list called `params` and `return`, to the list to the calling function. Add the following code:

```python
def getParameterInfo(self):

        """Define parameter definitions"""
        # First parameter
        param0 = arcpy.Parameter(displayName="ArcGIS Server Wildfire
URL",
                        name="url",
                        datatype="GPString",
                        parameterType="Required",
                        direction="Input")
        param0.value = "http://wildfire.cr.usgs.gov/arcgis/rest/
services/geomac_dyn/MapServer/0/query"

        # Second parameter
        param1 = arcpy.Parameter(displayName="Output Feature Class",
                        name="out_fc",
                        datatype="DEFeatureClass",
                        parameterType="Required",
                        direction="Input")

        params = [param0, param1]
        return params
```

Tool execution

The main work of a tool is done inside the `execute()` method. This is where the geoprocessing of your tool takes place. The `execute()` method, as shown in the following code snippet, can accept a number of arguments, including the tools `self`, `parameters`, and `messages`:

```
def execute(self, parameters, messages):
        """The source code of the tool."""
        return
```

To access the parameter values that are passed into the tool, you can use the `valueAsText()` method. The following steps will guide you through how to add and execute the `execute()` method:

1. Find the `execute()` function in the `USGSDownload` class and add the following code snippet to access the parameter values that will be passed into your tool. Remember from a previous step that the first parameter will contain a URL to a map service containing the wildfire data and the second parameter will be the output feature class where the data will be written:

    ```
    def execute(self, parameters, messages):
            inFeatures = parameters[0].valueAsText
            outFeatureClass = parameters[1].valueAsText
    ```

    At this point, you have created a Python Toolbox, added a tool, defined the parameters for the tool, and created variables that will hold the parameter values that the end user has defined. Ultimately, this tool will use the URL that is passed to the tool to connect to an ArcGIS Server map service, download the current wildfire data, create a feature class, and write the wildfire data to the feature class. However, we're going to save the geoprocessing tasks for later. For now, I just want you to print out the values of the parameters that have been entered so that we know that the structure of the tool is working correctly. Print this information using the following code:

    ```
    def execute(self, parameters, messages):
            inFeatures = parameters[0].valueAsText
            arcpy.AddMessage(inFeatures)
            outFeatureClass = parameters[1].valueAsText
            arcpy.AddMessage(outFeatureClass)
    ```

2. Execute the tool by double-clicking on **USGS Download** from the **InsertWildfires.pyt** toolbox that you've created. You should see the following dialog box:

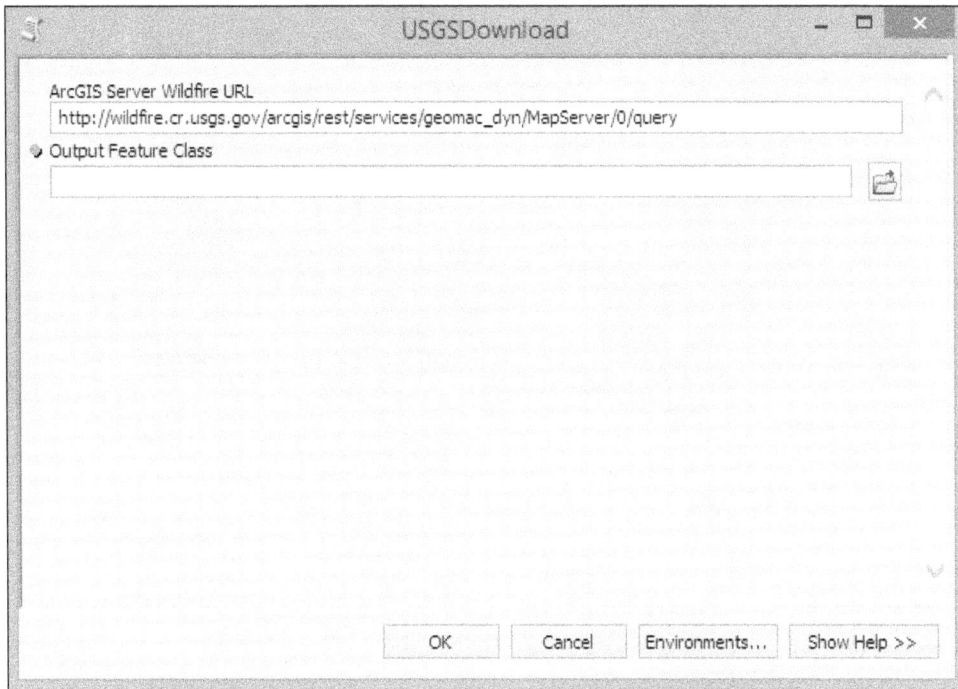

3. Leave the default URL parameter as it is and select an output feature class. You'll want to click on the **Browse** button and then navigate to the `C:\ArcGIS_Blueprint_Python\Data\WildfireData` folder and select the `WildlandFires` geodatabase. Inside, there is an empty feature class called `CurrentFires`. Select this feature class.

4. Now click on **OK**. The progress dialog box will contain the parameter values that you passed in. Your output feature class will probably be different than mine, as shown in the following screenshot:

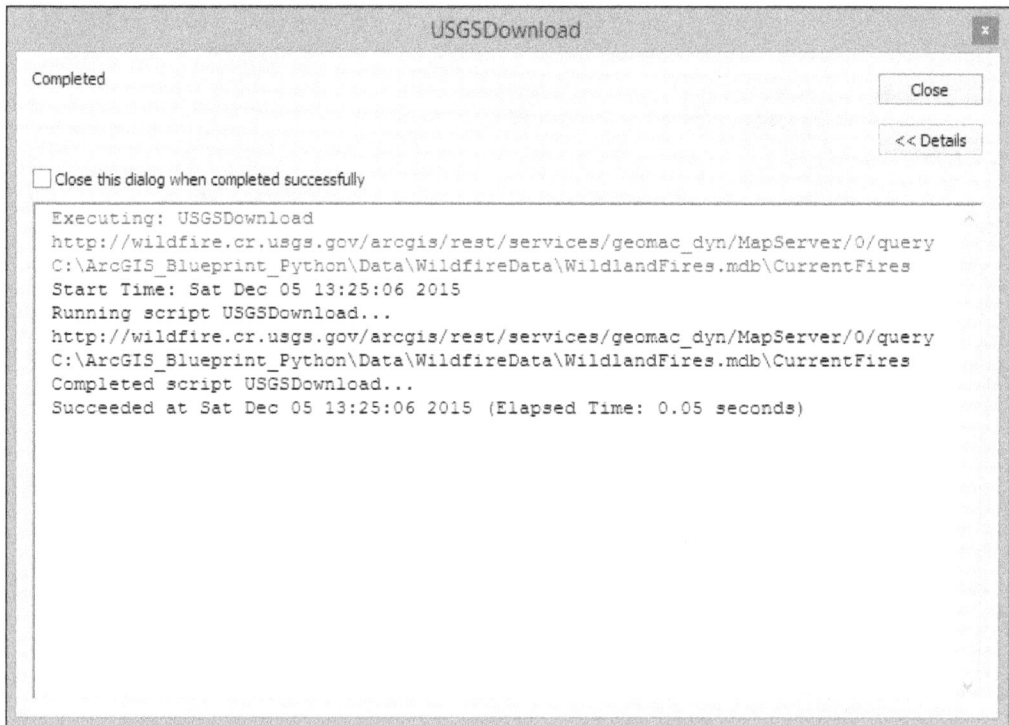

We'll complete the actual geoprocessing of this tool in the next section.

Populating the feature class

In the previous section, you learned how to create a Python Toolbox and add tools. You created a new toolbox called **InsertWildfires** and added a tool called **USGS Download**. However, in that exercise, you didn't complete the geoprocessing operations that connect to an ArcGIS Server map service, query the service for current wildfires, and populate the feature class from the data pulled from the map service query. You'll complete these steps in the following section.

Installing pip and the requests module

This section of the application uses the Python requests module. If you don't already have this module installed on your computer, you will need to do this at this time using pip.

The pip is a *package manager* that serves as a repository and installation manager for Python modules. It makes finding and installing Python modules much easier. There are several steps that you'll need to follow in order to install pip and the requests module. Instructions to install pip and the requests module are provided in the first few steps:

1. Open the **Environment Variables** dialog box in Windows. The easiest way to display this dialog box is to go to **Start** and then type **Environment Variables** in the search box. This should display an **Edit environment variables for your account** entry. Select this item.

2. If you don't see a variable called **PATH**, click on the **New** button to create one. If you already have a **PATH** variable, you can just add to the existing content. Select the PATH variable, and click on **Edit**, and then set the value to `C:\Python27\ArcGIS10.3;C:\Python27\ArcGIS10.3\Scripts`. The first path will provide a reference to the location of the Python executable and the second will reference the location of pip when it is installed. This makes it possible to run Python and pip from the Command Prompt in Windows.

3. Click on **OK** and then click on **OK** again to save the changes.

4. Next, we'll install pip if you haven't already done so in the past. You need to install pip before you can install requests. In a web browser, go to `https://pip.pypa.io/en/latest/installing.html` and scroll down to the install pip section. Right-click on `get-pip.py` and select **Save Link As** or something similar. This will vary depending upon the browser you are using. Save it to your `C:\ArcGIS_Blueprint_Python` folder.

5. Open the Command Prompt in Windows, type python `C:\ArcGIS_Blueprint_Python\get-pip.py`, and press *Enter* on your keyboard. This will install pip.

6. In the Command Prompt, type `pip install requests` and press *Enter* on your keyboard. This will install the requests module.

7. Close the Command Prompt.

Requesting data from ArcGIS Server

In the following steps, we will learn how to request data from ArcGIS Server:

1. Open ArcCatalog and navigate to the location where you've created your Python Toolbox, it would look like following screenshot:

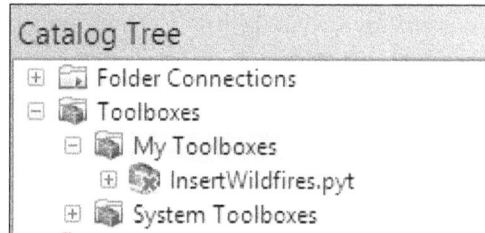

2. Right-click on **InsertWildfires.pyt** and select **Edit** to display the code for the toolbox.

3. First, we'll clean up a little by removing the **AddMessage()** functions. Clean up your **execute()** method so that it appears as follows:

```
def execute(self, parameters, messages):
        inFeatures = parameters[0].valueAsText
        outFeatureClass = parameters[1].valueAsText
```

4. Next, add the code that connects to the wildfire map service, to perform a query. In this step, you will also define the QueryString parameters that will be passed into the query of the map service. First, import the requests and json modules:

```
import arcpy
import requests, json


class Toolbox(object):
    def __init__(self):
        """Define the toolbox (the name of the toolbox is the name
of the
```

```
    .pyt file)."""
    self.label = "Toolbox"
    self.alias = ""

    # List of tool classes associated with this toolbox
    self.tools = [USGSDownload]
```

5. Then, create the `agisurl` and `json_payload` variables that will hold the `QueryString` parameters. Note that, in this case, we have defined a WHERE clause so that only wildfires where the acres are greater than 5 will be returned. The `inFeatures` variable holds the **ArcGIS Server Wildfire URL**:

```
def execute(self, parameters, messages):
    inFeatures = parameters[0].valueAsText
    outFeatureClass = parameters[1].valueAsText

    agisurl = inFeatures
    json_payload = { 'where': 'acres > 5', 'f': 'pjson',
    'outFields': 'latitude,longitude,incidentname,acres' }
```

6. Submit the request to the ArcGIS Server instance; the response should be stored in a variable called `r`. Print a message to the dialog box indicating the response as:

```
def execute(self, parameters, messages):
    inFeatures = parameters[0].valueAsText
    outFeatureClass = parameters[1].valueAsText

    agisurl = inFeatures
    json_payload = { 'where': 'acres > 5', 'f': 'pjson',
'outFields': 'latitude,longitude,incidentname,acres' }

    r = requests.get(agisurl, params=json_payload)
    arcpy.AddMessage("The response: " + r.text)
```

7. Test the code to make sure that we're on the right track. Save the file and refresh `InsertWildfires` in **ArcCatalog**. Execute the tool and leave the default URL. If everything is working as expected, you should see a JSON object output to the progress dialog box. Your output will probably vary from the following screenshot:

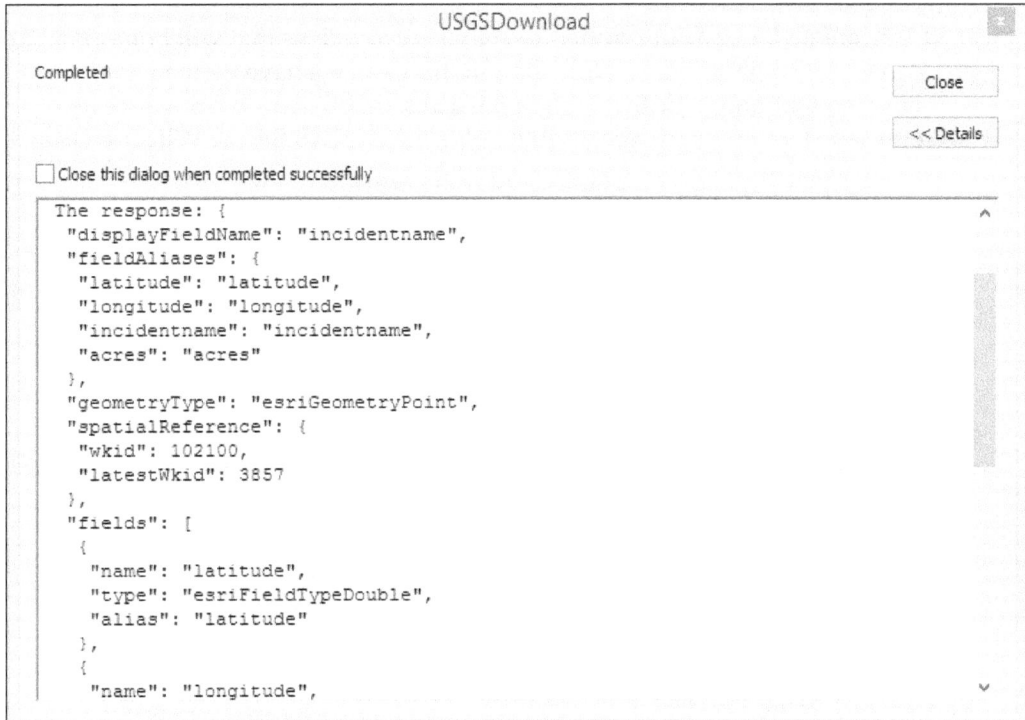

8. Return to the `execute()` method and convert the JSON object to a Python dictionary using the `json.loads()` method:

```python
def execute(self, parameters, messages):
        inFeatures = parameters[0].valueAsText
        outFeatureClass = parameters[1].valueAsText

        agisurl = inFeatures
        json_payload = { 'where': 'acres > 5', 'f': 'pjson',
'outFields': 'latitude,longitude,incidentname,acres' }

        r = requests.get(inFeatures, params=json_payload)
        arcpy.AddMessage("The response: " + r.text)

        decoded = json.loads(r.text)
```

Inserting data in a feature class with the ArcPy data access module

The following steps will guide you, to insert data in a feature class with the help of the ArcPy data access module:

1. Now, we'll use the ArcPy data access module, that is Arcpy.da, to create an InsertCursor object by passing the output feature class defined in the tool dialog box along with the fields that will be populated:

```
def execute(self, parameters, messages):
    inFeatures = parameters[0].valueAsText
    outFeatureClass = parameters[1].valueAsText

    agisurl = inFeatures
    json_payload = { 'where': 'acres > 5', 'f': 'pjson',
'outFields': 'latitude,longitude,fire_name,acres' }

    r = requests.get(inFeatures, params=json_payload)
    arcpy.AddMessage("The response: " + r.text)

    decoded = json.loads(r.text)
    cur = arcpy.da.InsertCursor(outFeatureClass, ("SHAPE@XY",
"NAME", "ACRES"))
```

2. Create a For loop that you can see in the following code, and then we'll discuss what this section of code accomplishes:

```
def execute(self, parameters, messages):
    inFeatures = parameters[0].valueAsText
    outFeatureClass = parameters[1].valueAsText

    agisurl = inFeatures
    json_payload = { 'where': 'acres > 5', 'f': 'pjson',
'outFields': 'latitude,longitude,fire_name,acres' }

    r = requests.get(inFeatures, params=json_payload)
    arcpy.AddMessage("The response: " + r.text)

    decoded = json.loads(r.text)
    cur = arcpy.da.InsertCursor(outFeatureClass, ("SHAPE@XY",
"NAME", "ACRES"))
    cntr = 1
    for rslt in decoded['features']:
        fireName = rslt['attributes']['incidentname']
        latitude = rslt['attributes']['latitude']
```

```
                longitude = rslt['attributes']['longitude']
                acres = rslt['attributes']['acres']
                cur.insertRow([(longitude,latitude),fireName, acres])
                arcpy.AddMessage("Record number: " + str(cntr) + "
        written to feature class")
                cntr = cntr + 1
            del cur
```

The first line simply creates a `counter` that will be used to display the progress information in the **Progress Dialog** box. We then start a `For` loop that loops through each of the features (wildfires) that have been returned. The decoded variable is a Python dictionary. Inside the `For` loop, we retrieve the wildfire name, latitude, longitude, and acres from the attributes dictionary. Finally, we call the `insertRow()` method to insert a new row into the feature class along with the wildfire name and acres as attributes. The progress information is written to the **Progress Dialog** box and the counter is updated.

3. Save the file and refresh your Python Toolbox.

4. Double-click on the **USGS Download** tool.

5. Leave the default URL and select the `CurrentFires` feature class in the `WildlandFires` geodatabase. The `CurrentFires` feature class is empty and has fields for `NAMES` and `ACRES`:

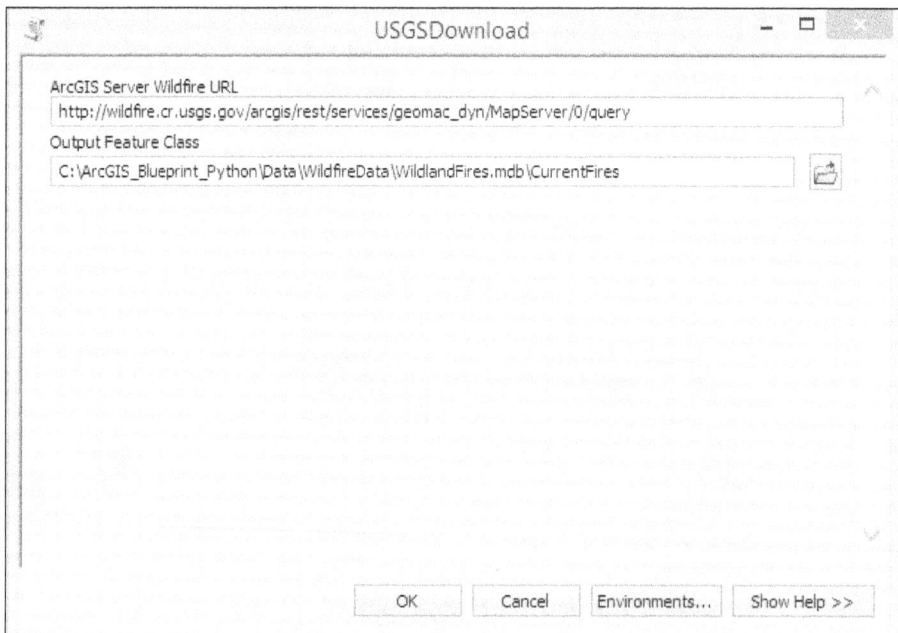

6. Click on **OK** to execute the tool. The number of features written to the feature class will vary depending upon the current wildfire activity. Most of the time, there is at least a little activity, but it is possible that there wouldn't be any wildfires in the U.S. as shown in the following screenshot:

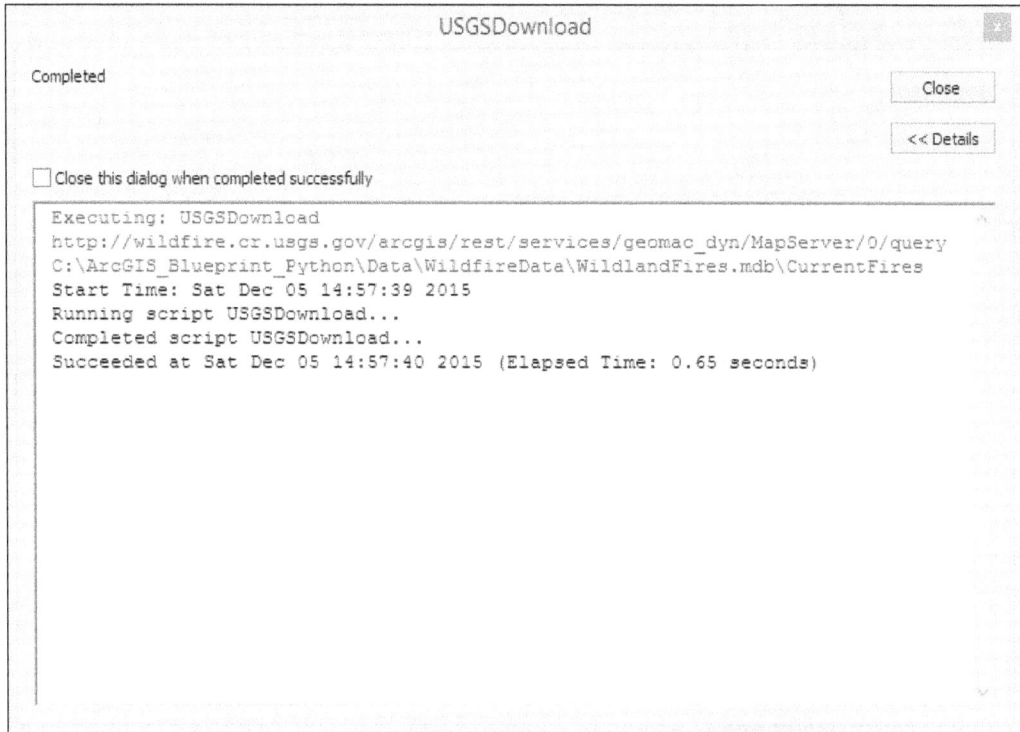

```
                                  USGSDownload

Completed                                                          [ Close ]

                                                                   [ << Details ]

☐ Close this dialog when completed successfully

Executing: USGSDownload
http://wildfire.cr.usgs.gov/arcgis/rest/services/geomac_dyn/MapServer/0/query
C:\ArcGIS_Blueprint_Python\Data\WildfireData\WildlandFires.mdb\CurrentFires
Start Time: Sat Dec 05 14:57:39 2015
Running script USGSDownload...
Completed script USGSDownload...
Succeeded at Sat Dec 05 14:57:40 2015 (Elapsed Time: 0.65 seconds)
```

7. View the feature class in **ArcMap**. To view the feature class in the following screenshot, I've plotted the points along with a **Basemap** topography layer. Your data will almost certainly be different than mine as we are pulling real-time data:

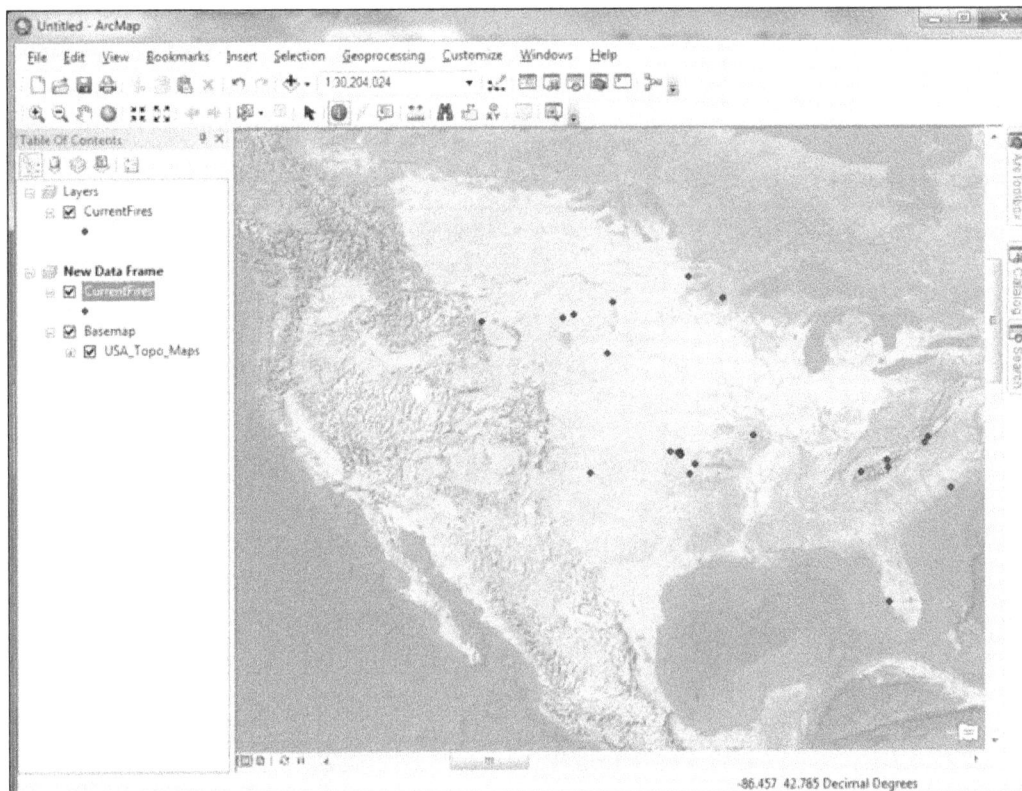

Summary

Integrating ArcGIS Desktop and ArcGIS Server is easily accomplished using the ArcGIS REST API and the Python programming language. In this chapter, we created an ArcGIS Python Toolbox containing a tool that connects to an ArcGIS Server map service containing real-time wildfire information and hosted by the **USGS**. The connection was accomplished through the use of the Python request module that we used in order to submit a request and handle the response. Finally, we used the ArcPy data access module to write this information to a local geodatabase.

In the next chapter, we'll continue working with ArcGIS Python toolboxes, and you'll also learn how to read CSV files with the Python CSV module, insert data into a feature class, and use the arcpy.mapping module to work with time-enabled data.

2
Tracking Elk Migration Patterns with GPS and ArcPy

In this chapter, we're going to build an application that imports a CSV file containing **Global Positioning System** (**GPS**) locations that depict elk migration patterns into a feature class that will be time-enabled to display migration patterns over time and space. We'll use the ArcPy data access (arcpy.da) module and the Python csv module to read the file containing GPS locations, and write the data to a new feature class. Next, we'll use the ArcPy mapping (archy.mapping) module to make the output feature class time-enabled, and then visualize the migration patterns of the elk over time and space. The application will be built as an ArcGIS Python Toolbox in much the same way as what we did in *Chapter 1, Extracting Wildfire Data from an ArcGIS Server Map Service with the ArcGIS REST API*.

In this chapter, we will cover the following topics:

- ArcGIS Desktop Python toolboxes
- Reading CSV files with the Python csv module
- Inserting data in a feature class using the ArcPy data access module
- Using the ArcPy mapping module to make a layer time-enabled
- Visualizing time-enabled data with ArcGIS Desktop

Design

Let's spend a little time going over the design of what we're going to build in this chapter. This application, like the one we built in the *Chapter 1, Extracting Wildfire Data from an ArcGIS Server Map Service with the ArcGIS REST API*, will include the creation of an ArcGIS Desktop Python Toolbox. The toolbox, `MigrationPatterns.pyt`, will include two tools: `ImportCollarData` and `VisualizeMigration`. The `ImportCollarData` tool will import GPS data from a collar that was attached to an elk in northern California. The GPS data will have been extracted to a comma-delimited text file (`csv` format), that will be read using the Python `csv` module and then imported into a local feature class stored in a file geodatabase using the `arcpy.da` which is a data access module. We'll then need to do a little manual work inside ArcMap. First, we'll make the `feature` class that was created as a result of the `ImportCollarData` tool time-enabled, and then we'll save the time-enabled data in a `map` document file. The `VisualizeMigration` tool will then enable the visualization of the migration patterns on a daily basis between a date range provided by the end user. This tool will also export a map of the layout view to a PDF file for each of the days.

The elk migration data provided by McCrea Cobb of the U.S. Fish and Wildlife Service is as follows:

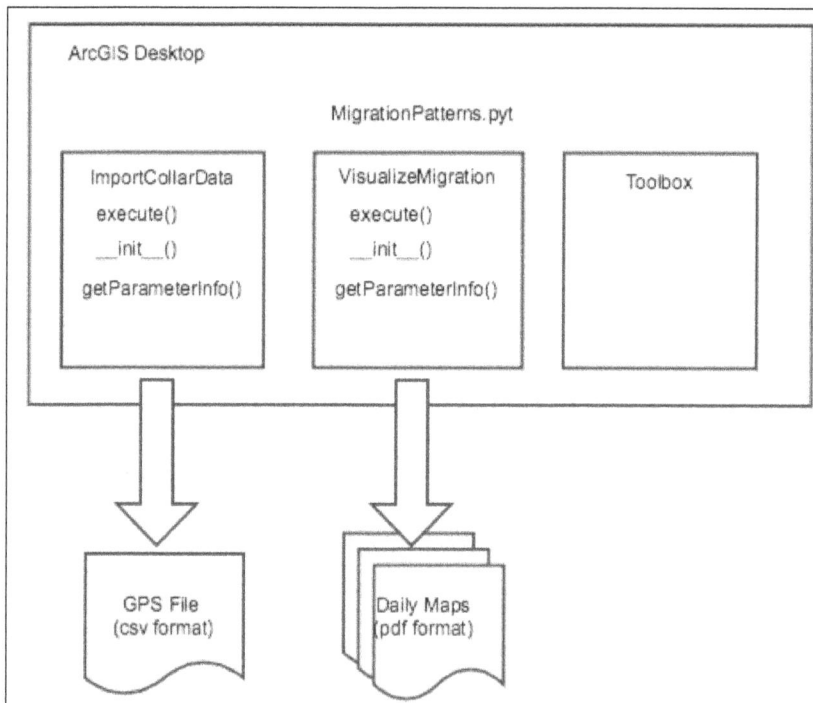

Let's get started and build the application.

Creating migration patterns for Python toolbox

Just like we did in the first chapter of the book, we'll build an **ArcGIS Python Toolbox** to hold the code for our application. I won't walk you through every single step like I did in the first chapter, but I will provide some general guidelines instead. If needed, refer to the first chapter for the specifics of how to create an **ArcGIS Python Toolbox**.

The Python toolboxes encapsulate everything in one place: parameters, validation code, and source code. A Python Toolbox functions like any other toolbox in `ArcToolbox`, but it is created entirely in Python and has a file extension of `.pyt`. As you learned in the last chapter, it is created programmatically as a class named `Toolbox`.

The following steps will help you to create migration patterns for Python toolbox:

1. Open **ArcCatalog**. You can create a Python Toolbox in a folder by right-clicking on the folder and navigating to **New | Python Toolbox**. In **ArcCatalog**, there is a folder called **Toolboxes**; inside the folder, there is a **My Toolboxes** folder, as shown in the following screenshot. Right-click on this folder and navigate to **New | Python Toolbox**:

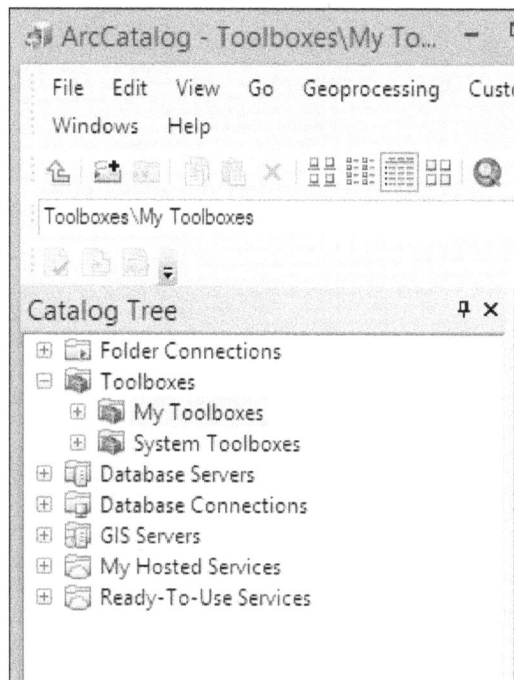

2. The name of the toolbox is controlled by the filename. Name the toolbox `MigrationPatterns.pyt`.

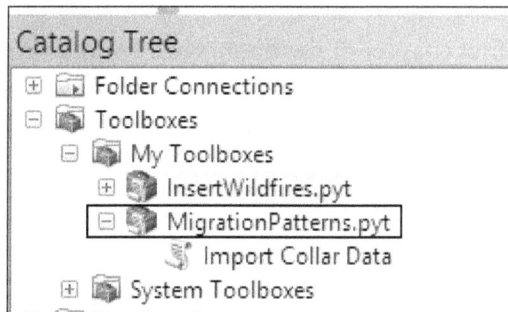

Creating the Import Collar Data tool

The following steps will help you to create **Import Collar Data** tool:

1. Right-click on `MigrationPatterns.pyt` and select **Edit**. This will open your development environment, as shown in the following screenshot. Your environment will vary depending upon the editor that you defined in *Chapter 1, Extracting Real-Time Wildfire Data from ArcGIS Server with the ArcGIS REST API*:

```
import arcpy


class Toolbox(object):
    def __init__(self):
        """Define the toolbox (the name of the toolbox is the name of the
        .pyt file)."""
        self.label = "Toolbox"
        self.alias = ""

        # List of tool classes associated with this toolbox
        self.tools = [ImportCollarData, VisualizeMigration]


class Tool(object):
    def __init__(self):
        """Define the tool (tool name is the name of the class)."""
        self.label = "Import Collar Data"
        self.description = "Import Elk Collar Data"
        self.canRunInBackground = False

    def getParameterInfo(self):

        return params

    def isLicensed(self):
        """Set whether tool is licensed to execute."""
        return True

    def updateParameters(self, parameters):
        """Modify the values and properties of parameters before internal
        validation is performed.  This method is called whenever a parameter
        has been changed."""
        return
```

2. Remember that you will not be changing the name of the class, which is
   `Toolbox`. However, you will rename the `Tool` class to reflect the name of the
   tool you want to create.

3. Find the class named `Tool` in your code and change the name of this tool to
   `ImportCollarData`, and set the label and description properties:

```
class ImportCollarData(object):
    def __init__(self):
        """Define the tool (tool name is the name of the class)."""
        self.label = "Import Collar Data"
        self.description = "Import Elk Collar Data"
        self.canRunInBackground = False
```

You can use the `Tool` class as a template for other tools you'd like to add to the toolbox by copying and pasting the class and its methods. We'll do this in a later step, when we create the tool that enables the display of the elk migration patterns over time.

4. You will need to add each tool to the tools property (the Python list) in the `Toolbox` class. Add the `ImportCollarData` tool, as shown in the following code:

```
def __init__(self):
        """Define the toolbox (the name of the toolbox is the name
of the
        .pyt file)."""
        self.label = "Toolbox"
        self.alias = ""

        # List of tool classes associated with this toolbox
        self.tools = [ImportCollarData]
```

5. Assuming that you don't have any syntax errors, you should see the following `Toolbox` or `Tool` structure:

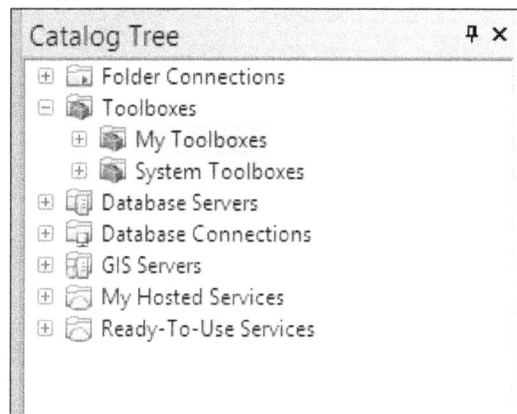

6. In this step, we'll set the parameters for the tool. The `ImportCollarData` tool will need parameters that accept the `csv` file to be imported along with an output feature class, where the data will be written, and an input feature class to be used for schema purposes. Use the `getParameterInfo()` method to define the parameters for your tool. Individual parameter objects are created as part of this process. Add the following parameters, and then we'll discuss what the code is doing:

```
def getParameterInfo(self):
        param0 = arcpy.Parameter(displayName = "CSV File to
Import", \
```

```
                          name="fileToImport", \
                          datatype="DEFile", \
                          parameterType="Required",\
                          direction="Input")


        param1 = arcpy.Parameter(displayName = "Output Feature
Class", \
                          name="out_fc", \
                          datatype="DEFeatureClass",\
                          parameterType="Required",\
                          direction="Output")


        param2 = arcpy.Parameter(displayName = "Schema Feature
Class", \
                          name="schema_fc", \
                          datatype="DEFeatureClass",\
                          parameterType="Required",\
                          direction="Input")
```

Each `Parameter` object is created using `arcpy.Parameter` and is passed a number of arguments that define the object.

For the first parameter object (`param0`), we are going to capture a file reference that, in this case, will be a reference to a `csv` file containing the elk migration data. We give it a display name (**CSV File to Import**), which will be displayed on the dialog box for the tool, a name for the parameter, a datatype, a parameter type (required), and a direction.

For the second parameter, we're going to define an output feature class, where the elk migration data that is read from the file will be written. Our tool will create this feature class.

The final parameter is also a feature class but has a direction of input and will be used to specify an existing feature class from which we'll pull the schema.

7. Next, we'll add both the parameters to a Python list called `params` and return the list to the calling function. Add the following code:

```
def getParameterInfo(self):
        param0 = arcpy.Parameter(displayName = "CSV File to
Import", \
                          name="fileToImport", \
                          datatype="DEFile", \
```

```
                                parameterType="Required",\
                                direction="Input")

        param1 = arcpy.Parameter(displayName = "Output Feature
    Class", \
                                name="out_fc", \
                                datatype="DEFeatureClass",\
                                parameterType="Required",\
                                direction="Output")

        param2 = arcpy.Parameter(displayName = "Schema Feature
    Class", \
                                name="schema_fc", \
                                datatype="DEFeatureClass",\
                                parameterType="Required",\
                                direction="Input")

        params = [param0, param1, param2]
        return params
```

Reading data from the CSV file and writing to the feature class

The following steps will help you to read and write data from CSV file to a write to feature class:

1. The main work of a tool is done inside the `execute()` method. This is where the geoprocessing of the tool takes place. The `execute()` method, as shown in the following code, can accept a number of arguments, including the tools `self`, `parameters`, and `messages`:

   ```
   def execute(self, parameters, messages):
       """The source code of the tool."""
       return
   ```

2. To access the parameter values that are passed into the tool, you can use the `valueAsText()` method. Add the following code to access the parameter values that will be passed into your tool. Remember from a previous step that the first parameter will contain a reference to a CSV file that will be imported and the second parameter is the output feature class where the data will be written:

   ```
   def execute(self, parameters, messages):
   ```

```
inputCSV = parameters[0].valueAsText
outFeatureClass = parameters[1].valueAsText
schemaFeatureClass = parameters[2].valueAsText
```

3. Import the `csv` and `os` modules at the top of the script:

```
import arcpy
import csv
import os
```

4. Inside the `execute()` method, create the `try`/`except` block that will wrap the code into an exception-handling structure:

```
def execute(self, parameters, messages):
    """The source code of the tool."""
    inputCSV = parameters[0].valueAsText
    outFeatureClass = parameters[1].valueAsText
    schemaFeatureClass = parameters[2].valueAsText

    try:
    except Exception as e:
        arcpy.AddMessage(e.message)
```

5. Create the output feature class by passing in the `output` path, `feature` class name, geometry type, schema feature class, and spatial reference:

```
try:
    #create the feature class
    outCS = arcpy.SpatialReference(26910)
    arcpy.CreateFeatureclass_management(os.path.
split(outFeatureClass)[0], os.path.split(outFeatureClass)[1],
"point", schemaFeatureClass, spatial_reference=outCS)
```

6. Create the `InsertCursor` object just below the line above where you created a new feature class:

```
#create the insert cursor
with arcpy.da.InsertCursor(outFeatureClass, ("SHAPE@XY", "season",
"date", "time", "hour", "temp", "bearing", "slope", "elevation"))
as cursor:
```

7. Open the `csv` file and read the data items that we need. There is a lot of data in the `csv` file, but we're not going to need everything. For our purposes, we are just going to pull out a handful of items, including the season, date, time, hour, coordinates, temperature, bearing, slope, and elevation. The following code block should be placed inside the `WITH` statement. In this block of code, we open the file in read mode, create a CSV reader object, skip the header row, and then loop through all the records in the file and pull out the individual items that we need:

```
with arcpy.da.InsertCursor(outFeatureClass,("SHAPE@XY", "season",
"date", "time", "hour", "temp", "bearing", "slope", "elevation"))
as cursor:

                #loop through the csv file and import the data
into the feature class
                cntr = 1
                with open(inputCSV, 'r') as f:
                    reader = csv.reader(f)
                    #skip the header
                    next(reader, None)
                    #loop through each of the rows and write to
the feature class
                    for row in reader:
                        season = row[4]   ## season
                        dt = row[6]   ## date
                        tm = row[7]   ## time
                        hr = row[8]   ## hour
                        lng = row[10]
                        lat = row[11]
                        temperature = row[13]   ## temperature
                        bearing = row[17]   ## bearing
                        slope = row[23]   ## slope
                        elevation = row[28]   ## elevation
```

8. In the final code block for this tool, we'll add statements that convert the coordinates to a float datatype, encapsulate all the values within a Python list object, insert each row into the feature class, and update the progress dialog:

```
if lng != 'NA':
    lng = float(row[10])
    lat = float(row[11])
```

```
    row_value = [(lng,lat),season,dt,tm,hr,temperature,bearing,slo
pe,elevation]
    cursor.insertRow(row_value)    # Inserts the row into the
feature class
    arcpy.AddMessage("Record number " + str(cntr) + " written to
feature class") # Adds message to the progress dialog
    cntr = cntr + 1
```

9. Check your code against the solution file found at `C:\ArcGIS_Blueprint_ Python\solutions\ch2\ImportCollarData.py` to make sure you have coded everything correctly.

10. Double-click on the **Import Collar Data** tool to execute your work. Fill in the parameters, as shown in the following screenshot. The data we're importing from the `csv` file is from a single elk, which we'll call `Betsy`:

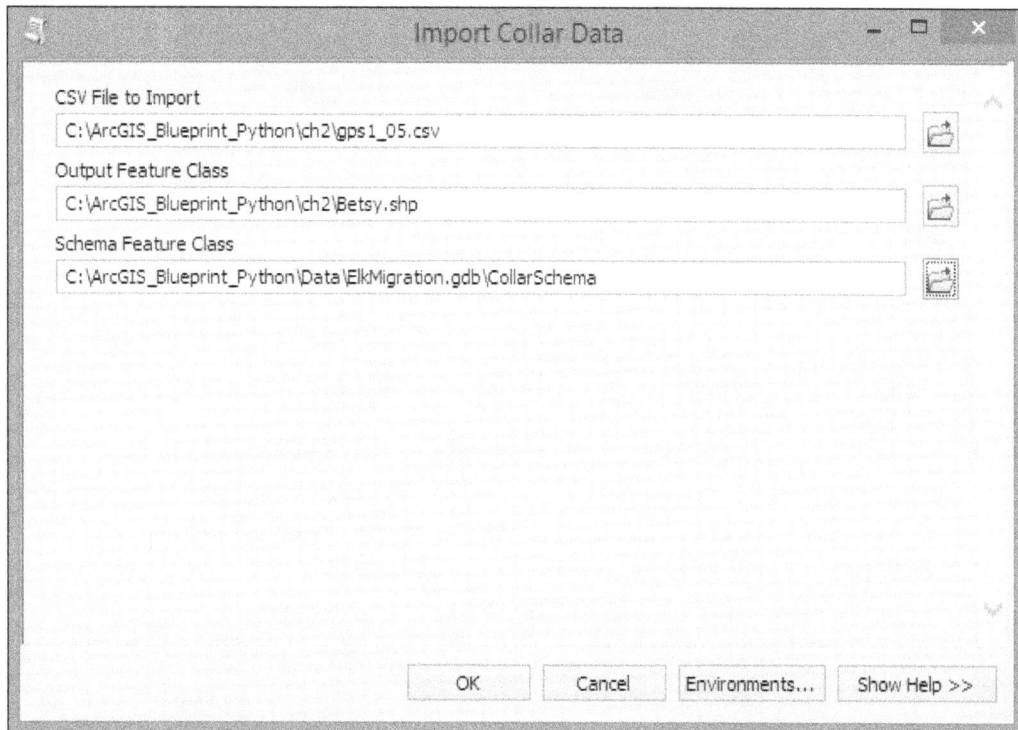

11. Click on **OK** to execute the tool, and if everything has gone correctly, your progress dialog should start getting updated as the data is read from the file and written to the feature class.

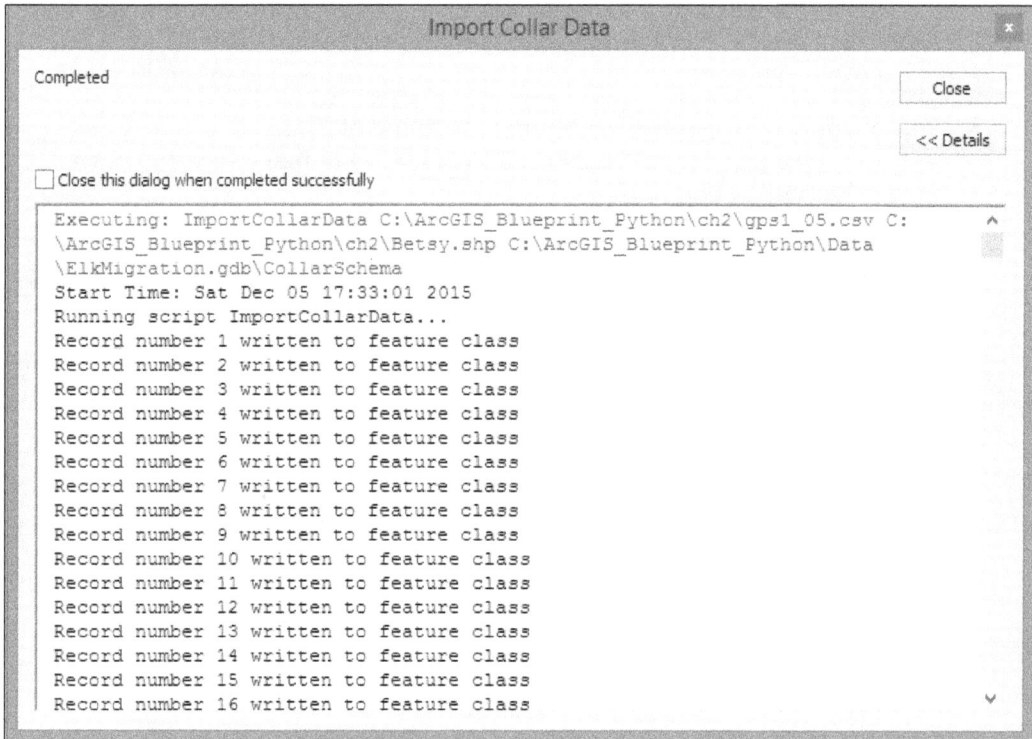

```
                          Import Collar Data                                    x

Completed                                                           Close

                                                                   << Details

 Close this dialog when completed successfully

Executing: ImportCollarData C:\ArcGIS_Blueprint_Python\ch2\gps1_05.csv C:  ^
\ArcGIS_Blueprint_Python\ch2\Betsy.shp C:\ArcGIS_Blueprint_Python\Data
\ElkMigration.gdb\CollarSchema
Start Time: Sat Dec 05 17:33:01 2015
Running script ImportCollarData...
Record number 1 written to feature class
Record number 2 written to feature class
Record number 3 written to feature class
Record number 4 written to feature class
Record number 5 written to feature class
Record number 6 written to feature class
Record number 7 written to feature class
Record number 8 written to feature class
Record number 9 written to feature class
Record number 10 written to feature class
Record number 11 written to feature class
Record number 12 written to feature class
Record number 13 written to feature class
Record number 14 written to feature class
Record number 15 written to feature class
Record number 16 written to feature class                                    v
```

12. Close the progress dialog and open **ArcMap** with an empty map document file.

13. In **ArcMap**, click on the **Add Data** from the drop-down list and select **Add Basemap**. Select the **Topographic** basemap and click on the **Add** button.

14. Click on the **Add Data** button again, and navigate to the `Betsy` feature class that you created using the tool that you just created. Add this feature class to **ArcMap**. The data should be displayed in a cluster north-west of San Francisco in the Point Reyes National Seashore area. You will need to zoom in on this area. You should see something similar to what is shown in the following screenshot:

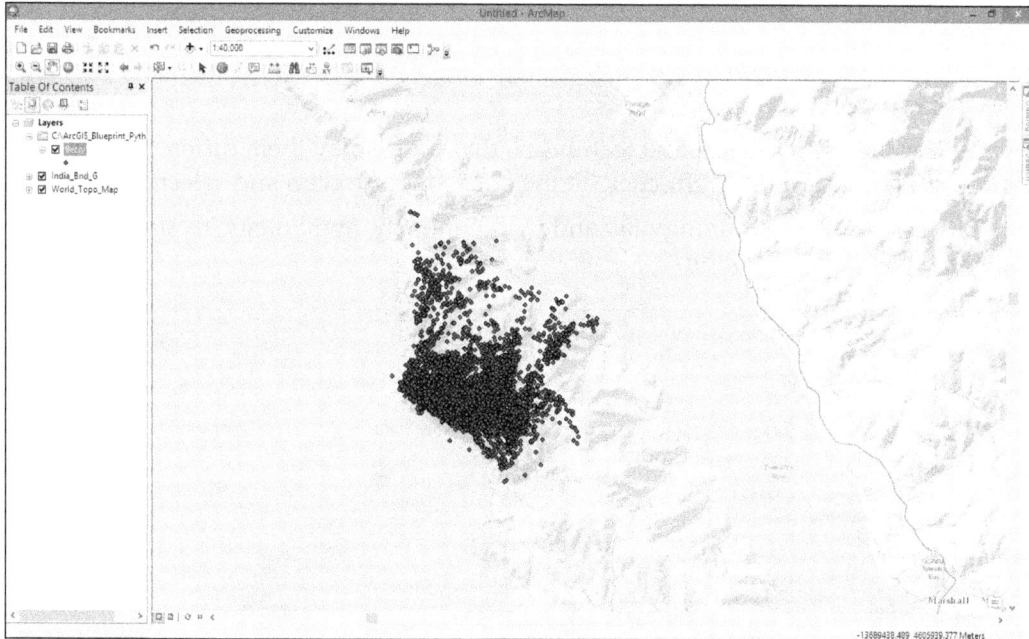

15. Save the map document as `C:\ArcGIS_Blueprint_Python\ch2\ ElkMigration.mxd`.

16. In the next section, we'll make this data time-enabled so that we can get a better understanding of how this elk moves through its environment over time. Right now, the data just looks like one big cluster but, by making our data time-enabled, we'll be able to better understand the data.

Making the data frame and layer time-enabled

In this section, you will learn how to make a layer and data frame time-enabled. You will then add a tool to the **Migration Patterns** toolbox that cycles through the time range for the layer and exports a PDF map showing the movement of the elk over time and space:

1. If necessary, open `C:\ArcGIS_Blueprint_Python\ch2\ElkMigration.mxd` in **ArcMap**.

2. First, we'll symbolize the features so that we display them differently for wet and dry seasons. Right-click on the `Betsy` feature class and select **Properties**.

3. Click on the **Symbology** tab and then define the symbology, as shown in the following screenshot:

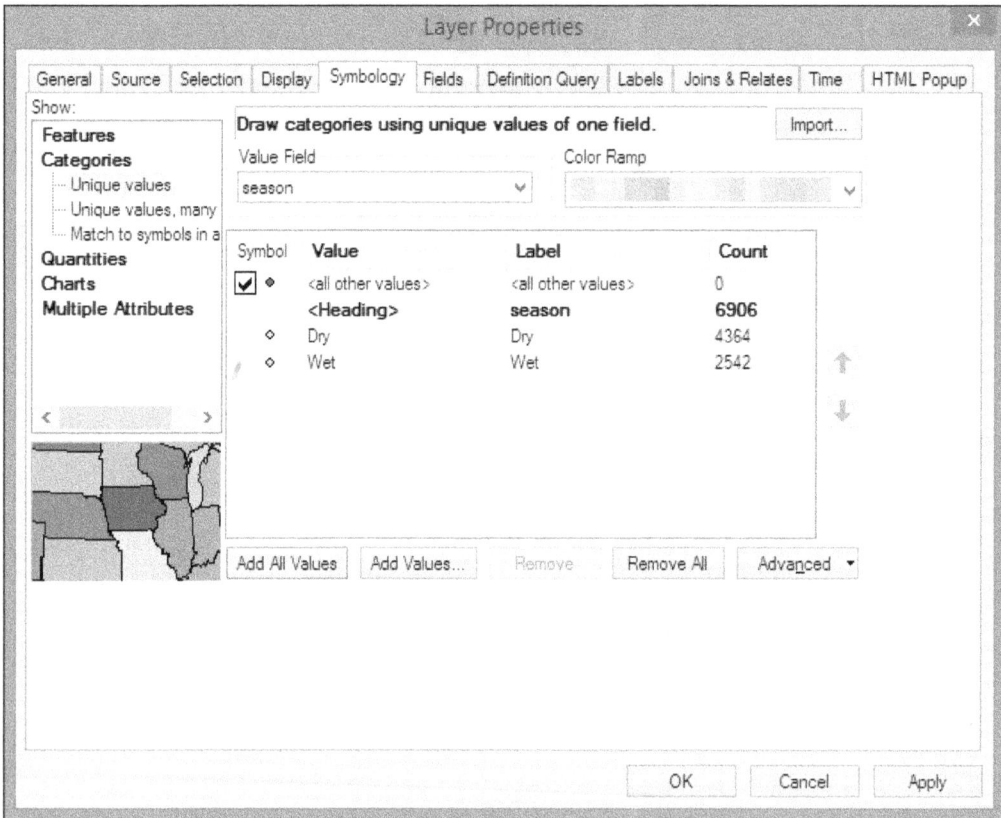

4. Now, select the **Time** tab, as shown in the following screenshot:

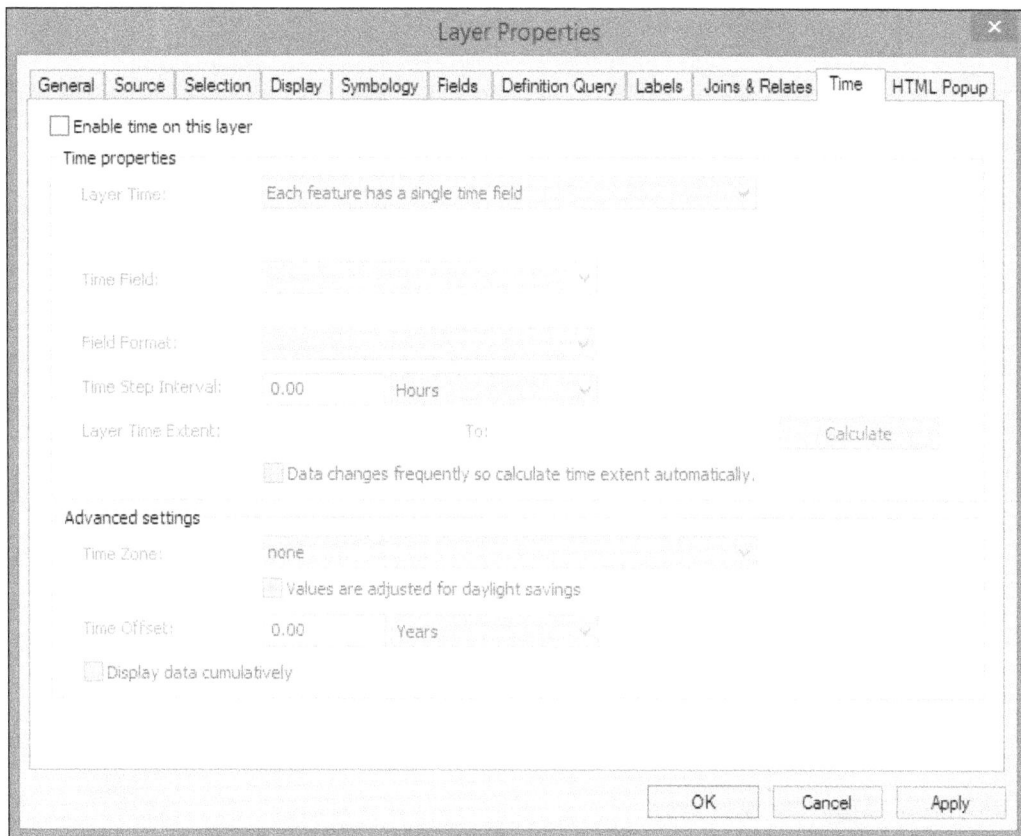

5. Enable the time for the layer by clicking on the **Enable time for this layer** checkbox.

6. Define **Layer Time Extent** by clicking on the **Calculate** button.

7. Under **Time** properties, select **Each feature has a single time field** for **Layer Time**. Select the date field for **Time Field**. Define a **Time Step Interval** of **1 Days**, as shown in the following screenshot:

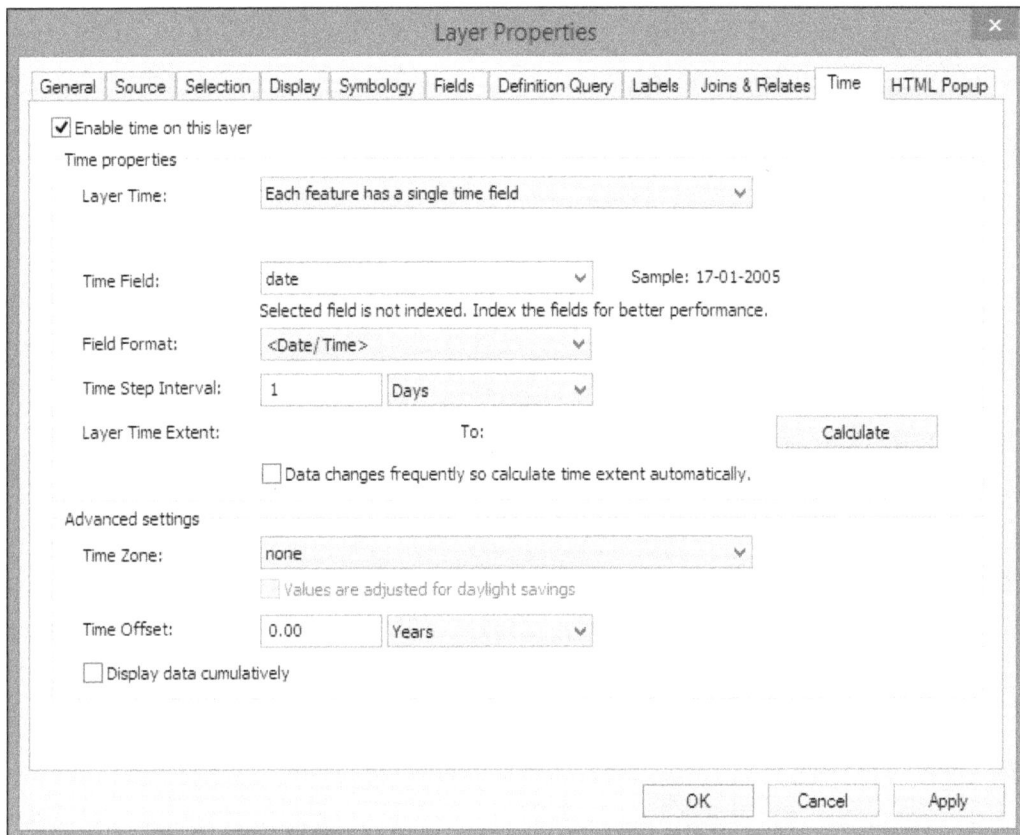

8. Define **Layer Time Extent** by clicking on the **Calculate** button circled in the following screenshot:

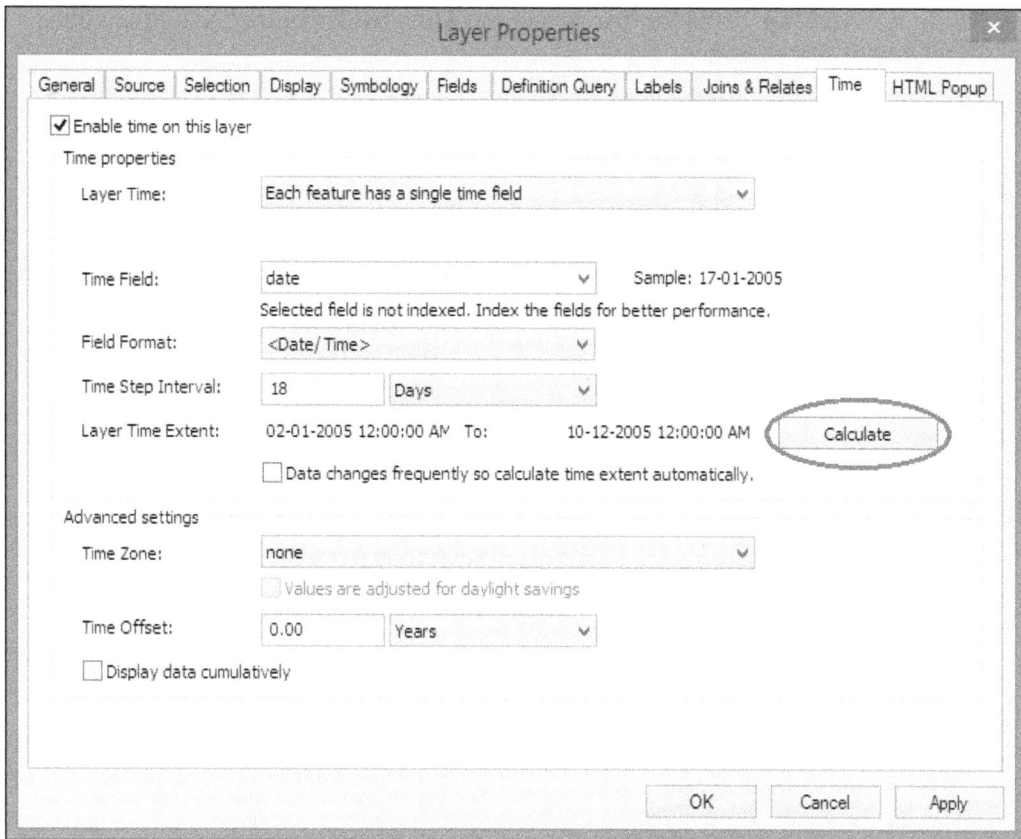

The Layer Properties dialog shows the following:

- Layer Properties (title bar)
- Tabs: General | Source | Selection | Display | Symbology | Fields | Definition Query | Labels | Joins & Relates | Time | HTML Popup
- ✓ Enable time on this layer
- Time properties
  - Layer Time: Each feature has a single time field
  - Time Field: date   Sample: 17-01-2005
    - Selected field is not indexed. Index the fields for better performance.
  - Field Format: <Date/ Time>
  - Time Step Interval: 18   Days
  - Layer Time Extent: 02-01-2005 12:00:00 AM To: 10-12-2005 12:00:00 AM   Calculate
    - ☐ Data changes frequently so calculate time extent automatically.
- Advanced settings
  - Time Zone: none
    - ☐ Values are adjusted for daylight savings
  - Time Offset: 0.00   Years
  - ☐ Display data cumulatively
- OK   Cancel   Apply

9. Select **Time Step Interval**. You may need to reset it to **1 Days**.

10. Click on **Apply** and then click on **OK**.

11. In the **ArcMap Tools** toolbar, select the **Time Slider** button to display the **Time Slider** dialog.

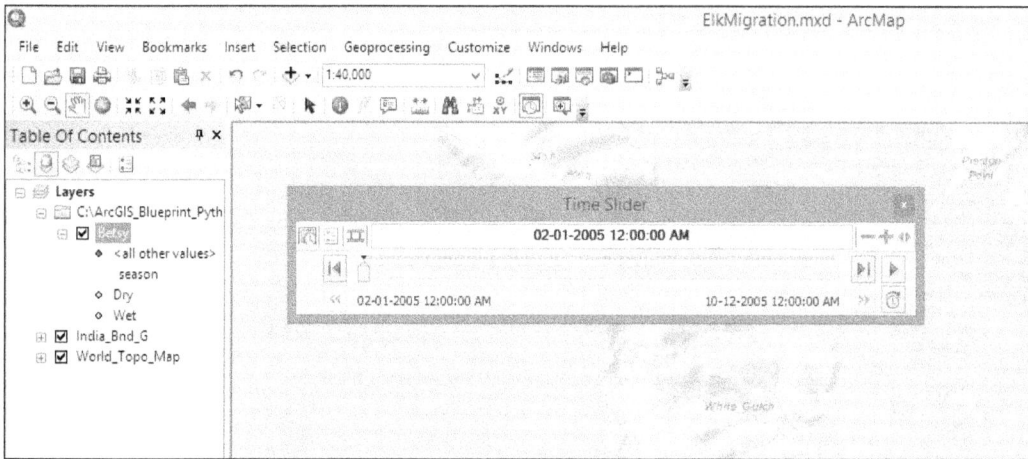

12. Click on the **Options** button to display the **Time Slider Options** dialog.

13. In the **Time Display** tab of the **Time Slider Options** dialog, make sure that **Time step interval** is set to **1 Days**. If not, set it to **1 Days**. Do this for the **Time window** option as well, as shown in the following screenshot:

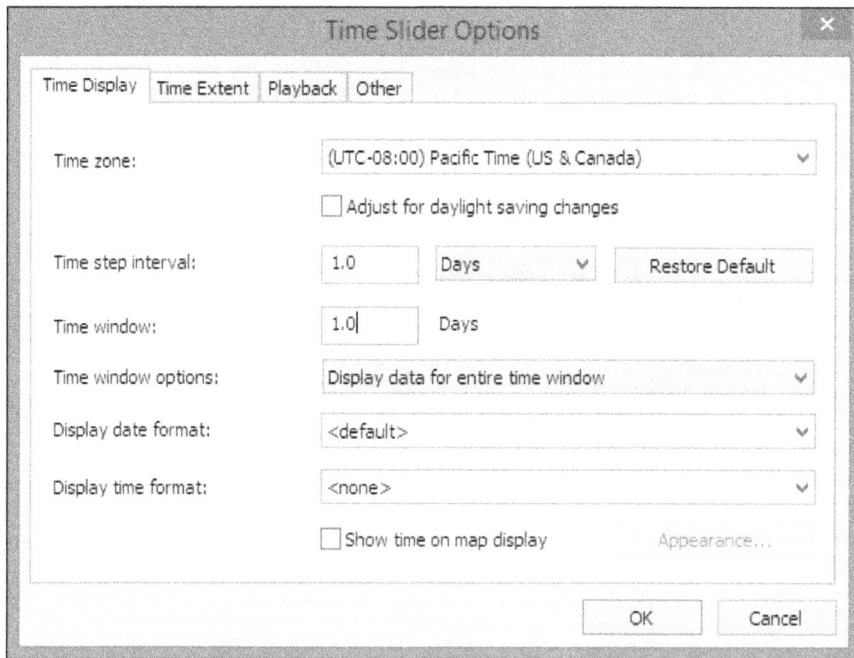

14. Click on **OK**.

15. Switch to the **Layout** view in **ArcMap**.

16. Add a **Title** text element to the layout, as shown in the following screenshot:

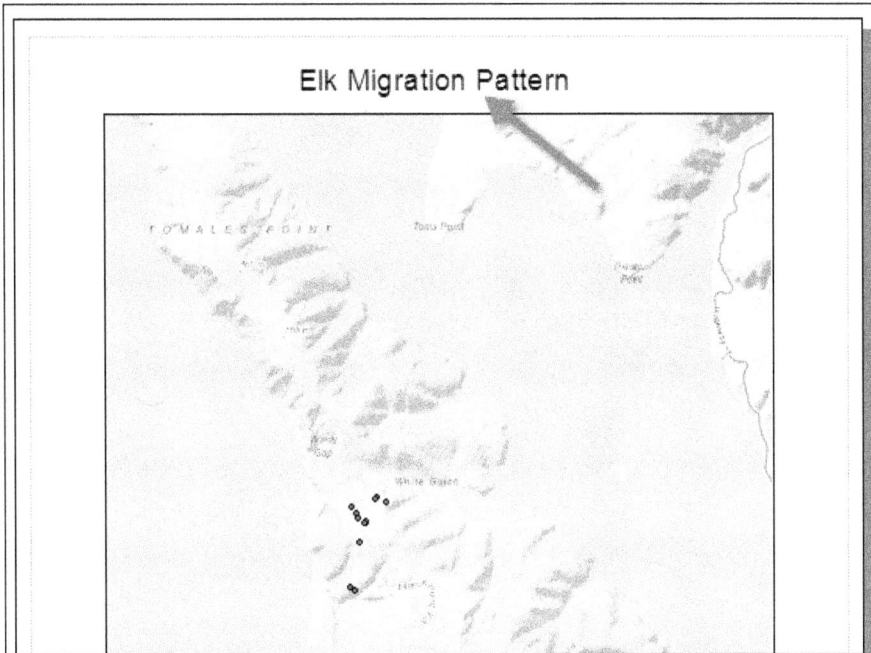

17. Right-click on the **New text element** and select **Properties**; then, select the **Size** and **Position** tab, as shown in the following screenshot. Add an **Element Name** called `title`. Adding **Element Name** is important because we'll reference it in the script that we write, which automatically updates the title to include the current date:

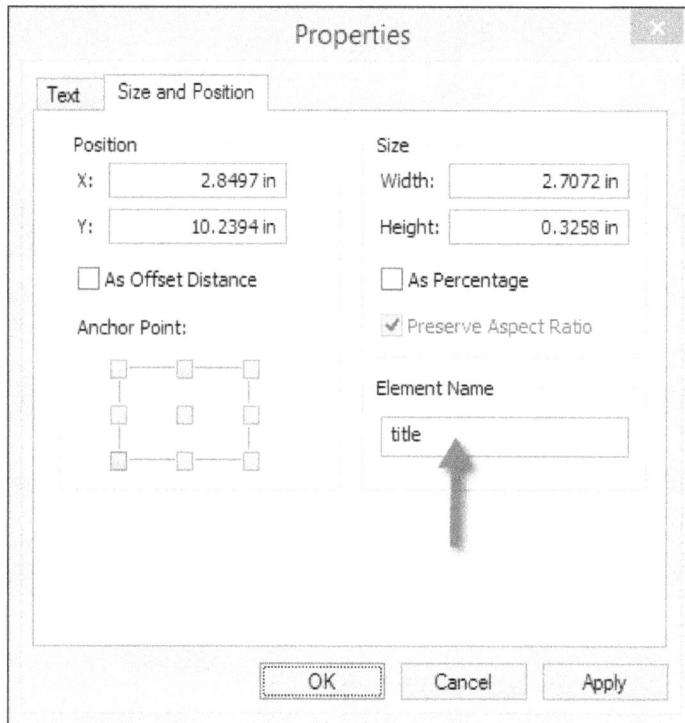

18. Save your map document. It's very important that you save the time-enabled data with your map document. The code you write next won't work unless you do this.

Coding the VisualizeMigration tool

In this final section, you'll create a new tool in the **Migration Patterns** toolbox that can be used to visualize elk migration patterns in one-week increments that have beginning and end dates, as specified through the tool. The tool will also export maps in the PDF format for each week:

1. Import the `arcpy.mapping` and `datetime` modules that will be used in this class:

```
import arcpy
```

```
import arcpy.mapping as mapping
import csv
import os
import datetime
```

2. Create a new tool called `VisualizeMigration` by copying and pasting the existing `ImportCollarData` code and then renaming the class to `VisualizeMigration`.

3. In the `VisualizeMIgration` class, set the `label` and `description` properties, as shown in the following code through the `__init__` method:

```
class VisualizeMigration(object):
    def __init__(self):
        """Define the tool (tool name is the name of the
class)."""
        self.label = "Visualize Elk Migration"
        self.description = "Visualize Elk Migration"
        self.canRunInBackground = False
```

4. You'll need two parameters to capture the `start` and `end` dates. Add the `Parameter` objects, as shown in the following code through the getParameterInfo() method:

```
def getParameterInfo(self):
        param0 = arcpy.Parameter(displayName = "Begin Date", \
                    name="beginDate", \
                    datatype="GPDate", \
                    parameterType="Required",\
                    direction="Input")

        param1 = arcpy.Parameter(displayName = "End Date", \
                    name="endDate", \
                    datatype="GPDate",\
                    parameterType="Required",\
                    direction="Input")

        params = [param0, param1]
        return params
```

5. Capture the `start` and `end` data parameter values in the `execute()` method:

```
def execute(self, parameters, messages):
        """The source code of the tool."""
        beginDate = parameters[0].valueAsText
        endDate = parameters[1].valueAsText
```

6. Split the day, month, and year values for the start and end dates:

```
def execute(self, parameters, messages):
        """The source code of the tool."""
        beginDate = parameters[0].valueAsText
        endDate = parameters[1].valueAsText

        #begin date
        lstBeginDate = beginDate.split("/")
        beginMonth = int(lstBeginDate[0])
        beginDay = int(lstBeginDate[1])
        beginYear = int(lstBeginDate[2])

        #end date
        lstEndDate = endDate.split("/")
        endMonth = int(lstEndDate[0])
        endDay = int(lstEndDate[1])
        endYear = int(lstEndDate[2])
```

7. Get the current MapDocument, DataFrame, and DataFrameTime objects:

```
def execute(self, parameters, messages):
        """The source code of the tool."""
        beginDate = parameters[0].valueAsText
        endDate = parameters[1].valueAsText

        #begin date
        lstBeginDate = beginDate.split("/")
        beginMonth = int(lstBeginDate[0])
        beginDay = int(lstBeginDate[1])
        beginYear = int(lstBeginDate[2])

        #end date
        lstEndDate = endDate.split("/")
        endMonth = int(lstEndDate[0])
        endDay = int(lstEndDate[1])
        endYear = int(lstEndDate[2])

        mxd = mapping.MapDocument("current")
        df = mapping.ListDataFrames(mxd, "Layers")[0]
        dft = df.time
```

8. Set the `currentTime` and `endTime` properties on the `DataFrameTime` object. This will set the boundaries of the visualization and map export:

```
mxd = mapping.MapDocument("current")
df = mapping.ListDataFrames(mxd, "Layers")[0]
dft = df.time
dft.currentTime = datetime.datetime(beginYear, beginMonth,
beginDay)
dft.endTime = datetime.datetime(endYear, endMonth, endDay)
```

9. In the last section of this method, you will create a loop that accomplishes several tasks. The loop will operate between the `start` and `end` dates, set the visible features to the current date, dynamically set the title to the current date, export a PDF file, and reset the `currentTime` property to the next day. Add the WHILE loop just below the last line of code you wrote in the last step. Note that you will have to hardcode a path to the output folder where the PDF files will be created. If you are so inclined, you may want to convert this into a parameter that is provided as input from the user:

```
while dft.currentTime <= dft.endTime:
    for el in mapping.ListLayoutElements(mxd, "TEXT_ELEMENT",
"*title*"):
        el.text = "Elk Migration Pattern: " + str(dft.
currentTime).split()[0]

    fileName = str(dft.currentTime).split(" ")[0] + ".pdf"
    mapping.ExportToPDF(mxd,os.path.join(r"c:\ArcGIS_Blueprint_
Python\ch2", fileName))
    arcpy.AddMessage("Exported " + fileName)
    dft.currentTime = dft.currentTime + dft.timeStepInterval
```

10. Check your code against the solution file found at `C:\ArcGIS_Blueprint_Python\solutions\ch2\VisualizeMigration.py` to make sure you have coded everything correctly.

11. Close your code editor.

12. In **ArcMap**, open `C:\ArcGIS_Blueprint_Python\ch2\ElkMigration.mxd`. Open the **Catalog** view and execute the **Visualize Elk Migration** tool. You will be prompted to enter the `start` and `end` dates, as shown in the following screenshot. The data for this particular elk spans the period between January 17, 2005 to November 4, 2005. To keep it simple, enter a fairly small time period, such as `1/18/2005` to `2/18/2005`.

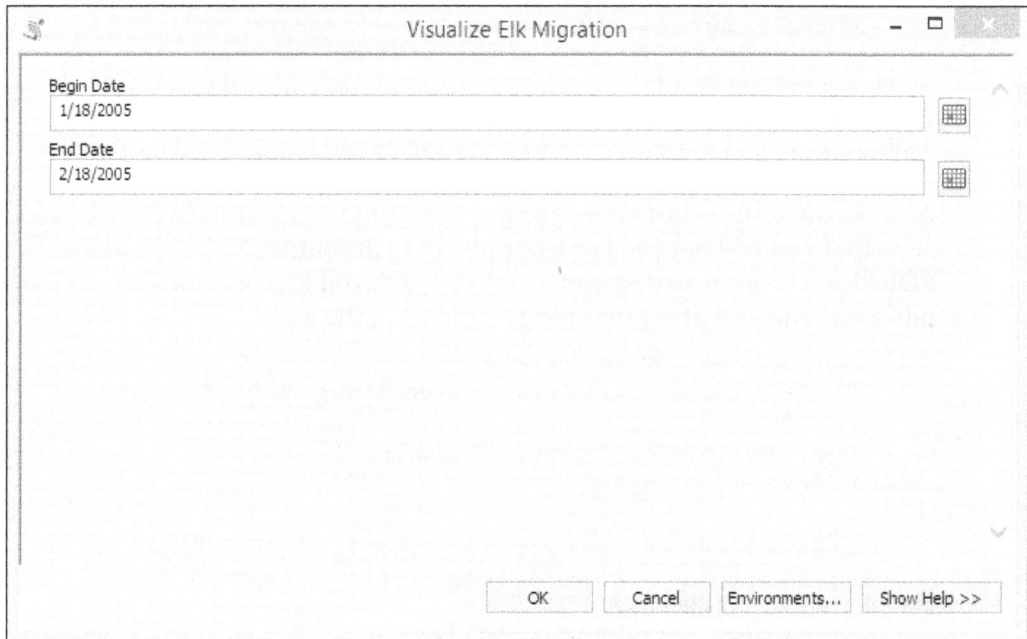

13. Click on **OK** to execute the tool. If everything has been coded correctly, the progress dialog should be updated as each day is visualized and exported to a PDF file.

14. After the execution, you can check the `C:\ArcGIS_Blueprint_Python\`
    `ch2` folder to see the output files. Each file should appear similar to what is
    shown in the following screenshot:

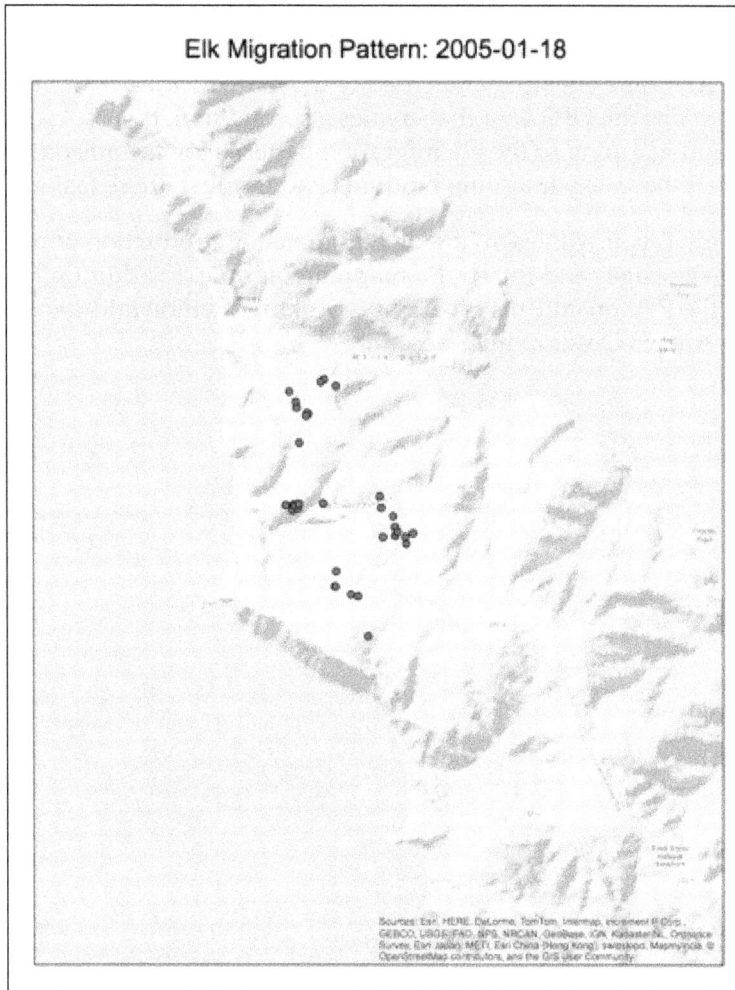

Summary

In this chapter, you created a new **ArcGIS Desktop Python** toolbox containing two tools that are used to support importing and visualizing of the GPS data extracted from a collar attached to an elk in northern California. The first tool used the Python csv and ArcPy data access modules to read the GPS data from a csv file into a local feature class. After time enabling and creating the map document and the feature class which contains the GPS data that creates a second tool, this tool controls the visualization and mapping of the elk migration patterns for specific dates. This second tool used the ArcPy mapping module to accomplish these tasks.

In the next chapter, you will learn how to automate the production of map books using Data-driven pages and the ArcPy mapping module. In addition, you'll be introduced to the Python add-ins for ArcGIS Desktop. Python add-ins allow you to customize the ArcGIS Desktop interface.

3

Automating the Production of Map Books with Data Driven Pages and ArcPy

Many organizations have a need to create map books that contain a series of individual maps that cover a larger geographic area. These map books contain a series of maps and some optional additional pages, including title pages, an overview map, and some other ancillary information, including reports and tables. For example, a utility company might want to generate a map book detailing their assets across a service area. A map book for this utility company could include a series of maps, each of a large scale, along with a title page and an overview map. These resources would then be joined in a single document that could be printed or distributed as a PDF file.

In this chapter, we will cover the following topics:

- The preparation of a map document to handle **Data Driven Pages**
- Using the data-driven pages toolbar in ArcGIS Desktop
- Using the `DataDrivenPages` object in the `ArcPy` mapping module
- Exporting map books with the `ArcPy` mapping module
- Creating a Python add-in for ArcGIS Desktop

Design

From a design perspective, ArcGIS Desktop isn't terribly complex. We'll spend a fair amount of time up front, preparing the map document. We'll add a `floodplain` layer to the `map` document file, create a grid index layer, enable data-driven pages, and prepare the layout view. After the map document has been prepared, we'll create a new Python add-in containing a button that will trigger the creation of the book map. The Python add-in will have a single button, which will contain an `onClick()` method. The `onClick()` method will append the individual map pages to a single output PDF file using the `PDFDocument` class in the `arcpy.mapping` module as shown in the following screenshot:

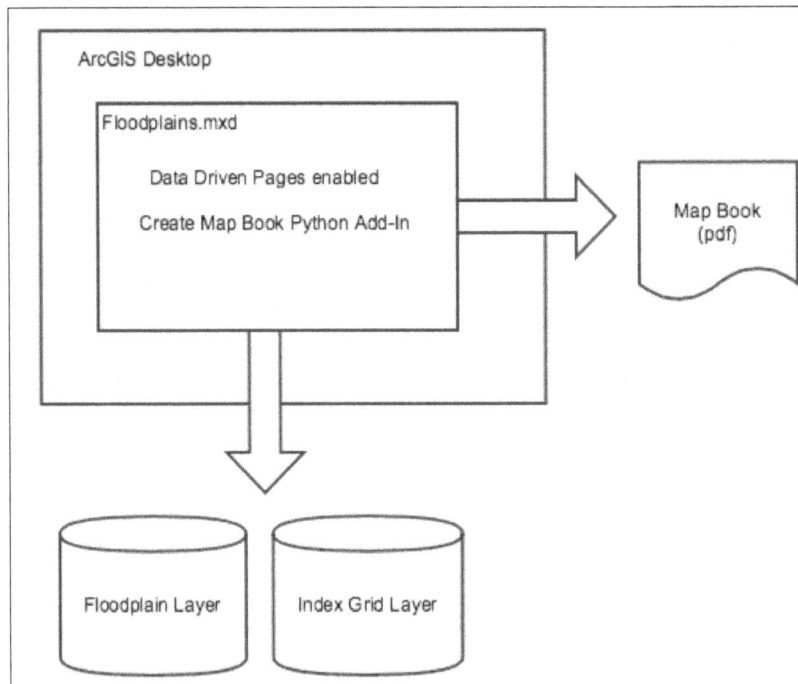

```
┌─────────────────────────────────────────────────────────────┐
│                                                              │
│  ArcGIS Desktop                                              │
│                                                              │
│  ┌────────────────────────────────────┐                     │
│  │ Floodplains.mxd                    │          ┌────────┐  │
│  │                                    │          │Map Book│  │
│  │    Data Driven Pages enabled       │ ──────▶  │ (pdf)  │  │
│  │                                    │          │        │  │
│  │    Create Map Book Python Add-In   │          └────────┘  │
│  │                                    │                     │
│  └────────────────────────────────────┘                     │
│                    │                                         │
│                    ▼                                         │
│      ┌─────────┐  ┌─────────┐                                │
│      │Floodplain│ │Index Grid│                               │
│      │ Layer    │ │ Layer    │                               │
│      └─────────┘  └─────────┘                                │
│                                                              │
└─────────────────────────────────────────────────────────────┘
```

Let's get started and build the application.

Setting up the Data Frame

The ArcGIS Desktop provides the ability to efficiently create a map book through a combination of the **Data Driven Pages** functionality along with an `arcpy.mapping` script. With a single map document file, you can use the **Data Driven Pages** toolbar to create a series of maps using the layout view along with your operational data and an index layer.

The index layer contains features that will be used to define the extent of each map in the series. It divides the map into sections, with each section representing a map that will be generated. These sections are sometimes called tiles or areas of interest, and they are often *rectangular* or *square* shapes.

If you need to include additional pages in the map book, including a title page, an overview map, and other ancillary pages, you'll need to combine the output from the **Data Driven Pages** toolbar with the functionality provided by the `arcpy.mapping` module.

In the following steps, we will learn how to use the **Data Driven Pages** toolbar, to set up a map document file for the **Data Driven Pages** functionality. We'll create a map book that will display a series of `floodplain` maps for a water-management district:

1. Open **ArcMap** with a **Blank Map** document and add the `World_Topo_Map` **Basemap** layer using the **Add Basemap** button.

2. Add the `floodplain_100yr_capcog` geodatabase and `reg_wtr_planning_dist.shp` layers from the `C:\ArcGIS_Blueprint_Python\data\Floodplain_100yr_capcog` folder.

3. Style the layers as shown in the following screenshot:

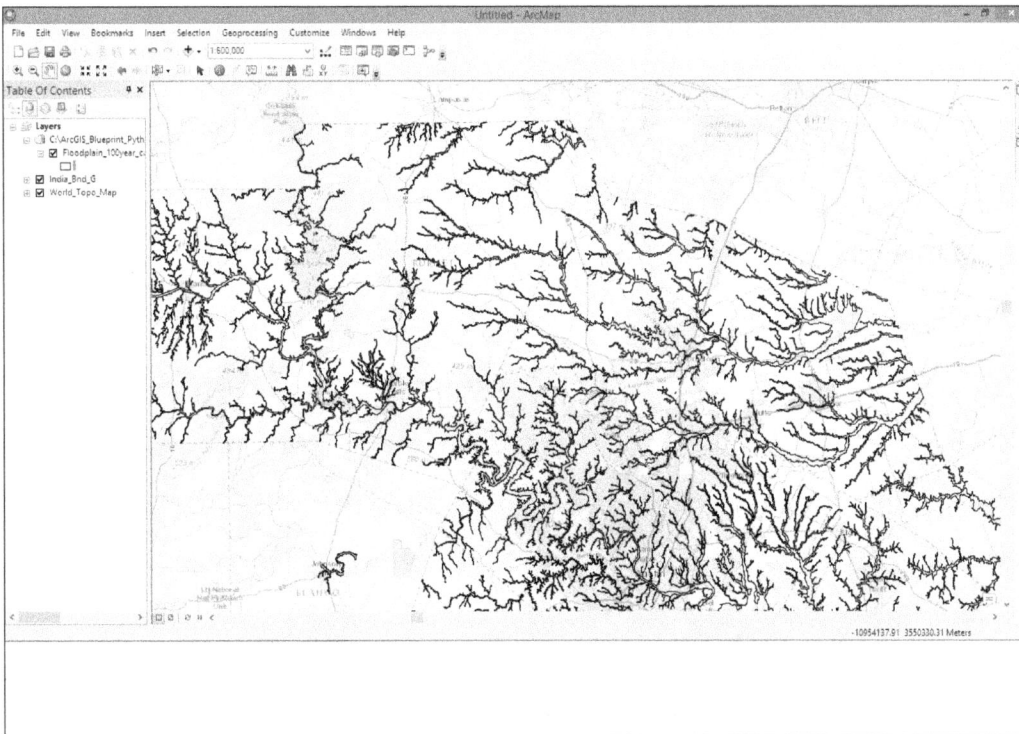

4. Save the map document as `C:\ArcGIS_Blueprint_Python\ch3\Floodplains.mxd`.

5. We need to define the map and page layout before creating the series. Let's create the reference series on a letter page size (8.5 by 11 inches) with a portrait orientation. The map scale will be 1:100,000. In addition, the map will have a title, labels for adjacent pages in the series, and some items in the map margin.

6. Double-click on the **Layers** data frame to display **Data Frame Properties** and select the **General** tab, as shown in the following screenshot:

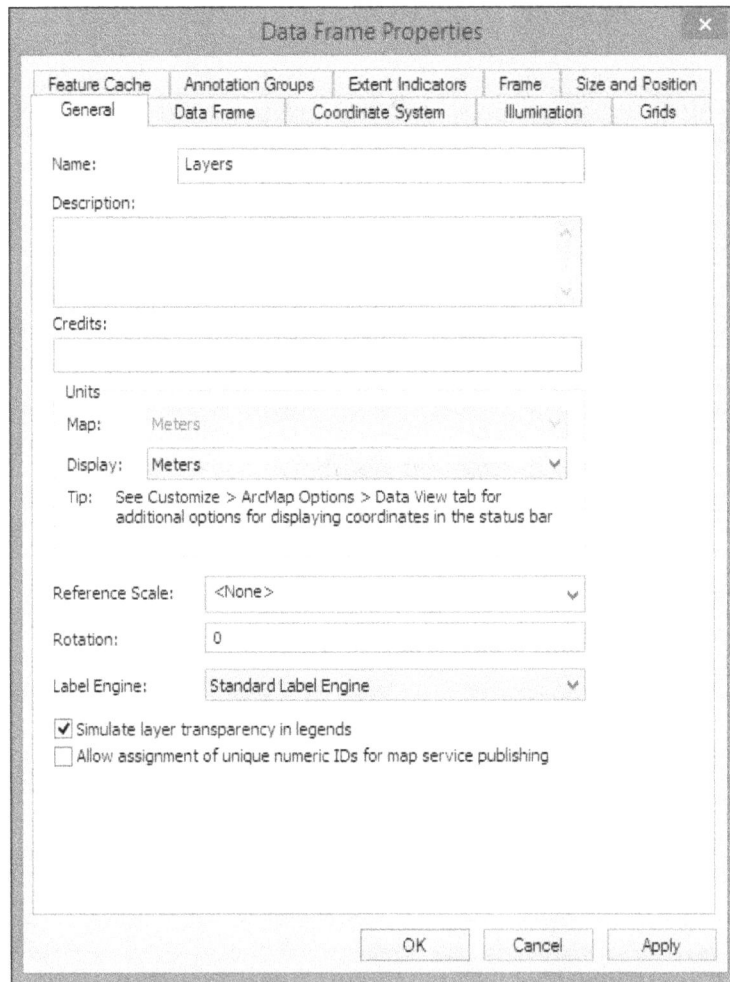

7. Rename the data frame to **Detail Map**.

8. Click on the **Size** and **Position** tab. Enter a value of 1 inch for *X* position and 2.5 inch. for *Y* position.

9. For size, enter 6.25 inch as **Width** and 7.5 inch as **Height**.

10. Click on the **Coordinate System** tab. The current coordinate system should be defined as NAD 1983 StatePlane Texas Central FIPS 4203 (US Feet). If not, change it now.

11. Click on **Apply** and then click on **OK**.

12. Switch to **Layout View** in ArcMap and make sure that you have left enough space for the title at the top and the footer area has ample space for text, scale bar, north arrow, and other marginalia, as shown in the following screenshot:

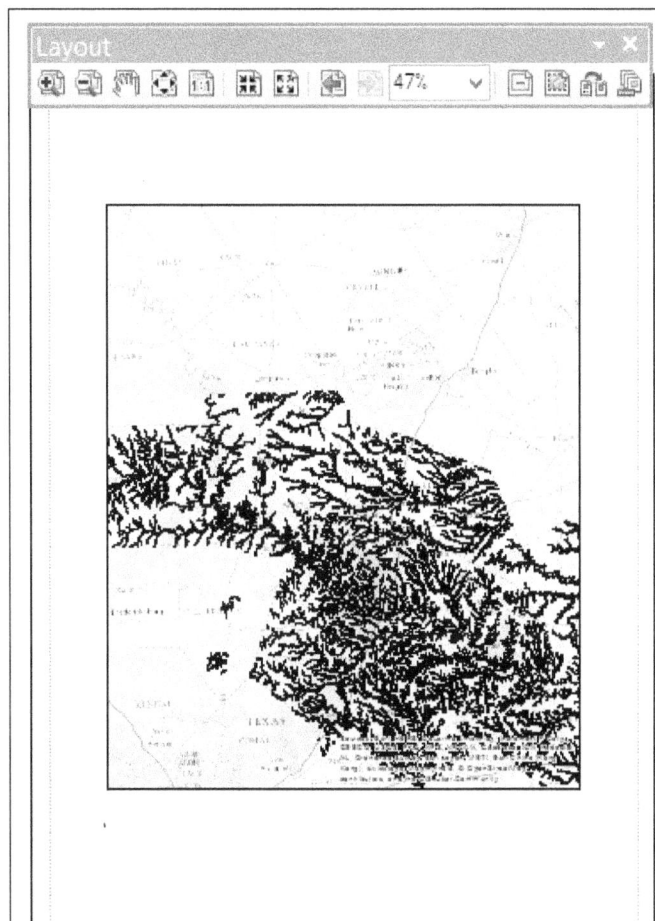

Creating the Grid Index Features

A grid index feature class can be used to set the map extent for each map in the series. We can use the **Grid Index Features** geoprocessing tool to create this layer, the following steps will guide you through, how to create **Grid Index Features**:

1. Using the selection tools in **ArcMap**, select the **Lower Colorado** region from the **Planning Districts** layer, as shown in the following screenshot. We're going to create an index grid for this region. The process will be as, though. if you decide to select other regions of the map:

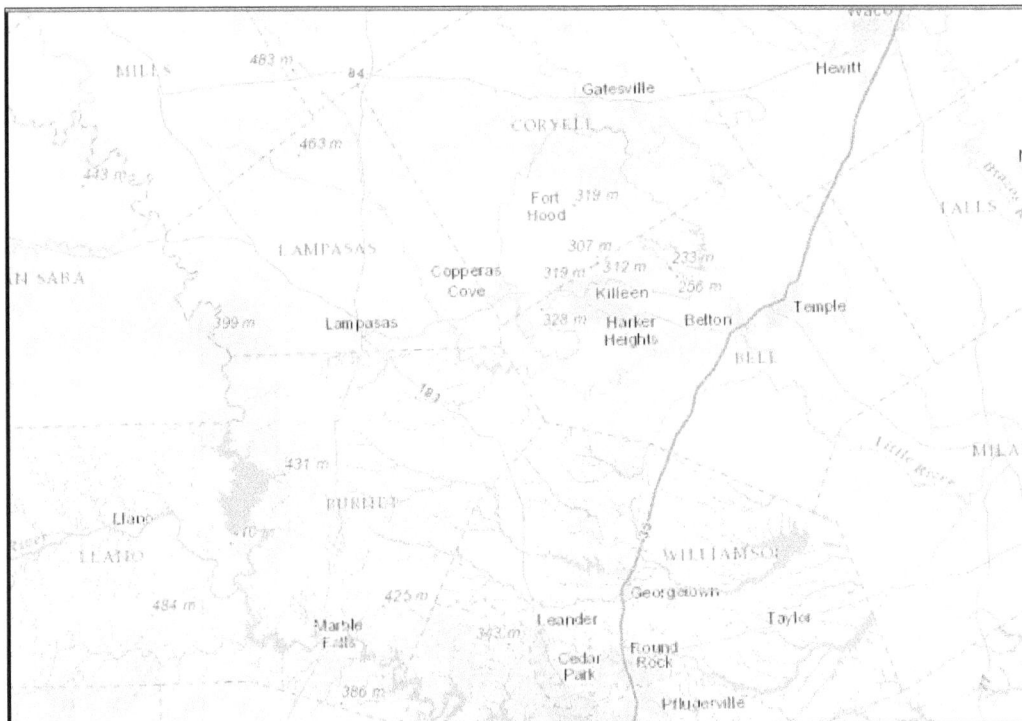

2. Open the **Grid Index Features** geoprocessing tool found in the **Data Driven Pages** toolset in the **Cartography Tools** toolbox.

3. Define the parameters seen in the following screenshot:

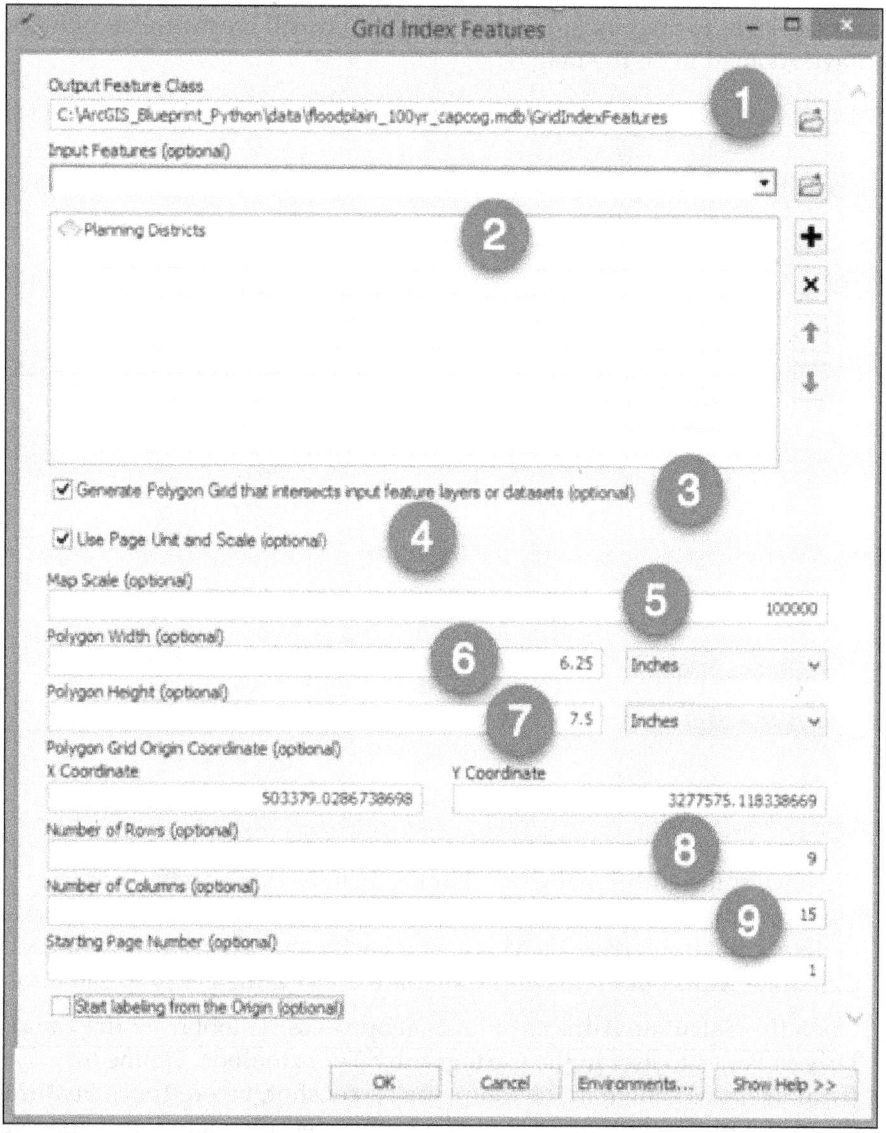

4. Click on **OK** to generate the grid index layer. It should look similar to what is shown in the following screenshot. Note that I have altered the zymology for the layer to only include an outline with no fill for the index polygons and have zoomed in on the map:

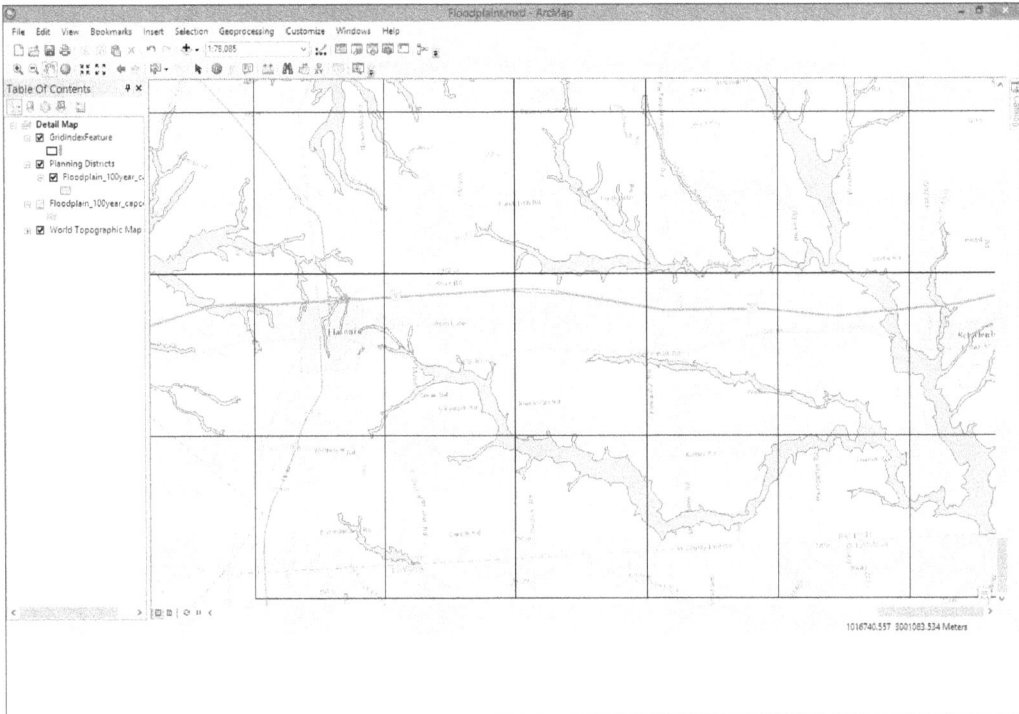

5. We have some additional work that needs to be done on the grid index layer, including the addition of a field for the labeling of adjacent pages and a field to determine the correct **UTM** zone for each page. To do this, we'll use the **Calculate Adjacent Fields** and **Calculate UTM Zone** geoprocessing tools.

6. Open the **Calculate Adjacent Fields** geoprocessing tool from the **Data Driven Pages** toolset in the **Cartography Tools** toolbox. Define the parameters, as shown in the following screenshot, where **Input Features** is the grid index layer that you created in the preceding steps and **PageName** is the default field that has already been added to the grid index layer:

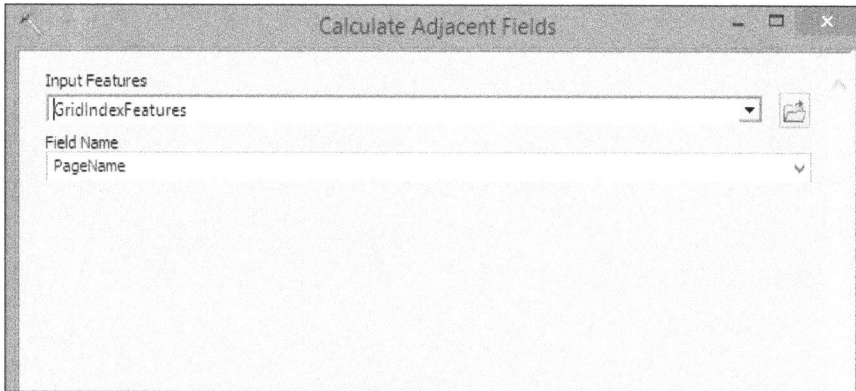

7. Click on **OK** to execute the tool. Open the attribute table for the `GridIndexFeatures` feature class to view the attribute data that has been created for the **PageName_*** fields. Attributes that define the adjacent index pages have been populated.

8. Open the **Add Field** geoprocessing tool found in the **Fields** toolset in the **Data Management** toolbox.

9. Add the field parameters shown in the following screenshot. This will add a field that we'll then populate with the **Calculate UTM Zone** geoprocessing tool:

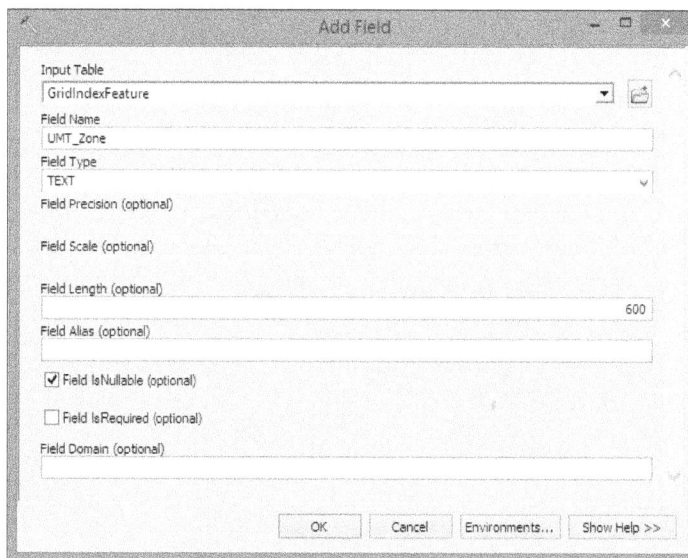

10. Open the **Calculate UTM Zone** tool found in the **Data Driven Pages** toolset in the **Cartography Tools** toolbox.

11. Define the parameters, as shown in the following screenshot:

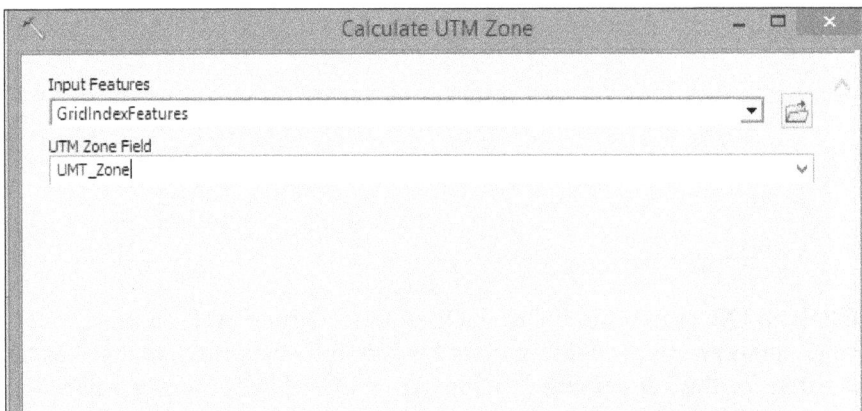

12. Click on **OK** to execute the tool. If you'd like, open the attribute table and view the contents of the UTM_Zone field.

Enabling Data Driven Pages

In this section, you'll enable the **Data Driven Pages** functionality for the map document. The following steps will guide you through, how to enable **Data Driven Pages**:

1. In **ArcMap**, click on the **Scale Control** on the **Standard** toolbar and type 1:100,000 as the map scale.

2. Open the **Data Driven Pages** toolbar by navigating to **Customize | Toolbars | Data Driven Pages** from the main **ArcMap** menu.

3. Click on the **Data Driven Pages Setup** button.

4. **Enable Data Driven Pages** by clicking on the checkbox shown in the following screenshot. Also, select `GridIndexFeatures` as index **Layer** and **PageName** as **Name Field**. Make sure all the other values are set, as shown in the following screenshot:

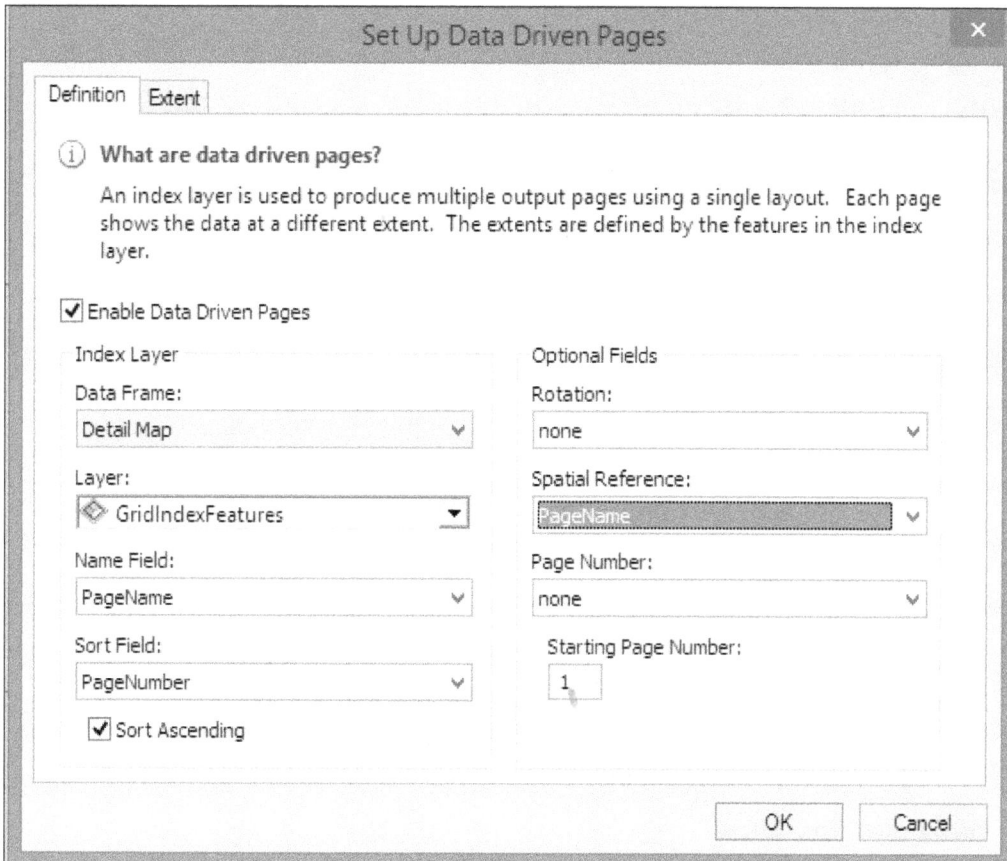

5. Click on the **Extent** tab, choose **Center** and **Maintain Current Scale**, and then click on **OK**, as shown in the following screenshot:

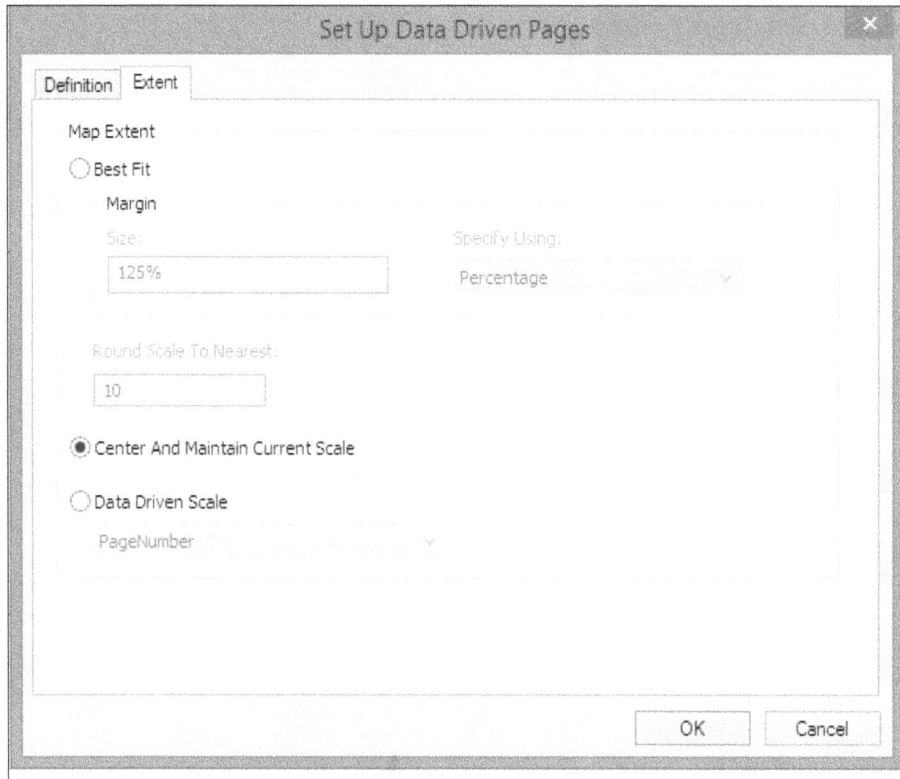

6. Save the map document file.

Creating the Locator Map

The **Locator Map** provide an overview of the spatial location of the current map within the context of a large geographic area. They provide an overview of the location of the current map in the series. In this section, you'll create a **Locator Map** for the layout view. We'll use a feature class copied from the grid index layer and edited to create a mask layer and a current page layer. The mask layer is used to *gray out* the features that are not in the current map while the current page layer highlights the current map:

1. In **ArcMap**, create a new data frame and name it **Locator Map**.

2. Copy the `Basemap`, `Floodplains`, and `GridIndexFeatures` layers from the **Detail Map** and paste them into the **Locator Map**. Rename the `GridIndexFeatures` layer to **Page Labels**.

3. Right-click on the **Page Labels** feature class and navigate to **Data | Export Data**. Save it to the same location as the grid index features and name it `LocatorMask`. Add the layer to the map. Your **ArcMap** table of contents should now appear, as shown in the following screenshot:

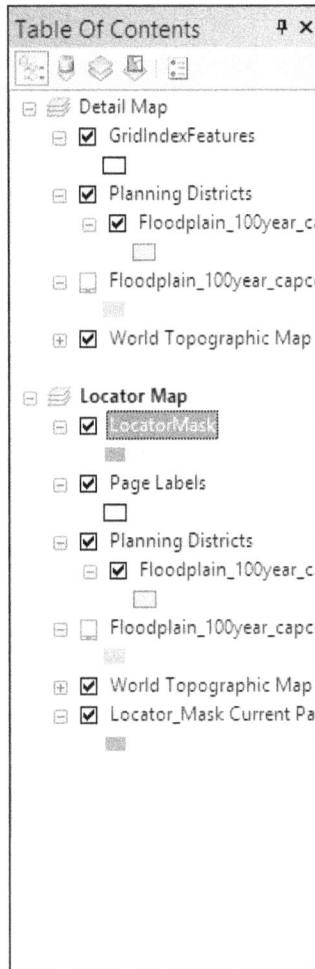

4. Right-click on the **Page Labels** feature class and select **Properties**.

5. Click on the **Labels** tab and choose **PageName** for the **Label** field and check the box next to **Label** features in this layer. Click on **OK**.

6. In the **Locator Map** data frame, click on the symbol for the LocatorMask feature class.

7. Using **Symbol Selector**, change **Fill Color** to black, **Outline Width** to 1, and **Outline Color** to white.

8. In the **Properties** dialog box for the LocatorMask feature class, click on the **Display** tab and type **60** for the transparency value. Click on **OK**.

9. Next, we'll create a layer that will serve as the highlight layer. Right-click on the LocatorMask feature class and select **Copy**.

10. Right-click on the **Locator Map** data frame and select **Paste Layer**.

11. Rename the layer to Locator_Mask Current Page.

12. Click on the symbol for Locator_Mask Current Page in the **Locator Map** data frame. From **Symbol Selector**, choose **Hollow**, set **Outline Width** to 1, and click on **OK**.

13. Now, we'll set the page definition queries for the LocatorMask and Locator_Mask Current Page layers so that they are displayed correctly.

14. Double-click on the LocatorMask layer and then click on the **Definition Query** tab.

15. Click on the **Page Definition Query** button and set the properties shown in the following screenshot. Click on **OK** when you're done:

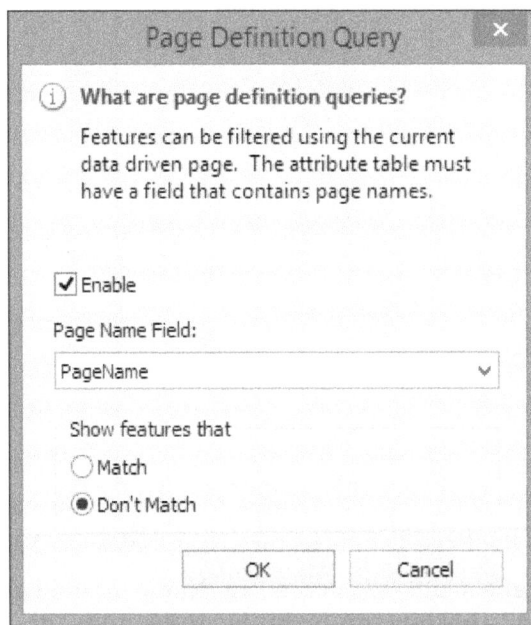

16. Click on **OK** to exit the **Layer Properties** dialog box.

17. Your view should now appear as shown in the following screenshot. Any features that do not match the current page are drawn so that the area outside the current page is displayed as a gray mask:

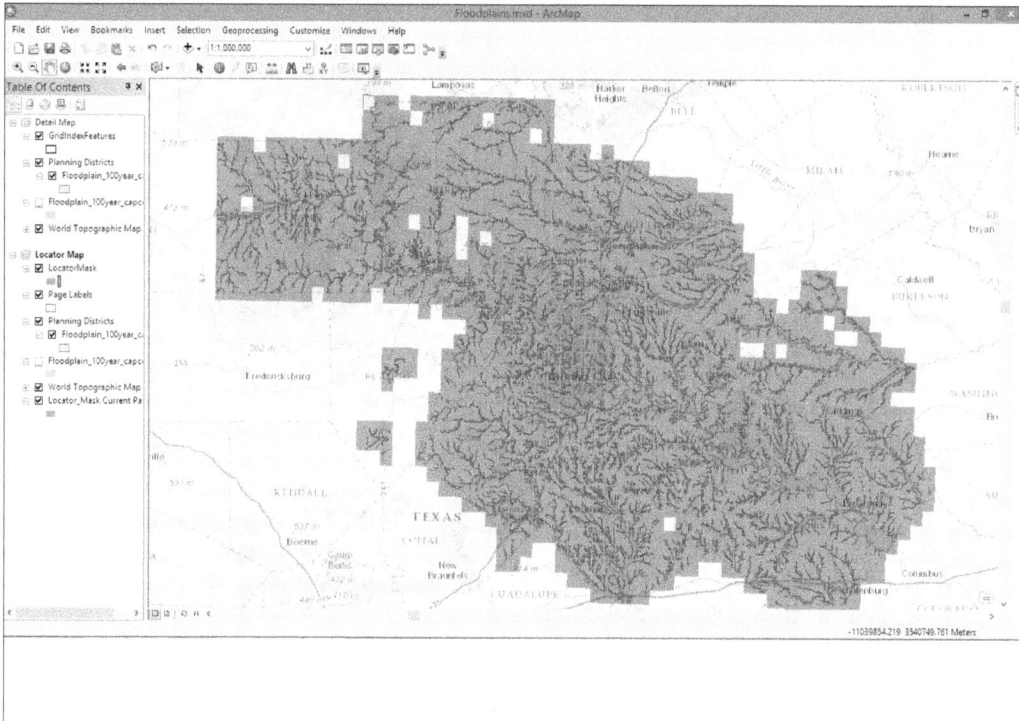

18. Open the **Properties** dialog box for the `Locator_Mask Current Page` layer and select the **Definition Query** tab and then select the **Page Definition Query** button.

19. Set the properties shown in the following screenshot. Click on **OK** when you're done and then click on **OK** again to dismiss the **Layer Properties** dialog:

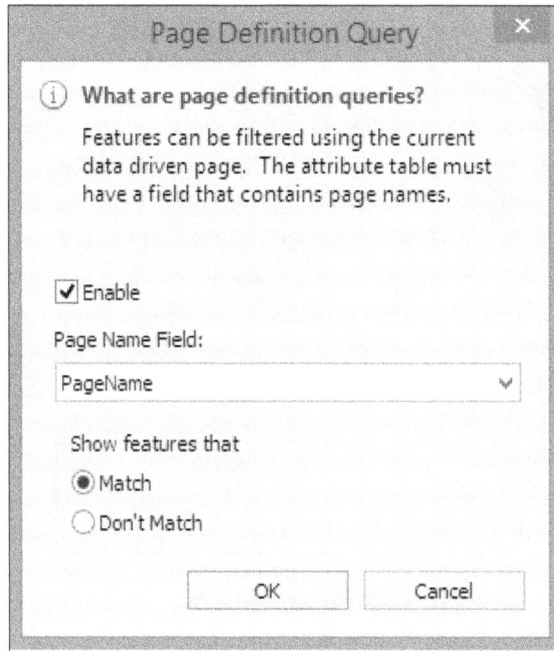

20. Switch to **Layout View** and resize the **Locator Map** data frame on the layout so that it appears just below the main data frame, as shown in the following screenshot:

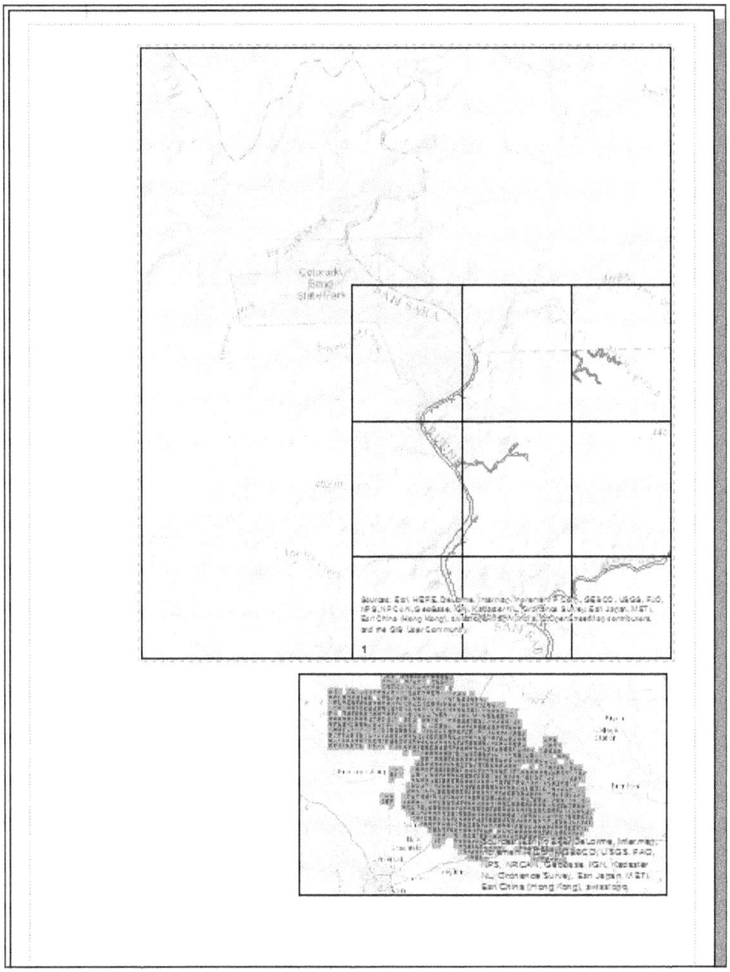

Adding dynamic text to the layout

The last thing that we need to do before writing our script to automate the process of generating the map book is add dynamic text to the layout. Dynamic text includes a title, a page number, a label for an adjacent page number, and other items added to the margins of the map. Dynamic text items are necessary when we have text items that will change for each map that is created. We'll also add a north arrow and scale bar:

1. First, we'll add the north arrow and scale bar. In **Layout View** inside **ArcMap**, add a north arrow and scale bars, as shown in the following screenshot. You don't have to select the same style as mine:

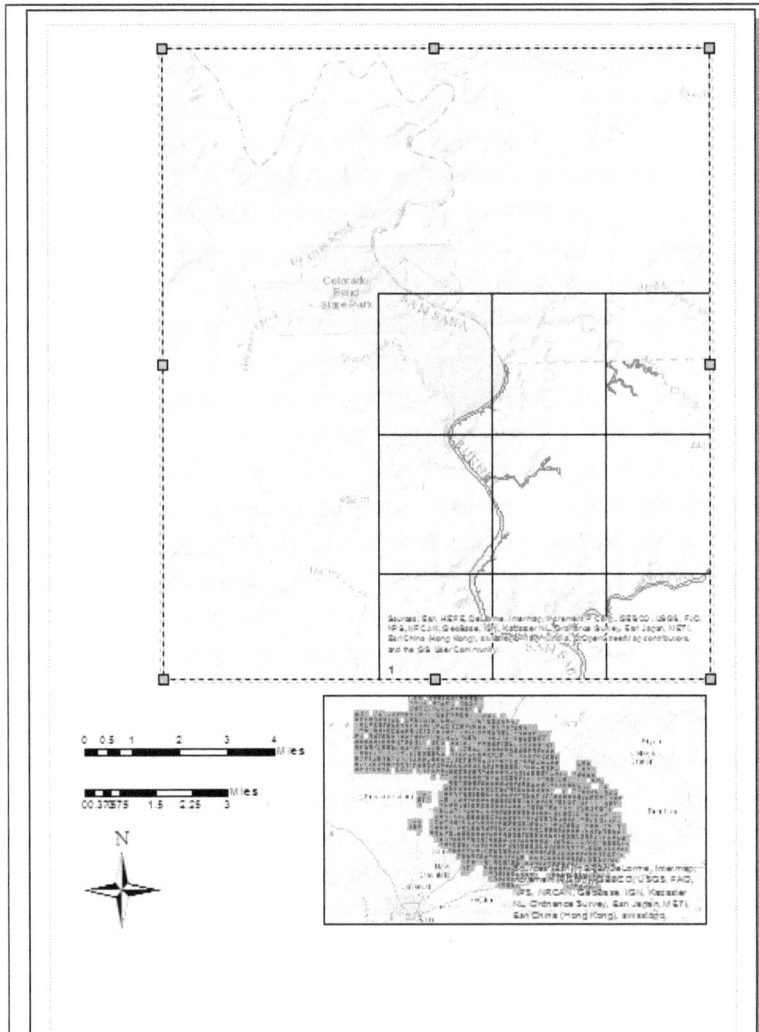

2. Now, we'll add the page number. If necessary, open the **Data Driven Pages** toolbar and navigate to **Page Text | Data Driven Page Name** from the toolbar.

3. The page number will be placed directly in the center of the main data frame for the map. Drag it just above the first scale bar, as shown in the following screenshot. You may want to make the text larger than the default font size of 10. I've changed mine to 16 by right-clicking on the **Text**, and selecting **Properties**, and then clicking on the **Change Symbol** button.

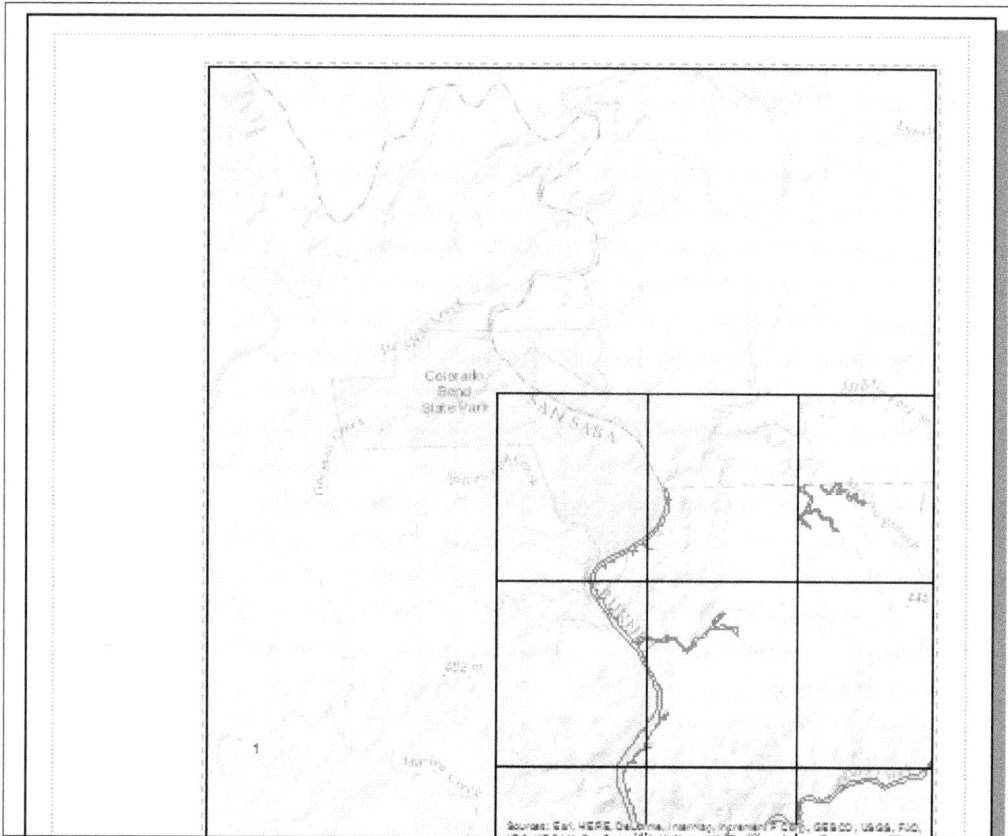

4. Next, we'll add the current page number. In the **Data Driven Pages** toolbar, navigate to **Page Text | Data Driven Page Count**. Like earlier, it will add a new text element to the center of the main data frame. Drag this text item to a new location somewhere in the margin.

5. You can add additional margin items, such as the date the map was saved, the author, the username, the coordinate system, and more by navigating to **Insert | Dynamic Text** from the main **ArcMap** menu. I've added the reference scale and date saved dynamic text items to my layout.

6. You'll also want to add a title to the map by navigating to **Insert | Title in ArcMap**. Call it `Floodplain Map for Lower Colorado River Planning District`.

7. Your map should look similar to what is shown in the following screenshot, though you may choose to add or change some of the items, as you see fit:

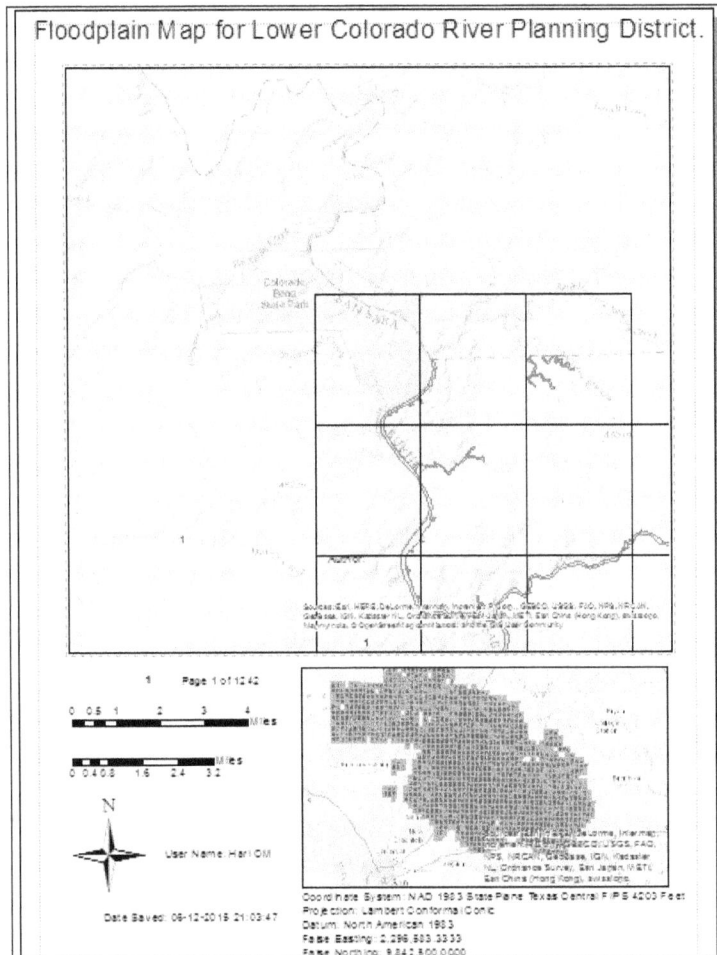

8. Save your map document file.

Exporting the map series with ArcPy mapping

In this final section, we'll use the `arcpy.mapping` module along with a Python add-in to automate the process of exporting the map series to PDF files. Python add-ins are an easy way to add user interface items to ArcGIS Desktop. The **Python Add-In Wizard** is a utility that greatly simplifies the creation of the user interface items. In this section of the chapter, you'll use the **Python Add-In Wizard** to create a toolbar containing a button that will trigger the export of your map book to a PDF file. The following steps will guide you through, how to export map series with **ArcPy** mapping:

1. If necessary, download and install the **Python Add-In Wizard** from `http://www.arcgis.com/home/item.html?id=5f3aefe77f6b4f61ad3e4c62f30bf f3b#!`.

2. In the folder where you unzipped the **Python Add-In Wizard**, find and double-click on the `addin_assistant.exe` file to start the wizard.

3. Choose or create a directory to be used as the add-in project root. Remember the name of the folder because you'll need it later. I'm going to use `C:\ MapBook`.

4. The **Python Add-In Wizard** has two tabs: **Project Settings** and **Add-In Contents**. In the **Project Settings** tab, define the parameters, as shown in the following screenshot. Your working folder may be different than mine depending on your action in the last step, and you'll obviously want to change the author and company:

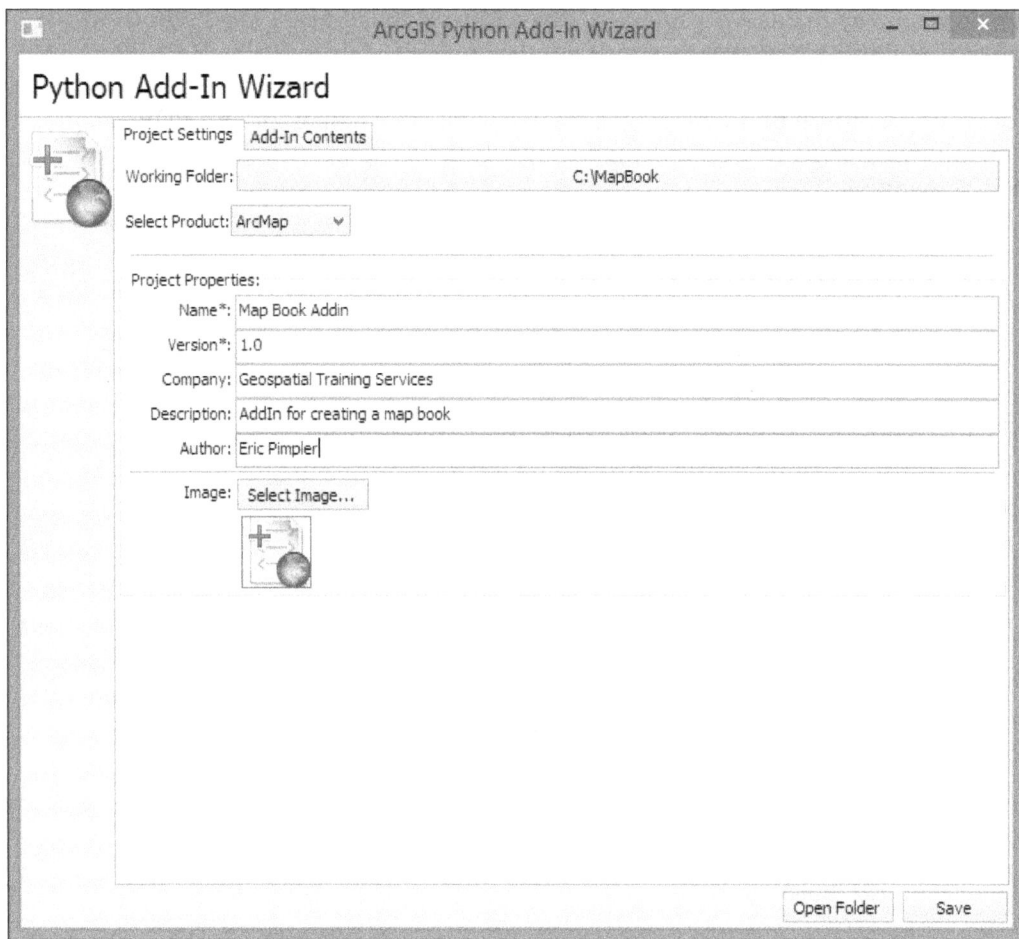

5. Select the **Add-In Contents** tab.

6. Create a new toolbar by right-clicking on **Toolbars** and selecting **New Toolbar**. Define the caption as **Create Map Book** and click on **Save**. The toolbar will serve as a container for the button that triggers the creation of the map book.

7. Right-click on the new toolbar that you just created, and click on **New** button. Fill in the parameters, as shown in the following screenshot:

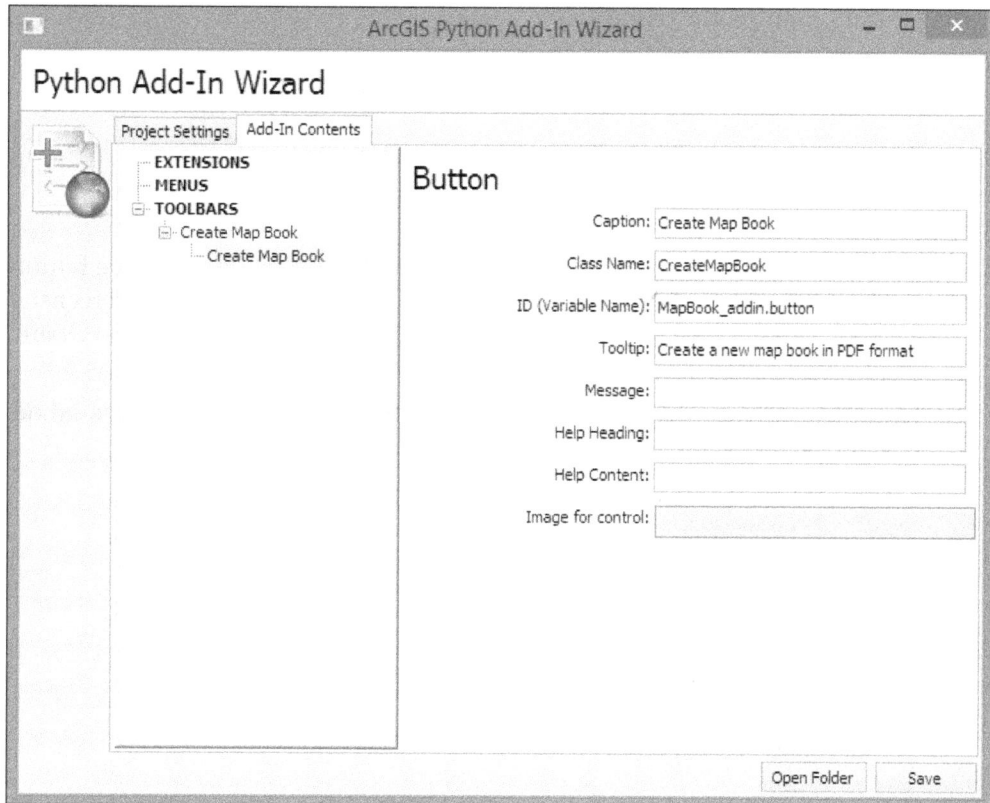

8. Click on **Save**.

9. Click on the **Open Folder** button to display the folders and files that the **Python Add-In Wizard** has created. You should see something very similar to what is shown in the following screenshot:

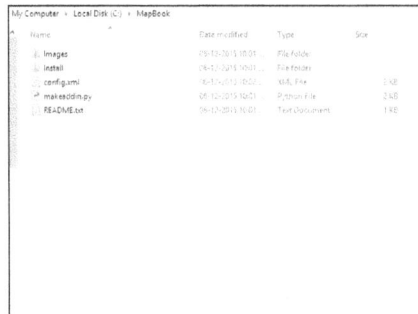

10. Now, it's time to add the Python code that will be executed when the button is clicked on. Go to the `Install` folder, as shown in the preceding screenshot, and you should see a single Python script called `MapBook_addin.py`. Open this file in your Python development environment.

11. Find the `onClick(self)` method shown in the following code. This method is executed when the button is clicked:

```
import arcpy
import pythonaddins

class CreateMapBook(object):
    """Implementation for MapBook_addin.button (Button)"""
    def __init__(self):
        self.enabled = True
        self.checked = False
    def onClick(self):
        pass
```

12. Remove the `pass` statement from the `onClick()` method.

13. Add an `import` statement for the `arcpy.mapping` and `os` modules. In this line of code, we're importing the mapping module and assigning it to a variable called `MAP`. It will make referencing the functions in the mapping module easier as we proceed:

```
import arcpy
import pythonaddins
import arcpy.mapping as MAP
import os
```

14. The rest of our code will go inside the `onClick()` method. Use the `pythonaddins.SaveDialog()` function to display a dialog box that will allow the end user to save the map book as a PDF file. Also, create an output directory variable:

```
def onClick(self):

        # Create an output directory variable
        finalpdf_filename = pythonaddins.SaveDialog("Save Map
Book", "MapBook.pdf", r"C:\ArcGIS_Blueprint_Python\ch3")
        outDir = os.path.split(finalpdf_filename)[0]
```

15. If you are using ArcGIS Desktop 10.3, create a new `pythonaddins.ProgressDialog` object using a `WITH` statement. This will display a progress dialog while the map book is being created. Set the title, description, and animation properties. If you are using ArcGIS Desktop 10.1 or 10.2, please skip this step and proceed to step 16:

```
def onClick(self):

        # Create an output directory variable
        finalpdf_filename = pythonaddins.SaveDialog("Save Map
Book", "MapBook.pdf", r"C:\ArcGIS_Blueprint_Python\ch3")
        outDir = os.path.split(finalpdf_filename)[0]

        with pythonaddins.ProgressDialog as dialog:
            dialog.title = "Progress Dialog"
            dialog.description = "Creating a map book....this
will take awhile!"
            dialog.animation = "File"
```

16. Create a new, empty PDF document in the specified output directory with `pythonaddins.ProgressDialog` as dialog:

```
dialog.title = "Progress Dialog"
dialog.description = "Creating a map book....this will take
awhile!"
dialog.animation = "File"

# Create a new pdf document in the output directory
if os.path.exists(finalpdf_filename):
    os.remove(finalpdf_filename)
finalPdf = MAP.PDFDocumentCreate(finalpdf_filename)
```

17. To simplify the process, a map title page (`TitlePage.pdf`), and overview map page (`IndexMap.pdf`), have also been created for you. These files are located in your `C:\ArcGIS_Blueprint_Python\ch3` folder. Add the title and index pages:

```
# Create a new, empty pdf document in the specified output
directory
if os.path.exists(finalpdf_filename):
    os.remove(finalpdf_filename)
finalPdf = MAP.PDFDocumentCreate(finalpdf_filename)

# Add the title page to the pdf
finalPdf.appendPages(r"C:\ArcGIS_Blueprint_Python\ch3\TitlePage.
pdf")

# Add the index map to the pdf
finalPdf.appendPages(r"C:\ArcGIS_Blueprint_Python\ch3\IndexMap.
pdf")
```

18. Export the **Data Driven Pages** to a temporary PDF file, and then add it to the final PDF:

```
# Add the title page to the pdf
finalPdf.appendPages(r"C:\ArcGIS_Blueprint_Python\ch3\TitlePage.
pdf")

# Add the index map to the pdf
finalPdf.appendPages(r"C:\ArcGIS_Blueprint_Python\ch3\IndexMap.
pdf")

# Export the Data Driven Pages to a temporary pdf and then add it
#to the final pdf.
mxd = MAP.MapDocument("CURRENT")
ddp = mxd.dataDrivenPages

temp_filename = outDir + r"\tempDDP.pdf"

if os.path.exists(temp_filename):
    os.remove(temp_filename)
ddp.exportToPDF(temp_filename, "ALL")
finalPdf.appendPages(temp_filename)
```

19. Update the properties of the final PDF and save it:

```
if os.path.exists(temp_filename):
    os.remove(temp_filename)
ddp.exportToPDF(temp_filename, "ALL")
```

```
finalPdf.appendPages(temp_filename)
# Update the properties of the final pdf.
finalPdf.updateDocProperties(pdf_open_view="USE_THUMBS", pdf_
layout="SINGLE_PAGE")

# Save your result
finalPdf.saveAndClose()
```

20. Remove the temporary **Data Driven Pages** file:

```
# Update the properties of the final pdf.
finalPdf.updateDocProperties(pdf_open_view="USE_THUMBS", pdf_
layout="SINGLE_PAGE")

# Save your result
finalPdf.saveAndClose()

# remove the temporary data driven pages file
if os.path.exists(temp_filename):
    os.remove(temp_filename)
```

21. Check your script against the solution file found in `C:\ArcGIS_Blueprint_Python\ch3\scripts\CreateMapBook.py` for accuracy.

22. Save your script and close the file.

23. Now, it's time to install and test your Python add-in. Inside the main folder where you created the add-in (`C:\MapBook`, in my case), you will find a Python script called `makeaddin.py`. Double-click on this file to open.

24. A new `.esriaddin` file called `MapBook.esriaddin` will be created in the same folder, as shown in the following screenshot:

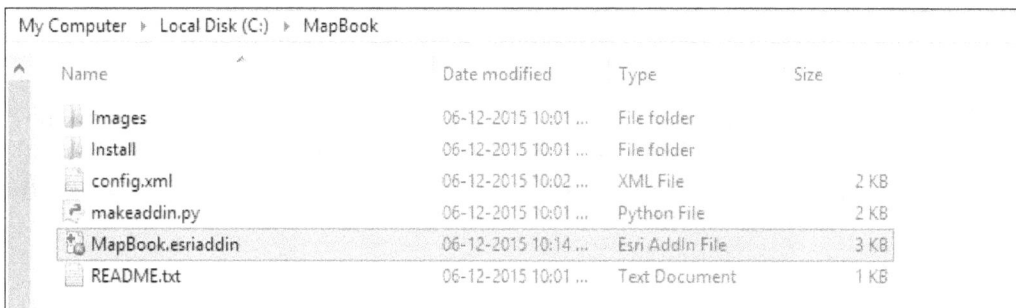

My Computer ▶ Local Disk (C:) ▶ MapBook

Name	Date modified	Type	Size
Images	06-12-2015 10:01 ...	File folder	
Install	06-12-2015 10:01 ...	File folder	
config.xml	06-12-2015 10:02 ...	XML File	2 KB
makeaddin.py	06-12-2015 10:01 ...	Python File	2 KB
MapBook.esriaddin	06-12-2015 10:14 ...	Esri Addin File	3 KB
README.txt	06-12-2015 10:01 ...	Text Document	1 KB

25. To install your new add-in using ArcGIS Desktop, double-click on the `MapBook.esriaddin` file to launch the **Esri ArcGIS Add-In Installation Utility** window seen in the following screenshot:

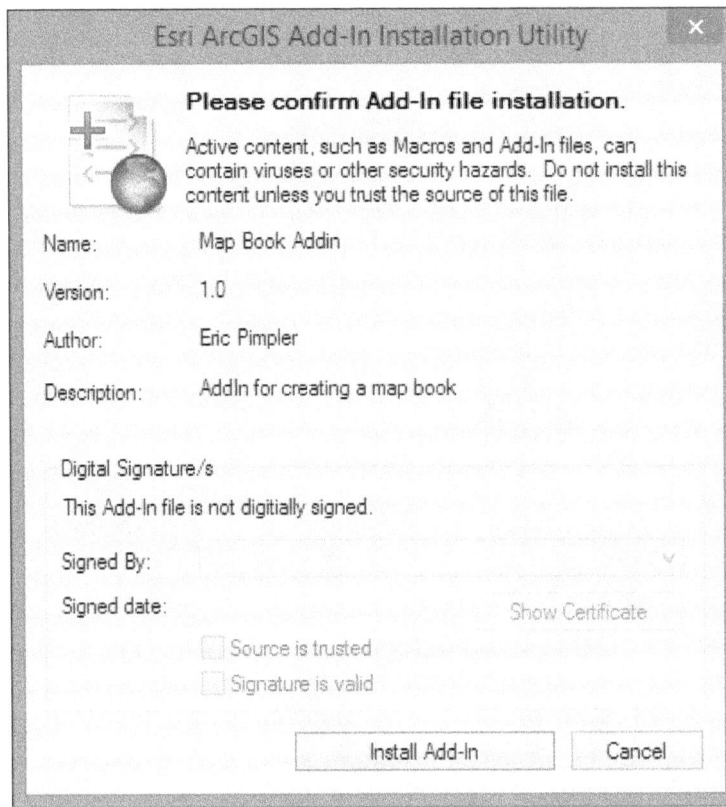

26. Click on **Install Add-In**. If everything is successful, you should see a success message.

27. Open ArcMap with the `Floodplains.mxd` file that you created earlier in the chapter, to test the add-in. Navigate to **Customize | Create Map Book** from the **ArcMap** menu. This will display the **Create Map Book** add-in, as shown in the following screenshot:

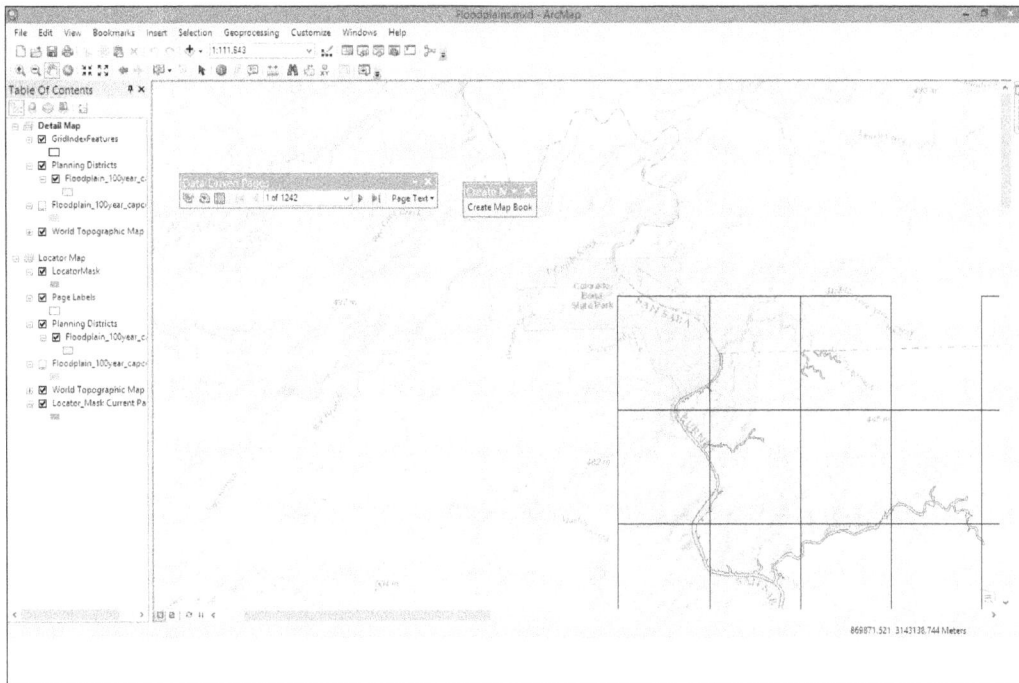

28. Click on the **Create Map Book** button to execute the code.

Summary

ArcGIS Desktop provides the ability to efficiently create a map book through a combination of **Data Driven Pages** along with an `arcpy.mapping` script. With a single map document file, you can use the **Data Driven Pages** toolbar to create a basic series of maps using the layout view along with your operational data and an index layer. The index layer contains features that will be used to define the extent of each map in the series. However, if you need to include additional pages in the map book, including a title page, an overview map, and other ancillary pages, you'll need to combine the output from the **Data Driven Pages** toolbar with the functionality provided by the `arcpy.mapping` module. With the `arcpy.mapping` module, you can automate the export of the map series and append the ancillary files into a single map book document. While it is certainly programmatically to generate the entire map book using only Python and the `arcpy.mapping` module, it is more efficient to use a combination of programming and the `Data Driven Pages` toolbar.

4
Analyzing Crime Patterns with ArcGIS Desktop, ArcPy, and Plotly (Part 1)

This is the first of two chapters that will cover the creation of crime analysis tools, using a combination of ArcGIS Desktop with `arcpy`, `arcpy.mapping`, and `arcpy.da` along with the Python requests and `plotly` modules. Data for the application will be pulled from the Seattle Open Data initiative, which contains crime data, among many other datasets. The `Socrata` API will be used to request the crime data that will be used in our analysis tools.

Three tools will be built in this chapter and added to a custom ArcGIS Python Toolbox. The initial focus of this chapter will be the construction of a tool that connects to the open database using the Python `requests` module with the `Socrata` API to request and receive data. The data will be written to a local geodatabase feature class. A second tool will take the imported records and aggregate to boundary datasets, such as census block groups, police precincts, and neighborhood boundaries. Finally, to automate the process of creating, printing, and exporting maps, we'll create a tool that makes this process easier.

In this chapter, we will cover the following topics:

- Creating ArcGIS Desktop Python toolboxes
- Using the Python requests module
- Accessing an open source database using the Socrata API
- Inserting data in a feature class using the ArcPy data access module
- Automating the process of creating, exporting, and printing maps using ArcPy mapping
- Using Spatial Statistics tools

Design

Let's spend a little time going over the design of what we're going to build in this chapter. This application will be contained within an ArcGIS Python Toolbox called CrimeAnalysis.pyt. Inside the toolbox, three tools will be created, including ImportRecords, AggregateCrimes, and CreateMap. The ImportRecords tool will use the Python requests module to request crime data from the Seattle Police Department open database using the Socrata API. Crime data will be returned to the tool and then written to a local SeattleCrimes geodatabase using the arcpy.da module. The AggregateCrimes tool will use these imported point feature classes and aggregate them to polygon boundary layers, including census block groups, police precincts, and neighborhood boundaries. Finally, the CreateMap tool will allow the end user to select one of the boundary files that include aggregated crime data and automate the process of creating, exporting, and printing maps, as shown in the following screenshot:

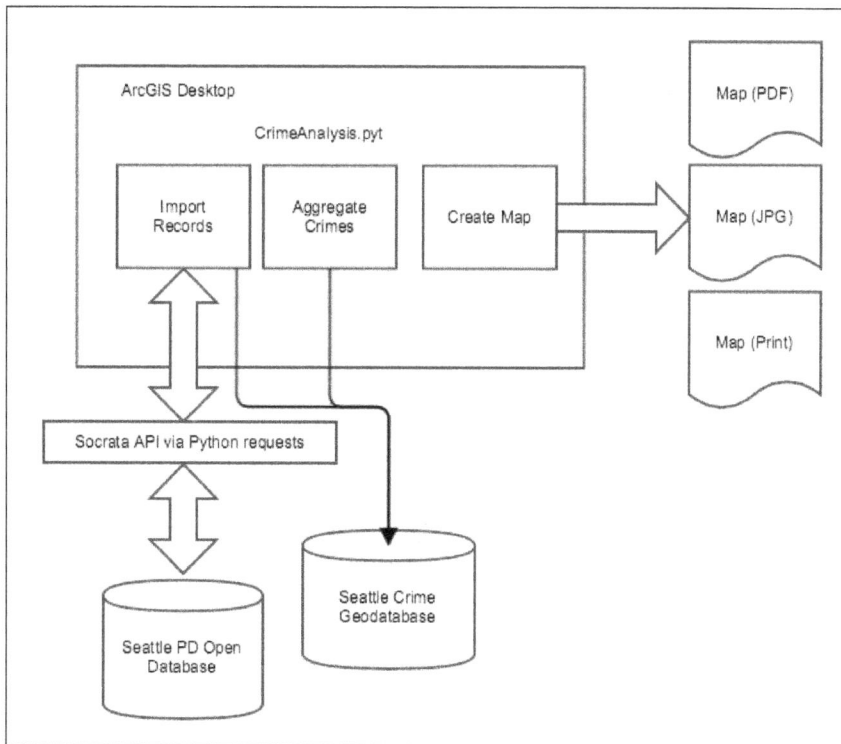

Let's get started and build the application.

Creating the Import Records tool

In this first section, we'll build a Python Toolbox for the crime analysis tools that we'll create over the course of the next two chapters, and we'll create the **Import Records** tool. By now, you should be comfortable with the basic process of creating an ArcGIS Python Toolbox, so I will provide only a minimum set of instructions to create the toolbox. If needed, refer to the first chapter for the specifics of how to create an ArcGIS Python Toolbox.

The **Import Records** tool, which will be created in this section, will dynamically **Import Records** from an online, open records dataset provided by the city of Seattle, WA. This dataset will be accessed through the `Socrata Open Data` API using the Python `requests` module. For this tool, we'll include several parameters, including start and end dates to filter the records, a filter for the crime type, an output feature class where the records will be written, and an optional parameter to filter by police district. The following steps will guide you through the creation of the **Import Records** tool:

1. We will be using open records that are provided by the city of Seattle, WA, and are accessible through their website, `https://data.seattle.gov`, using the `Socrata Open Data` API shown in the following screenshot. Open a browser and navigate to the city of Seattle, WA:

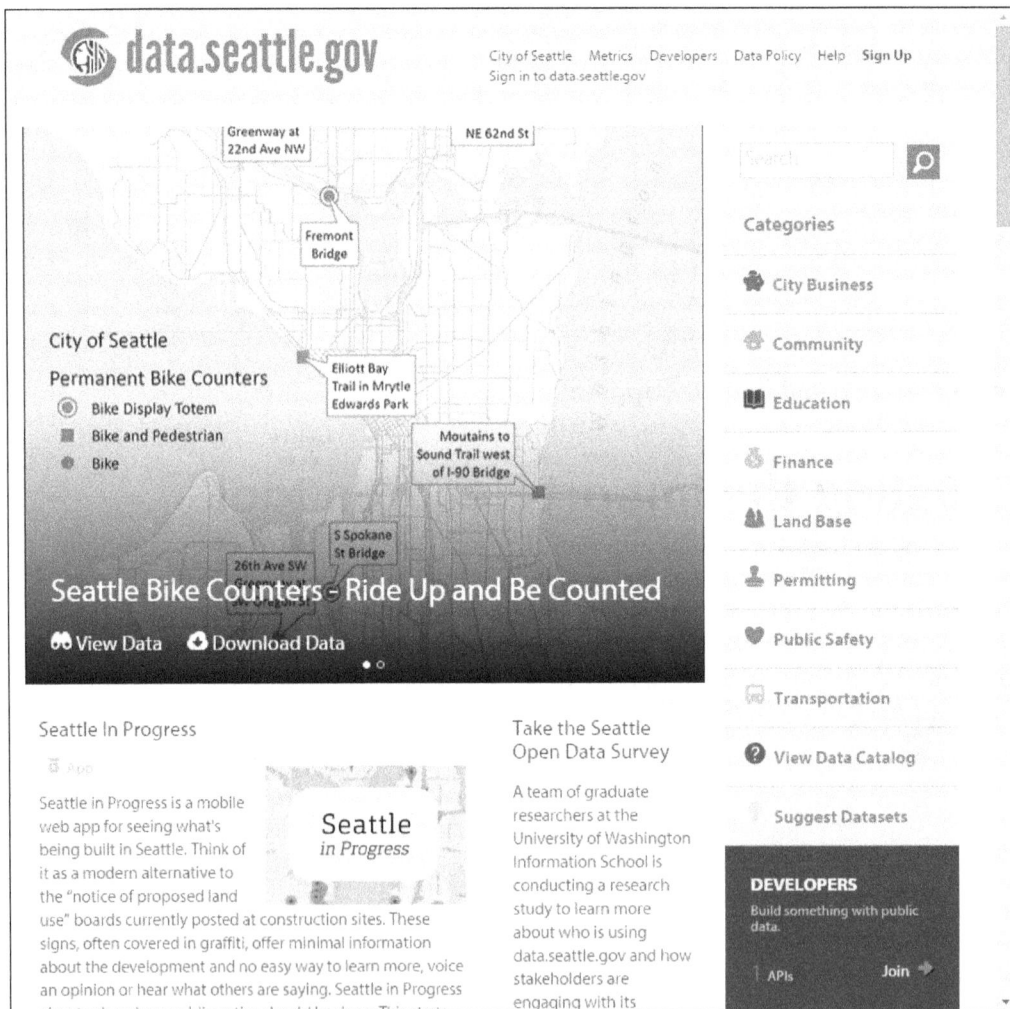

2. Before you can access the `Socrata Open Data` API, you will need to create a developer account. Click on the **Sign Up** link near the top-right of the webpage and fill out the details, as shown in the following screenshot:

3. The `Seattle Police Department Police Report Incident` API will be used to access the data, and it can be found at `http://dev.socrata.com/foundry/#/data.seattle.gov/y7pv-r3kh`.

4. An **App Token** will be needed when you submit requests through the API. A button with the **Sign up for App Token!** text is located about halfway down the page. This button will open the page shown in the following screenshot. You'll need to log in with the **Socrata ID** you created in the preceding few steps:

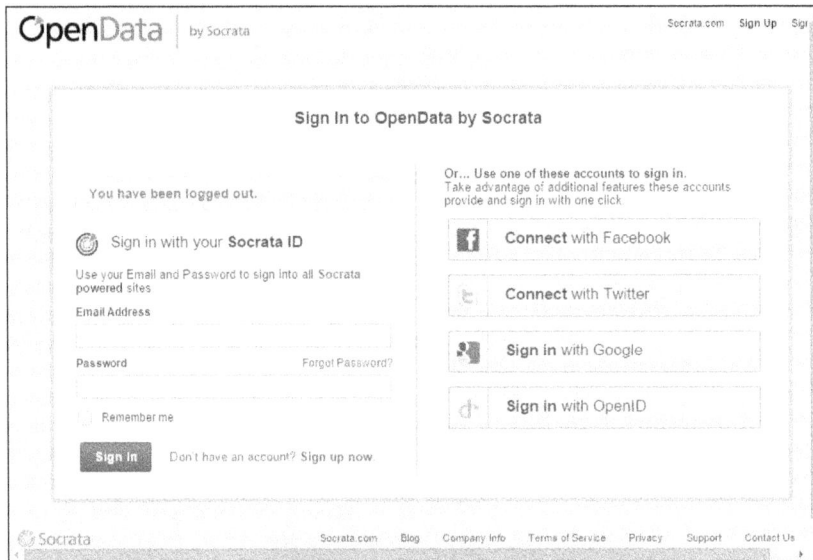

5. Click on **Edit Account Settings**, as shown in the following screenshot:

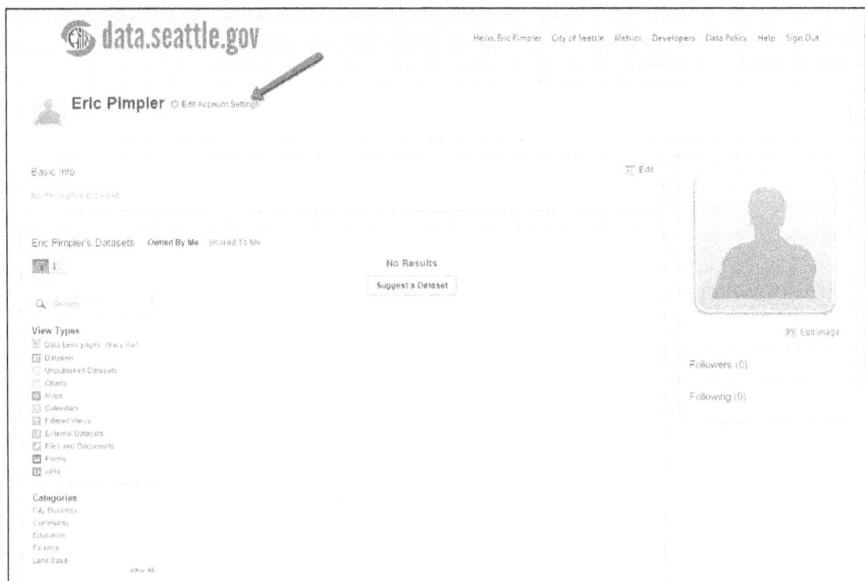

6. Navigate to **App Tokens | Create New Application**, as shown in the following screenshot, and enter a name and description for the application. You will need to enter a unique application name that has not already been created by someone else. For example, you might want to call the application **Crime Analysis** `<your name>`:

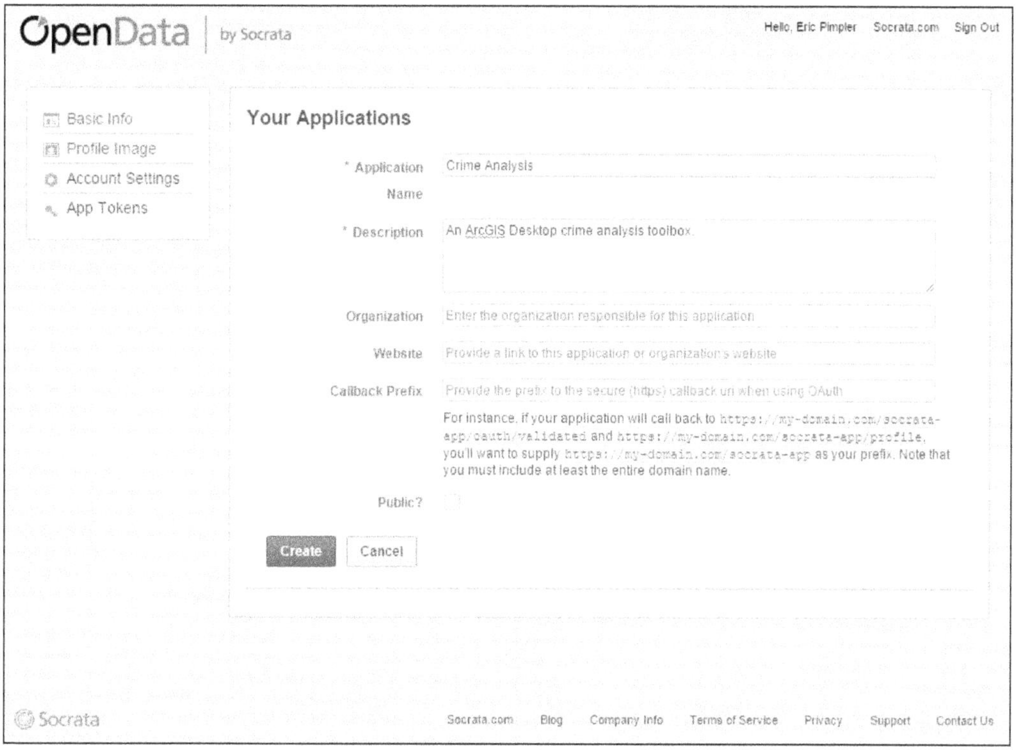

7. Click on the **Create** button to create the application.

8. Click on **App Tokens** to see the application token, as shown in the following screenshot. Note that your **App Token** will not be the same as mine. We'll use this **App Token** value in a later step, when we construct the query that imports records. Your **App Token** might look like the following screenshot:

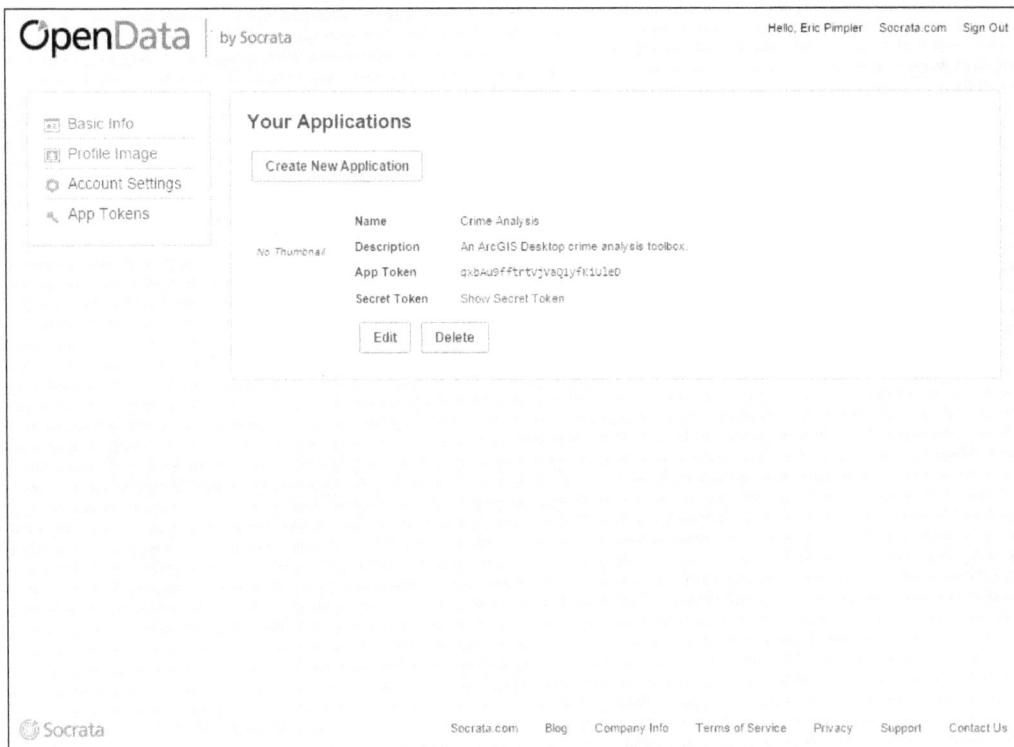

9. Open **ArcMap** and display the **ArcCatalog** pane. In the **Toolboxes** folder under **My Toolboxes**, create a new **Python Toolbox** and call it `CrimeAnalysis.pyt`, as shown in the following screenshot:

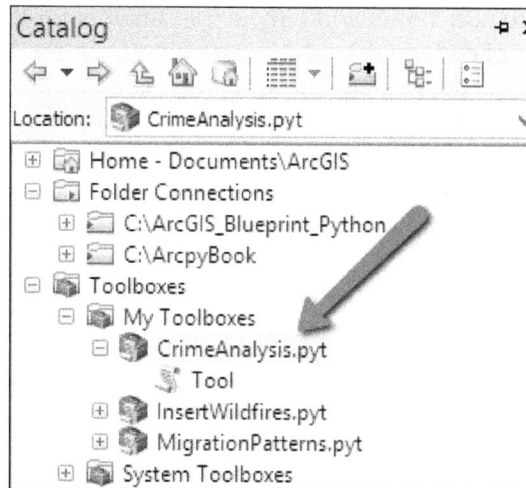

10. Open the code for the toolbox in your Python development environment.

11. Rename the `Tool` class `ImportRecords`. This tool will be used to import Seattle Police Department records using a REST API. Also, update the `self.label` and `self.description` properties, as shown in the following code:

```
class ImportRecords(object):
    def __init__(self):
        """Define the tool (tool name is the name of the
class)."""
        self.label = "Import Records"
        self.description = "Imports police records from Seattle PD
REST API"
        self.canRunInBackground = False
```

12. Add import statements for the `requests`, `json`, `datetime`, and `os` modules, as shown in the following code:

```
import arcpy
import requests
import json
import datetime
import os
```

13. Now it's time to add `input` and `output` parameters for the tool. Add the following lines of code to the `getParameterInfo()` method. The first parameter, `outFC`, defines the output feature class that will be created. The second parameter, `schemaFC`, defines an input feature class that can be used to extract schema information, as shown in the following code:

```
def getParameterInfo(self):
        """Define parameter definitions"""

        outFC = arcpy.Parameter(displayName = "Output Feature
Class",
            name="outFC",
            datatype="DEFeatureClass",
            parameterType="Required",
            direction="Output")

        schemaFC = arcpy.Parameter(displayName = "Schema Feature
Class",
            name="schemaFC",
            datatype="DEFeatureClass",
            parameterType="Required",
            direction="Input")
```

14. Two input parameters need to be created in order to define the `start` and `end` dates for the import parameter. Together, these two parameters will define a data range that can be used as a filter for the returned records. Add the following lines of code just below the parameters you created in the last step. The first parameter is `beginDate`, which captures the start date to be used for the filter; a default value `01/01/2014` is set. The second parameter is `endDate`, which captures the end date for the filter, and the initial value for this parameter is set to the current date:

```
## begin date for import
beginDate = arcpy.Parameter(
    displayName="Begin Crime Date",
    name="beginDate",
    datatype="GPDate",
    parameterType="Required",
    direction="Input")
beginDate.value = "01/01/2014"
```

```
## end date for import
endDate = arcpy.Parameter(
    displayName="End Crime Date",
    name="endDate",
    datatype="GPDate",
    parameterType="Required",
    direction="Input")
endDate.value = str(datetime.date.today())
```

15. The next parameter will be used to filter by crime type. To keep things simple, we'll limit the list of crime types to a small subset of the possibilities. This will be an input parameter presented as a combobox of values that can be selected. Add the following code just below the parameters you have already created. The final line of code for this parameter defines the list of possible values for the parameter. These values will populate the combobox:

```
## crime type
crimeType = arcpy.Parameter(
    displayName="Crime Type",
    name="crimeType",
    datatype="String",
    multiValue="False",
    parameterType="Required",
    direction="Input")
crimeType.filter.list = ["ASSAULT", "BURGLARY", "DUI",
"PROSTITUTION", "VEHICLE THEFT"]
```

16. The final parameter will be an optional parameter that will allow the end user to filter by police district. It will be similar to the parameter created in the last step in that it will be a combobox of possible values. However, this will be an optional parameter. Add the following code just below the parameter you created in the previous step:

```
district = arcpy.Parameter(
    displayName="Filter by District",
    name="district",
    datatype="String",
    parameterType="Optional",
    direction="Input",
    multiValue = False)
district.filter.list = ["B","C","D","E","F","G","J","K","L","M","N
","O","Q","R","S","U","W"]
```

17. Finally, add each of the parameters to the `params` list:

```
params = [outFC, schemaFC, beginDate, endDate, crimeType,
district]
return params
```

18. Add the tool to the `self.tools[]` list inside the `Toolbox` class:

```
self.tools[ImportRecords]
```

19. Find the `execute()` method inside the `ImportRecords()` class. This method will contain the functionality of the tool.

20. Add the following lines of code to capture the input variables submitted to the tool:

```
def execute(self, parameters, messages):
        """The source code of the tool."""
        outFC = parameters[0].valueAsText
        schemaFC = parameters[1].valueAsText
        beginDate = parameters[2].valueAsText
        endDate = parameters[3].valueAsText
        crimeType = parameters[4].valueAsText
        policeDistrict = parameters[5].valueAsText
```

21. Add the try/except exception-handling structures just below the code you just added. The rest of the code in this section should go inside the try block:

```
try:
except Exception as e:
    arcpy.AddMessage(e.message)
```

22. Create a new feature class that will be the container for the crime records. The end user in one of the input parameters defines the feature class location and name:

```
try:
    outCS = arcpy.SpatialReference(4326)
    arcpy.CreateFeatureclass_management(os.path.split(outFC)
[0], os.path.split(outFC)[1], "point", schemaFC, spatial_
reference=outCS))
```

23. Format the start and end dates:

```
try:
    #format the dates
    beginDate = datetime.datetime.strptime(beginDate, '%m/%d/%Y'
strftime('%Y-%m-%d')
    endDate = datetime.datetime.strptime(endDate, '%m/%d/%Y').
strftime('%Y-%m-%d')
```

24. Following highlighted lines of code, add the following line of code that defines the URL request that we'll submit through the API:

```
socrataURL = "https://data.seattle.gov/resource/y7pv-r3kh.
json?$$app_token=<your app token>&$where=occurred_date_or_
date_range_start between '" + str(beginDate) + "' and '" +
str(endDate) + "'&summarized_offense_description=" + crimeType +
"&$limit=10000"
```

The URL can be broken down into several parts that make it easier to understand. The first part of the URL (`https://data.seattle.gov/resource/y7pv-r3kh.json?`) is the API endpoint for the resource being accessed along with the requested data format (`json`). In this case, this is the Seattle Police Department's Police Report Incident database. What follows the `?` character is a sequence of parameters, with the first parameter, (`?$$app_token=<your app token>`), being the application token. This is where you'll enter the application token that you created earlier in this section. Next is the where clause parameter (`&$where=occurred_date_or_date_range_start between '" + str(beginDate) + "' and '" + str(endDate) + "'&summarized_offense_description=" + crimeType +`), which filters records between the start and end dates along with the crime type. Parameters in a URL query string are always separated by an `&` character. The final parameter, (`"&$limit=10000"`), sets the maximum number of records that can be returned by the query.

25. There is also an optional input parameter that allows the end user to filter records by the police beat. Because it's optional, we'll need to conditionally add this as a parameter to the URL. Do this by adding the following lines of code just below the last line:

```
if policeDistrict is not None:
    socrataURL = socrataURL + "&district_sector=" + policeDistrict
```

26. Next, use the Python `requests` module to submit the URL and the `json` module to convert the returned data from `json` format to a Python dictionary. The `json.loads()` method performs the conversion to a Python dictionary. The `crimes` variable holds the returned records in a dictionary:

```
r = requests.get(socrataURL)
crimes = json.loads(r.text)
```

27. The next section of code will process the returned records. Create an `InsertCursor` object to handle the insertion of the returned records in the feature class. The `InsertCursor()` constructor is passed a reference to the output feature class (`outFC`), along with a list of fields to be included, and is saved to a variable called `cursor`. The object is assigned to the `cursor` variable and contains the following:

```
with arcpy.da.InsertCursor(outFC,("SHAPE@XY", "DISTRICT_SECTOR",
"BEAT", "DATE_REPORTED", "MONTH", "YEAR")) as cursor:
```

28. Inside the `WITH` statement, create a variable that will serve as a counter, loop through each of the records in the Python dictionary, and pull out the geometry and attribute information that will be written to the feature class:

```
with arcpy.da.InsertCursor(outFC,("SHAPE@XY", "DISTRICT_SECTOR",
"BEAT", "DATE_REPORTED", "MONTH", "YEAR")) as cursor:
    cntr = 1
    for crime in crimes:

        if 'latitude' in crime:
            latitude = float(crime['latitude'])
        else:
            break

        if 'longitude' in crime:
            longitude = float(crime['longitude'])else:
            break

        if 'district_sector' in crime:
            district_sector = crime['district_sector']
        else:
            break

        if 'zone_beat' in crime:
            zone_beat = crime['zone_beat']
        else:
            break

        if 'date_reported' in crime:
            date_reported = crime['date_reported']
        else:
            break

        if 'month' in crime:
```

```
        month = crime['month']

else:
    break

if 'year' in crime:
    year = crime['year']
else:
    break
```

29. Define a new `row_value` variable that holds the geometry and attribute information for the new row. Insert the row into the `cursor` variable using the `insertRow()` method, add a message indicator to the progress dialog box, and update the counter. These lines should go just below the last `if-else` statement and line up exactly with those statements:

```
row_value = [(longitude,latitude),district_sector, zone_beat,
date_reported, month, year]
cursor.insertRow(row_value)
arcpy.AddMessage("Record number " + str(cntr) + " written to
feature class")
cntr = cntr + 1
```

30. The entire code for the `execute()` method should appear as follows:

```
def execute(self, parameters, messages):
        """The source code of the tool."""
        outFC = parameters[0].valueAsText
        schemaFC = parameters[1].valueAsText
        beginDate = parameters[2].valueAsText
        endDate = parameters[3].valueAsText
        crimeType = parameters[4].valueAsText
        policeDistrict = parameters[5].valueAsText

        try:
            outCS = arcpy.SpatialReference(4326)
            arcpy.CreateFeatureclass_management(os.path.
split(outFC)[0], os.path.split(outFC)[1], "point", schemaFC,
spatial_reference=outCS)

            #format the dates
            beginDate = datetime.datetime.strptime(beginDate,
'%m/%d/%Y').strftime('%Y-%m-%d')
            endDate = datetime.datetime.strptime(endDate,
'%m/%d/%Y').strftime('%Y-%m-%d')
```

```
        #arcpy.AddMessage("https://data.seattle.gov/resource/
y7pv-r3kh.json?$where=occurred_date_or_date_range_start between
'" + str(beginDate) + "' and '" + str(endDate) + "'&summarized_
offense_description=" + crimeType + "&district_sector=" +
policeDistrict)

        socrataURL = "https://data.seattle.gov/resource/y7pv-
r3kh.json?$$app_token=qxbAu9fftrtVjVaQ1yfKiUleD&$where=occurred_
date_or_date_range_start between '" + str(beginDate) + "' and '"
+ str(endDate) + "'&summarized_offense_description=" + crimeType +
"&$limit=10000"
        if policeDistrict is not None:
            socrataURL = socrataURL + "&district_sector=" +
policeDistrict

        r = requests.get(socrataURL)
        crimes = json.loads(r.text)
        with arcpy.da.InsertCursor(outFC,("SHAPE@XY",
"DISTRICT_SECTOR", "BEAT", "DATE_REPORTED", "MONTH", "YEAR")) as
cursor:
            cntr = 1
            for crime in crimes:

                if 'latitude' in crime:
                    latitude = float(crime['latitude'])
                else:
                    break

                if 'longitude' in crime:
                    longitude = float(crime['longitude'])
                else:
                    break

                if 'district_sector' in crime:
                    district_sector = crime['district_sector']
                else:
                    break

                if 'zone_beat' in crime:
                    zone_beat = crime['zone_beat']
                else:
                    break
```

```
                if 'date_reported' in crime:
                    date_reported = crime['date_reported']
                else:
                    break

                if 'month' in crime:
                    month = crime['month']
                else:
                    break

                if 'year' in crime:
                    year = crime['year']
                else:
                    break

                row_value = [(longitude,latitude),district_
sector, zone_beat, date_reported, month, year]

                cursor.insertRow(row_value)    # Inserts the
row into the feature class
                arcpy.AddMessage("Record number " + str(cntr)
+ " written to feature class") # Adds message to the progress
dialog
                cntr = cntr + 1
        except Exception as e:
            arcpy.AddMessage(e.message)
```

31. You can check your work by examining the `C:\ArcGIS_Blueprint_ Python\solutions\ch4\CrimeAnalysis.py` solution file. Refer to the `getParameterInfo()` and `execute()` methods.

32. Save the file and exit your Python development environment.

33. Now it's time to test the tool. In **ArcMap**, add a **basemap** by selecting the **Add Basemap** tool and selecting the **Streets** layer, as shown in the following screenshot:

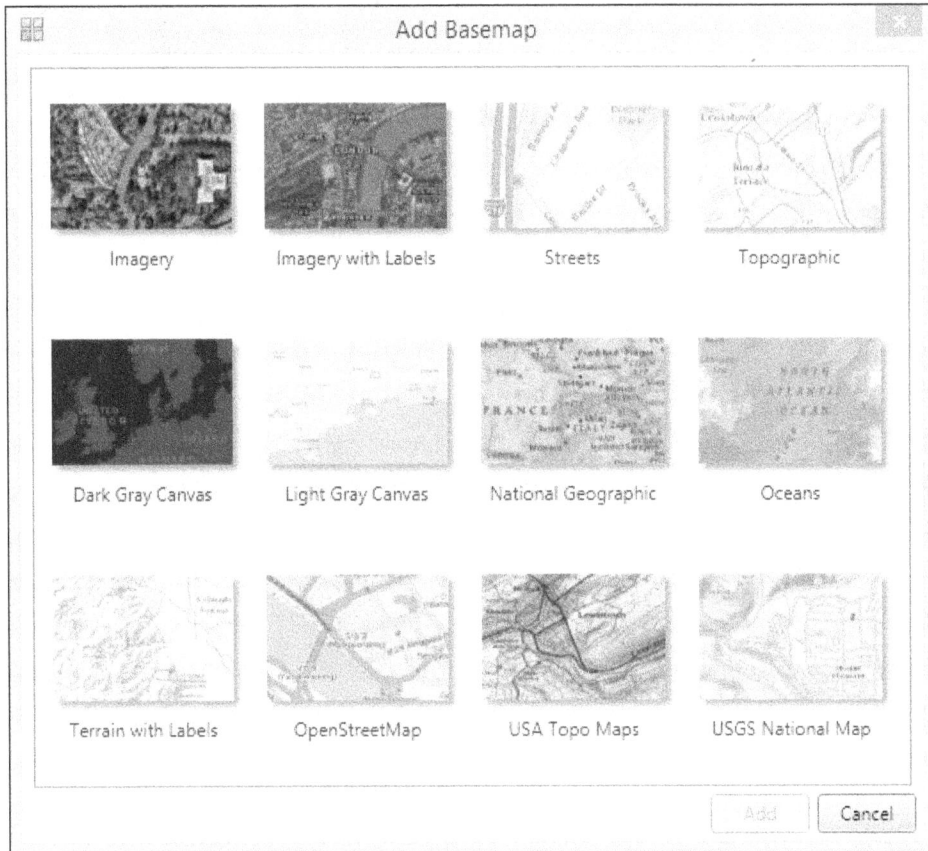

34. Zoom in to **Seattle, WA**.

35. In your C:\ArcGIS_Blueprint_Python\data\crime folder is a shape file called Seattle_BG. This layer contains census block groups. Add this dataset to the map document. Symbolize the layer if you'd like.

36. Double-click on the **Import Records** tool from the `CrimeAnalysis.pyt` toolbox to display the tool, as shown in the following screenshot:

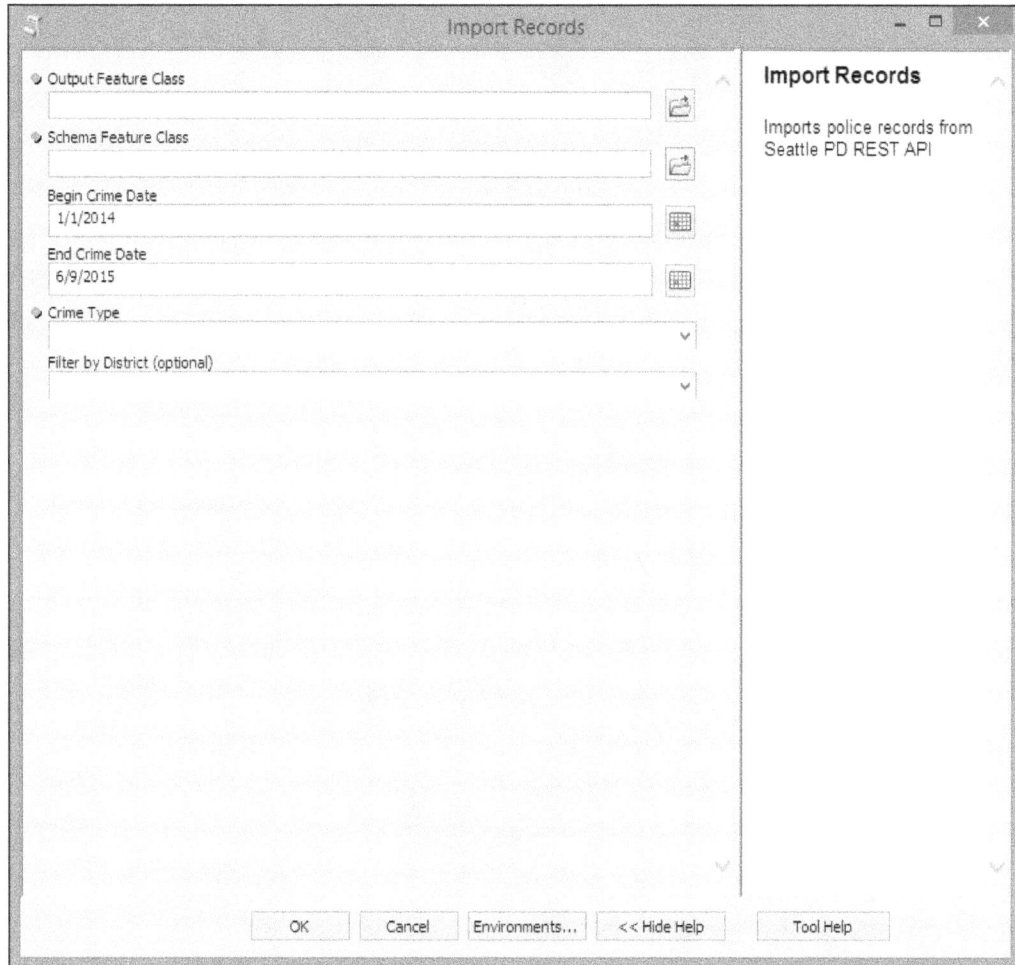

37. For this test, the tool will be used to import **Burglary** records for the first few months of 2015. In the `C:\ArcGIS_Blueprint_Python\data\crime` folder, there is a geodatabase file called `SeattleCrimeAnalysis`. Add a feature class called `Burglary_2015` to this geodatabase. There is a pre-created feature class called `CrimeSchema` in the same geodatabase. Use this for **Schema Feature Class**. Define a start date of **1/1/2014** and an end date of **6/9/2015**. Finally, select `Burglary` as the **Crime Type**. Don't define the optional filter **Filter by District (optional)** parameter for this test.

38. Your tool should now appear as follows:

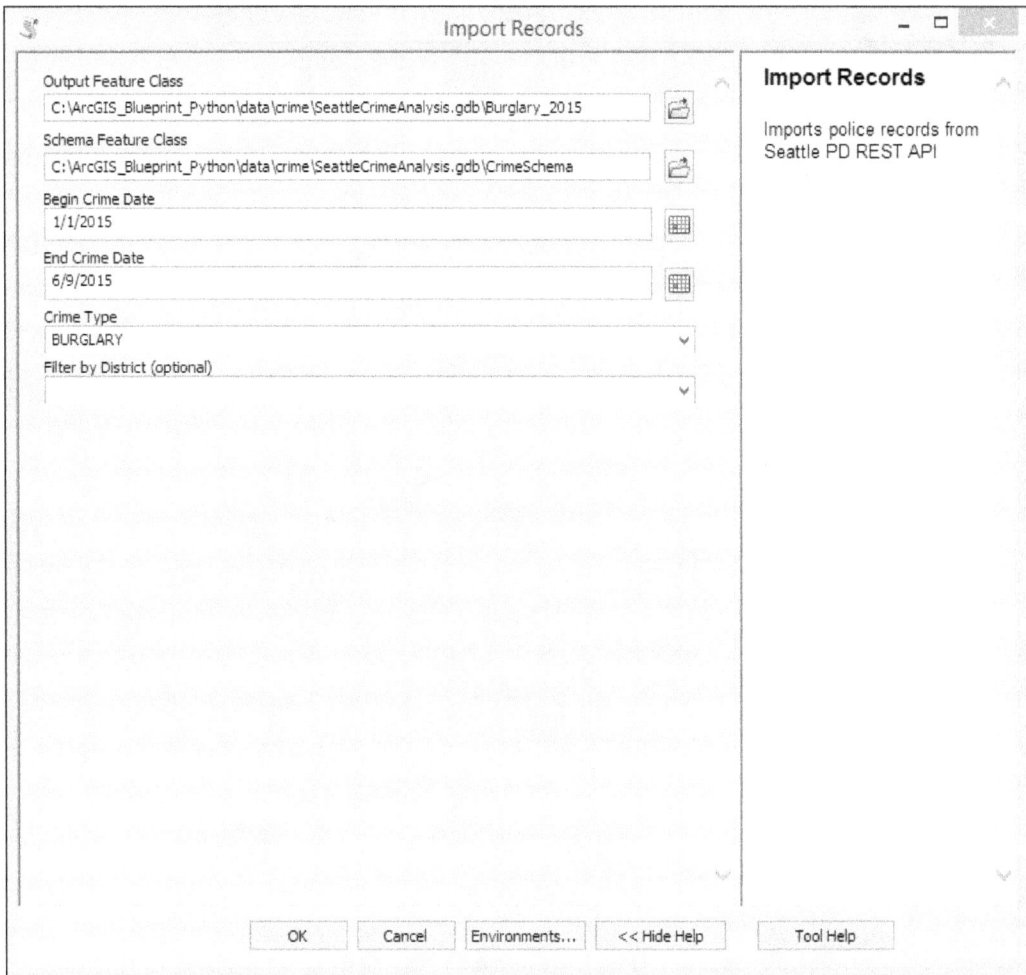

39. Click on **OK**.

40. If everything goes as expected, the data should be imported into the feature class and added to the table of contents, as shown in the following screenshot:

41. Save your map document as `C:\ArcGIS_Blueprint_Python\ch4\SeattleCrimes.mxd`.

In the next section, we'll create a tool that aggregates these points.

Creating the Aggregate Crimes tool

The **Aggregate Crimes** tool will aggregate the crimes to a polygon layer, such as census block groups or police precinct boundaries. This tool will use the existing SpatialJoin tool found in the **Analysis Tools** toolbox to summarize the total number of crimes in each polygon. Two polygon feature classes have been provided for you: Seattle_BG and Seattle_Merge_Precincts. The former contains census block groups for Seattle, while the later contains police precincts for Seattle. Both datasets are in the C:\ArcGIS_Blueprint_Python\data\crime folder. The **Aggregate Crimes** tool will prompt the user to select a polygon layer, crime dataset, and output feature class; it will then perform a spatial join:

1. Open C:\ArcGIS_Blueprint_Python\ch4\SeattleCrimes.mxd in **ArcMap**.

2. Duplicate the code that you have already created for the ImportRecords class by copying and pasting this class into the same CrimeAnalysis.pyt file.

3. Rename the duplicated ImportRecords class as **AggregateCrimes**.

4. Remove the code inside the getParameterInfo() and execute() methods for the new AggregateCrimes class.

5. Alter the self.label and self.description properties in the __init__ method, as shown in the following code:

    ```
    def __init__(self):
            """Define the tool (tool name is the name of the
    class)."""
            self.label = "Aggregate Crimes"
            self.description = "Aggregates crime points to a polygon
    layer"
            self.canRunInBackground = False
    ```

6. This tool will include three parameters. The first two will be input parameters that allow the end user to select the polygon feature class where the crimes will be aggregated, as well as the point feature class containing the crime information. The final parameter will be the output feature class where the aggregated records containing the summarized information will be written. Add the following code block to the getParameterInfo() method. By now, you should have a good understanding of how to create the parameters. One thing that may be new, though, is the use of the filter.list property that was used in this case to filter the feature classes so that only specific feature types are displayed for a particular parameter:

```
def getParameterInfo(self):
        """Define parameter definitions"""

        aggregateFC = arcpy.Parameter(displayName = "Boundary
Layer",
            name="aggregateFC",
            datatype="DEFeatureClass",
            parameterType="Required",
            direction="Input")
        aggregateFC.filter.list = ['Polygon']

        crimeFC = arcpy.Parameter(displayName = "Crime Point
Locations",
            name="crimeFC",
            datatype="DEFeatureClass",
            parameterType="Required",
            direction="Input")
        crimeFC.filter.list = ['Point']

        outputFC = arcpy.Parameter(displayName = "Output Feature
Class",
            name="outputFC",
            datatype="DEFeatureClass",
            parameterType="Required",
            direction="Output")
        outputFC.filter.list = ['Polygon']

        params = [aggregateFC, crimeFC, outputFC]
        return params
```

7. In the `execute()` method, add the following lines of code to accept the parameter information submitted by the user:

```
def execute(self, parameters, messages):
        """The source code of the tool."""
        polygonFC = parameters[0].valueAsText
        pointFC = parameters[1].valueAsText
        outputFC = parameters[2].valueAsText
```

8. Add the following code block to call the `SpatialJoin` tool:

```
def execute(self, parameters, messages):
        """The source code of the tool."""
        polygonFC = parameters[0].valueAsText
        pointFC = parameters[1].valueAsText
        outputFC = parameters[2].valueAsText

        try:
             arcpy.SpatialJoin_analysis (polygonFC, pointFC,
outputFC)
        except Exception as e:
             arcpy.AddMessage(e.message)
```

9. In the `Toolbox` class, add the `AggregateCrimes` tool to the `self.tools` list, as shown in the following code:

```
self.tools = [ImportRecords, AggregateCrimes]
```

10. You can check your work by examining the `C:\ArcGIS_Blueprint_Python\solutions\ch4\AggregateCrimes.py` solution file. Refer to the `AggregateCrimes` class.

11. Save the file and exit your Python development environment.

12. Now, it's time to test the tool. If necessary, open **ArcMap** and display the contents of the `CrimeAnalysis` toolbox. You should now see the **Aggregate Crimes** tool, as shown in the following screenshot:

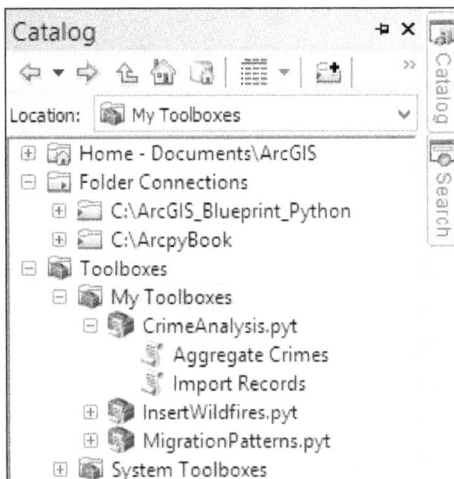

13. Double-click on the tool to display the tool dialog, which is shown as follows:

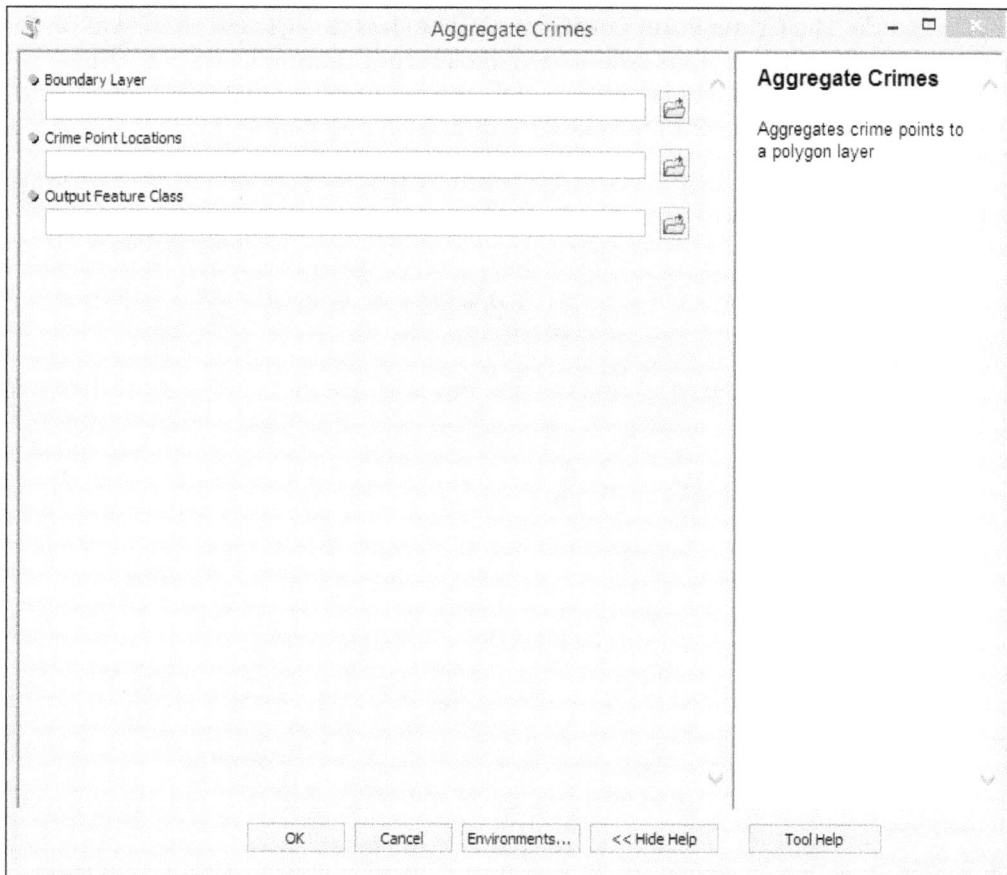

14. Add the references to the layers shown in the following screenshot.
    Boundary Layer (Seattle_BG.shp) contains block groups for the city of
    Seattle. The **Crime Point Locations** feature class (Burglary_2015) was
    created in the previous section, and the **Output Feature Class** (Seattle_BG_
    Burglary_2015) will be the new polygon feature class that will contain a
    count of the number of burglaries within each polygon.

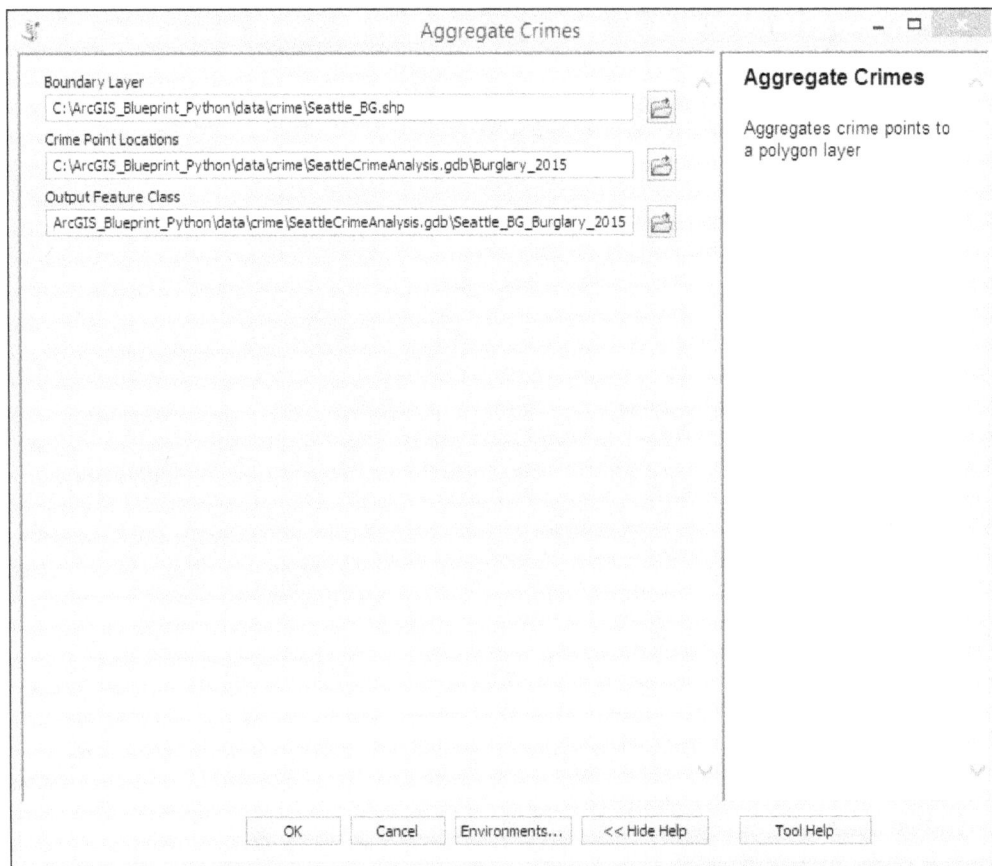

15. Click on **OK** to execute the tool. The new feature class (Seattle_BG_
    Burglary_2015) will be added to the table of contents. If you open the
    attribute table, you should see a new column called Join_Count. This is the
    count of the number of burglaries that are within each block group.

16. Symbolize the `Seattle_BG_Burglary_2015` layer by double-clicking on the layer and selecting the **Symbology** tab. Select **Graduated Colors** from the **Quantities** item and then select `Join_Count` as the **Value** field and `Shape_Area` as the **Normalization** field. Normalization is the process of dividing one numeric attribute value by another to minimize the differences in values based on the size of areas or the number of features in each area. In our case, normalizing (dividing) total crimes by the total polygon area yields crimes per unit area or density. The following screenshot illustrates how the symbolization should be applied:

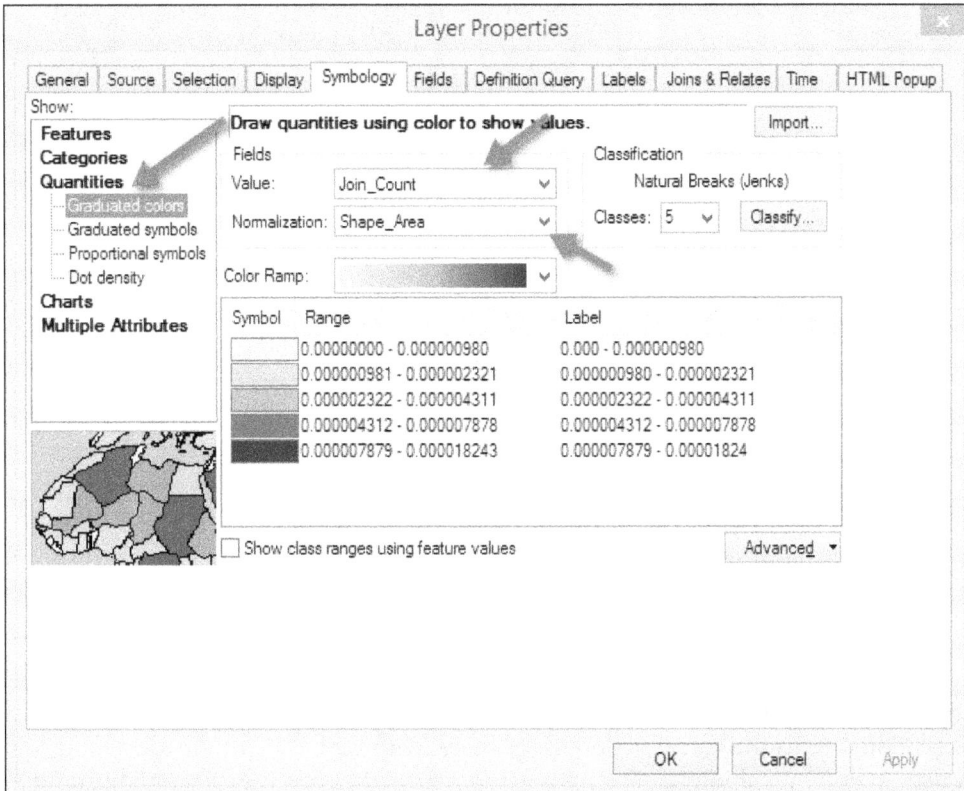

17. The resulting graduated color map should appear as shown in the following screenshot. Save your map document file.

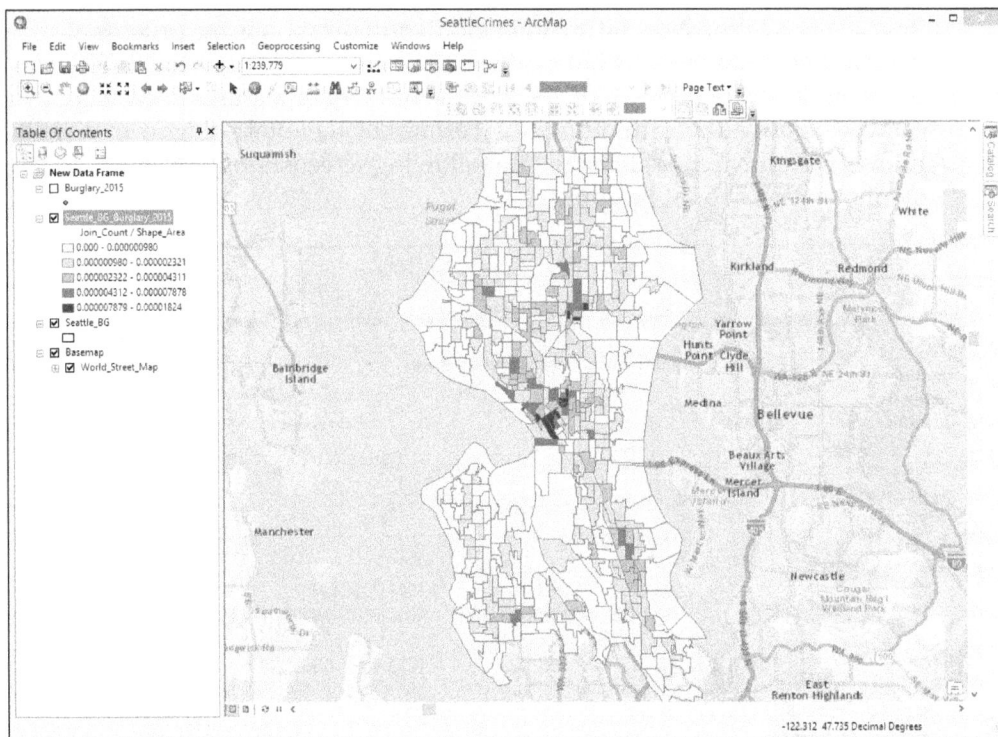

We'll use this dataset in the upcoming sections of the chapter, when we create a tool to automate the creation of maps and more advanced spatial statistical analysis.

Building the Create Map tool

This section details the construction of a **Create Map** tool, which will allow the end user to generate and export a map using a predefined layer file containing the symbology and a selected polygon feature class, which in turn will contain a count field (`Join_Count`) that was created with the **Aggregate Crimes** tool:

1. If necessary, open `C:\ArcGIS_Blueprint_Python\ch4\SeattleCrimes.mxd` in **ArcMap**.

2. Switch to the **Layout** view so that we can build the structure of the map that will be exported.

3. Build the layout so that it appears similar to what is shown in the following screenshot:

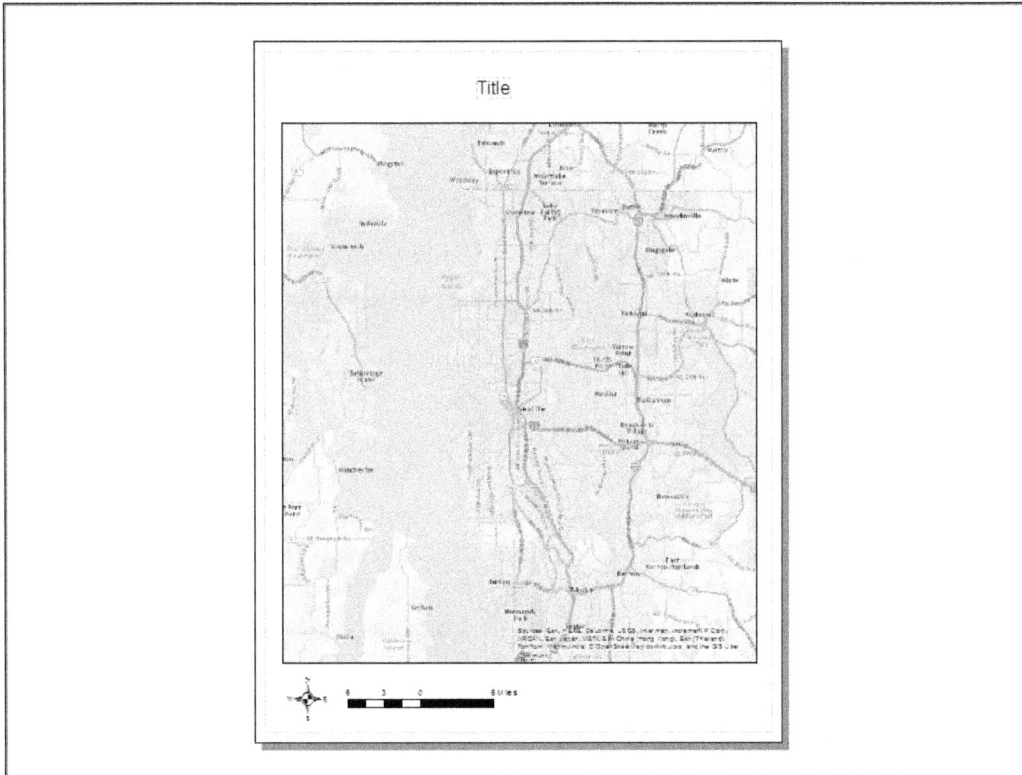

4. Double-click on the **Title** text element to display the properties.

5. Click on the **Size** and **Position** tab and set **Element Name** to CrimeTitle, as shown in the following screenshot. Setting **Element Name** will enable us to access this element through a Python script that will dynamically set the name based on the user input.

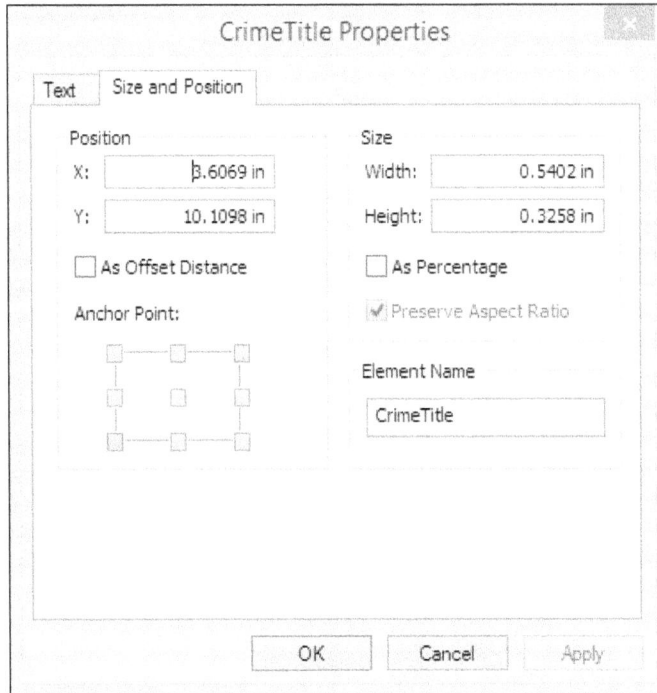

6. Save the map document.

7. From the **Catalog** window, open the Python development environment for the CrimeAnalysis.pyt toolbox.

8. Create a new tool by copying and pasting one of the existing tools. Rename the tool CreateMap and remove any existing code from the getParameterInfo() and execute() methods.

9. Update the __init__ method, as shown in the following code:

```
class CreateMap(object):
    def __init__(self):
        """Define the tool (tool name is the name of the
class)."""
```

```
        self.label = "Create Map"
        self.description = "Creates a map in layout view and
exports"
        self.canRunInBackground = False
```

10. This tool will have four parameters, which are used to capture the input polygon crime layer, the map type to be created, the export/print type, and the map title. Add the parameters to capture the input polygon, as shown in the following code:

```
def getParameterInfo(self):
        """Define parameter definitions"""

    crimeFC = arcpy.Parameter(displayName = "Input Crimes to Map",
        name="crimeFC",
        datatype="GPFeatureLayer",
        parameterType="Required",
        direction="Input")
    crimeFC.filter.list = ['Polygon']
```

Setting datatype to GPFeatureLayer will allow the end user to either select a layer from a drop-down box containing the polygon feature classes already in the table of contents, or click on a button to navigate to a polygon feature class on the computer but not in the map document. This feature class will be the class that was created using the **Aggregate Crimes** tool.

11. Next, add the map type parameter to allow the end user to select either a graduated color or a graduated symbol map type. Two layer files containing the symbology for both the types have already been created for you. They are located in your C:\ArcGIS_Blueprint_Python\data\ crime folder and are called SeattleCrimeGraduatedColor.lyr and SeattleCrimeGraduatedSymbol.lyr:

```
        ## map type
        mapType = arcpy.Parameter(
            displayName="Map Type",
            name="mapType",
            datatype="String",
            multiValue="False",
            parameterType="Required",
            direction="Input")
        mapType.filter.list = ["GRADUATED COLOR", "GRADUATED
SYMBOL"]
```

12. Add a parameter that will allow the user to select the type of export or print. The options will be PDF, JPEG, or Print. If the user selects Print, the geoprocessing script will send the layout view to the default printer associated with the computer in which the map document file resides:

```
## export type
exportType = arcpy.Parameter(
    displayName="Export Type",
    name="exportType",
    datatype="String",
    multiValue="False",
    parameterType="Required",
    direction="Input")
exportType.filter.list = ["PDF", "JPEG", "PRINT"]
```

13. Finally, add a parameter that will allow the end user to create a title for the map:

```
## map title
mapTitle = arcpy.Parameter(
    displayName="Map Title",
    name="mapTitle",
    datatype="String",
    multiValue="False",
    parameterType="Required",
    direction="Input")
```

14. Add each of the parameters to the params variable:

```
params = [crimeFC, mapType, exportType, mapTitle]
```

15. Find the execute() method inside the CreateMap class.

16. Add the following code block to accept the input parameters:

```
def execute(self, parameters, messages):
    """The source code of the tool."""
    crimeFC = parameters[0].valueAsText
    mapType = parameters[1].valueAsText
    exportType = parameters[2].valueAsText
    mapTitle = parameters[3].valueAsText
```

17. Add a try/except block, as we've done with the other tools.

18. Inside the `try` block, set the workspace environment variable and path to the layer files:

```
def execute(self, parameters, messages):
        """The source code of the tool."""
        crimeFC = parameters[0].valueAsText
        mapType = parameters[1].valueAsText
        exportType = parameters[2].valueAsText
        mapTitle = parameters[3].valueAsText

        try:
            arcpy.env.workspace = 'C:/ArcGIS_Blueprint_Python/ch4'

            #path to the layer file
            if mapType == "GRADUATED COLOR":
                lyrFile = r"C:\ArcGIS_Blueprint_Python\data\crime\
SeattleCrimeGraduatedColor.lyr"
            else:
                lyrFile = r"C:\ArcGIS_Blueprint_Python\data\crime\
SeattleCrimeGraduatedSymbol.lyr"
```

19. Get a reference to the `map` document file and `data` frame:

```
if mapType == "GRADUATED COLOR":
    lyrFile = r"C:\ArcGIS_Blueprint_Python\data\crime\
SeattleCrimeGraduatedColor.lyr"
else:
    lyrFile = r"C:\ArcGIS_Blueprint_Python\data\crime\
SeattleCrimeGraduatedSymbol.lyr"

mxd = arcpy.mapping.MapDocument("CURRENT")
df = arcpy.mapping.ListDataFrames(mxd)[0]
```

20. The first parameter of the tool allows the end user to either select a polygon feature class from the table of contents or click on a button and navigate to the feature class to be selected. The block of code that you add in this step will handle situations where the user elects to click on the button, navigate to the feature class somewhere on the computer or network, and add that feature class as a layer to the map document. Add the code block just below the lines of code that you added in the previous step:

```
layerList = []
for lyr in arcpy.mapping.ListLayers(mxd, "", df):
    layerList.append(lyr.name)
if not crimeFC in layerList:
```

```
arcpy.AddMessage("Adding the layer to the map document")
addLayer = arcpy.mapping.Layer(crimeFC)
arcpy.mapping.AddLayer(df, addLayer, "TOP")
crimeFC = os.path.split(crimeFC)[1]
```

21. Execute some checks to make sure that the `feature` class selected by the user has a `Join_Count` field:

```
#make sure the feature class has a join_count field
fldList = []
for fld in arcpy.ListFields(crimeFC):
    fldList.append(fld.name)
if "Join_Count" in fldList:
else:
 arcpy.AddMessage("Feature class does not contain Join_Count
field...can't create map")
```

22. The rest of the code for this method will go inside the `IF` block that you just created. We have verified that the `Join_Count` field exists in the selected `feature` class. Use the `arcpy.mapping UpdateLayer()` function to update the symbology of the map to the appropriate layer file:

```
updateLayer = arcpy.mapping.ListLayers(mxd, crimeFC, df)[0]
sourceLayer = arcpy.mapping.Layer(lyrFile)
arcpy.AddMessage("Updating the symbology")
arcpy.mapping.UpdateLayer(df, updateLayer, sourceLayer, True)
```

23. Set the title in the layout view to the value input by the user:

```
#set the layout map title
for elm in arcpy.mapping.ListLayoutElements(mxd, "TEXT_ELEMENT",
"CrimeTitle"):
    elm.text = mapTitle
```

24. Add the code that will export or print the map based on the user input:

```
#export or print the map
arcpy.AddMessage("Exporting/Printing the map")
if exportType == "PDF":
    arcpy.mapping.ExportToPDF(mxd, mapTitle + ".pdf")
elif exportType == "JPEG":
    arcpy.mapping.ExportToJPEG(mxd, mapTitle + ".jpg")
else:
    arcpy.mapping.PrintMap(mxd)
```

25. Add a `message` to the progress dialog box and make sure the except block is complete:

```
arcpy.AddMessage("Processing complete")
except Exception as e:
    arcpy.AddMessage(e.message)
```

26. Add the tool to the `self.tools` property of the `Toolbox` class:

```
class Toolbox(object):
    def __init__(self):
        """Define the toolbox (the name of the toolbox is the name
of the
        .pyt file)."""
        self.label = "Crime Analysis"
        self.alias = "crimeanalysis"

        # List of tool classes associated with this toolbox
        self.tools = [ImportRecords, AggregateCrimes, CreateMap]
```

27. The entire `execute()` method should appear as follows:

```
def execute(self, parameters, messages):
    """The source code of the tool."""
    crimeFC = parameters[0].valueAsText
    mapType = parameters[1].valueAsText
    exportType = parameters[2].valueAsText
    mapTitle = parameters[3].valueAsText

    try:
        arcpy.env.workspace = 'C:/ArcGIS_Blueprint_Python/ch4'

        #path to the layer file
        if mapType == "GRADUATED COLOR":
            lyrFile = r"C:\ArcGIS_Blueprint_Python\data\crime\
SeattleCrimeGraduatedColor.lyr"
        else:
            lyrFile = r"C:\ArcGIS_Blueprint_Python\data\crime\
SeattleCrimeGraduatedSymbol.lyr"

        mxd = arcpy.mapping.MapDocument("CURRENT")
        df = arcpy.mapping.ListDataFrames(mxd)[0]
```

```
            layerList = []
            for lyr in arcpy.mapping.ListLayers(mxd, "", df):
                layerList.append(lyr.name)
            if not crimeFC in layerList:
                arcpy.AddMessage("Adding the layer to the map
document")

                addLayer = arcpy.mapping.Layer(crimeFC)
                arcpy.mapping.AddLayer(df, addLayer, "TOP")
                crimeFC = os.path.split(crimeFC)[1]

            #make sure the feature class has a join_count field
            fldList = []
            for fld in arcpy.ListFields(crimeFC):
                fldList.append(fld.name)
            if "Join_Count" in fldList:

                updateLayer = arcpy.mapping.ListLayers(mxd,
crimeFC, df)[0]
                sourceLayer = arcpy.mapping.Layer(lyrFile)
                arcpy.AddMessage("Updating the symbology")
                arcpy.mapping.UpdateLayer(df, updateLayer,
sourceLayer, True)

                #set the layout map title
                for elm in arcpy.mapping.ListLayoutElements(mxd,
"TEXT_ELEMENT", "CrimeTitle"):
                    elm.text = mapTitle

                #export or print the map
                arcpy.AddMessage("Exporting/Printing the map")
                if exportType == "PDF":
                    arcpy.mapping.ExportToPDF(mxd, mapTitle +
".pdf")
                elif exportType == "JPEG":
                    arcpy.mapping.ExportToJPEG(mxd, mapTitle +
".jpg")
```

```
            else:
                arcpy.mapping.PrintMap(mxd)

            arcpy.AddMessage("Processing complete")
        else:
            arcpy.AddMessage("Feature class does not contain
Join_Count field...can't create map")

    except Exception as e:
        arcpy.AddMessage(e.message)
```

28. You can check your work by examining the `C:\ArcGIS_Blueprint_Python\solutions\ch4\CreateMap.py` solution file. Refer to the `CreateMap` class.

29. Save the file and exit your Python development environment.

30. Now, it's time to test the tool. If necessary, open **ArcMap** and display the contents of the `CrimeAnalysis` toolbox. You should now see the **Create Map** tool.

31. For this test, we're going to aggregate crimes to a polygon feature class containing neighborhood boundaries before creating the map. Double-click on the **Aggregate Crimes** tool.

32. Select `Seattle_Neighborhoods.shp` from the `C:\ArcGIS_Blueprint_Python\data\crime` folder as `Boundary Layer`.

33. Select `Burglary_2015` from the `C:\ArcGIS_Blueprint_Python\data\crime\SeattleCrimeAnalysis` geodatabase as the crime point locations parameter.

34. Define `Seattle_Neighborhood_Burglary_2015` inside the `SeattleCrimeAnalysis` geodatabase as output feature class.

35. Click on **OK** to execute the tool and generate the `feature` class. The `feature` class should be added to the table of contents. Open the attribute table and make sure a `Join_Count` field has been created.

36. Double-click on the **Create Map** tool to display the parameters shown in the following screenshot:

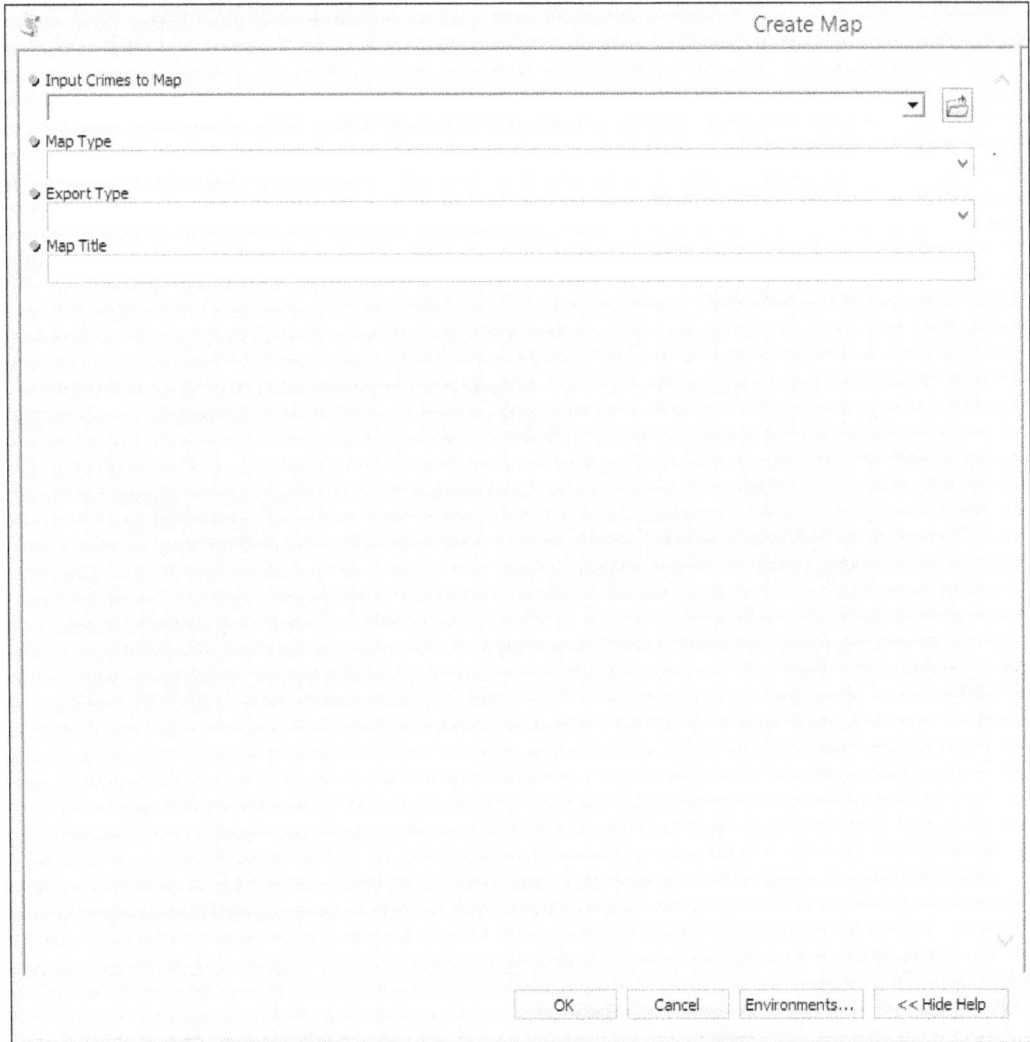

37. The `Seattle_Neighborhood_Burglary_2015` layer should already be present in your table of contents, so click on the drop-down arrow and select the layer.

38. Select **Graduated Color** for **Map** Type.

39. Select **PDF** as **Export Type**.

40. Set **Map Title** to something like **Seattle Burglaries by Neighborhood in 2015**.

41. If all goes as expected, a file called `Seattle Burglaries by Neighborhood in 2015.pdf` should be created in the `C:\ArcGIS_Blueprint_Python\ch4` folder and should look like what is shown in the following screenshot:

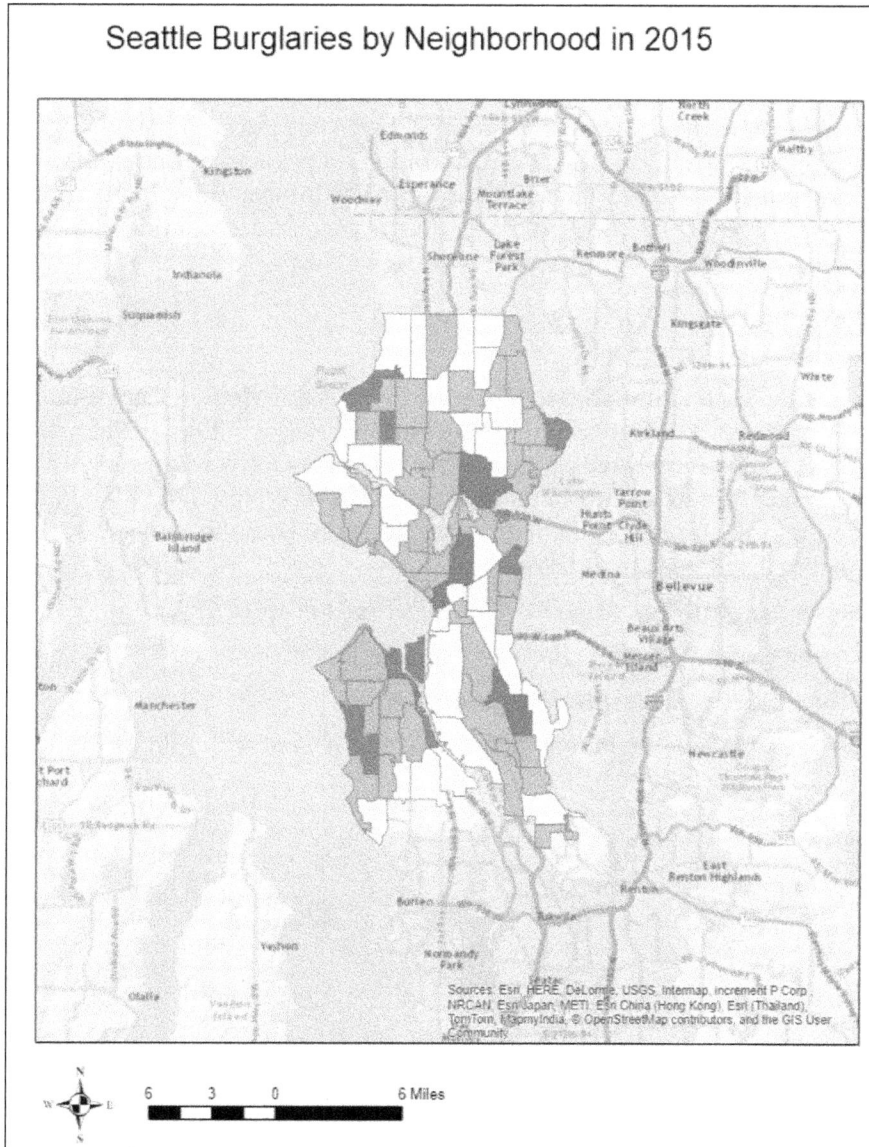

Feel free to experiment with the other parameters of this tool to print and export maps.

Performing Spatial Statistical Analysis

This section will cover the use of several spatial statistical tools in the analysis of crime data. Specifically, the Median Center, Directional Distribution, and Optimized Hot Spot Analysis Tools from the **Spatial Statistics** Tools toolbox will be used to create various analysis layers. Each of these tools is a **Python Script** tool, which means that you can view the code for the tool and even make changes if required. You can use the knowledge you gain in this section in the next chapter of the book as we continue to create **ArcGIS** tools for crime analysis.

In an earlier section, the **Aggregate Crimes** tool was used to aggregate burglaries by census block groups for Seattle, WA. A graduated color map was created from the resulting output. This provides some descriptive information about crimes and where they occur. In this section, we'll dig deeper to get a better understanding of the spatial characteristics of the data:

1. If necessary, open `C:\ArcGIS_Blueprint_Python\ch4\SeattleCrimes.mxd` in **ArcMap**.

2. The first **Spatial Statistics** tool that we'll run is the **Median Center** tool found in the **Measuring Geographic Distributions** toolset in the **Spatial Statistics Tools** toolbox. The **Median Center** tool identifies the median geographic location of a dataset and can include an optional Weight field that we'll use to find the geographic center of our burglary data. Find the tool and double-click on it to display the parameters.

3. Select `Seattle_BG_Burglary_2015` as **Input Feature Class**.

4. Define **Output Feature Class** as `Seattle_BG_Burglary_2015_MeanCenter`.

5. Select `Join_Count` as the **Weighted** field.

6. Click on **OK** to execute the tool. The **Median Center** should be located as shown in the following screenshot:

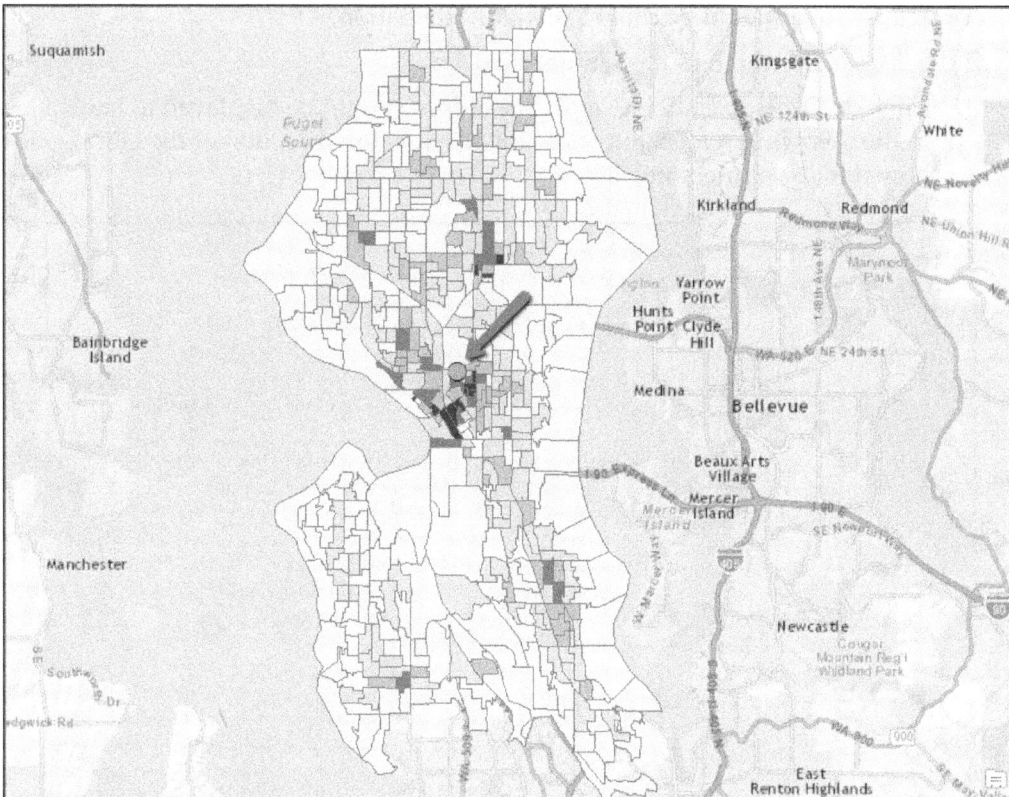

7. Next, the **Directional Distribution** tool will be used to create a standard deviational ellipse to summarize the spatial characteristics of the burglary data, including the central tendency, dispersion, and directional trends.

8. Find the **Directional Distribution** tool and double-click on it to display the parameters.

9. For **Input Feature Class,** select `Seattle_BG_Burglary_2015`.

10. Define `Seattle_BG_Burglary_2015_Directional` as **Output Ellipse Feature Class**.

11. Ellipse **Size** should be set to 1 standard deviation.

12. Select Join_Count as the **Weight** field.

13. Click on **OK** to execute the tool. The ellipse should be displayed as shown in the following screenshot. Note the north-south direction of the ellipse and how it follows **Interstate 5**.

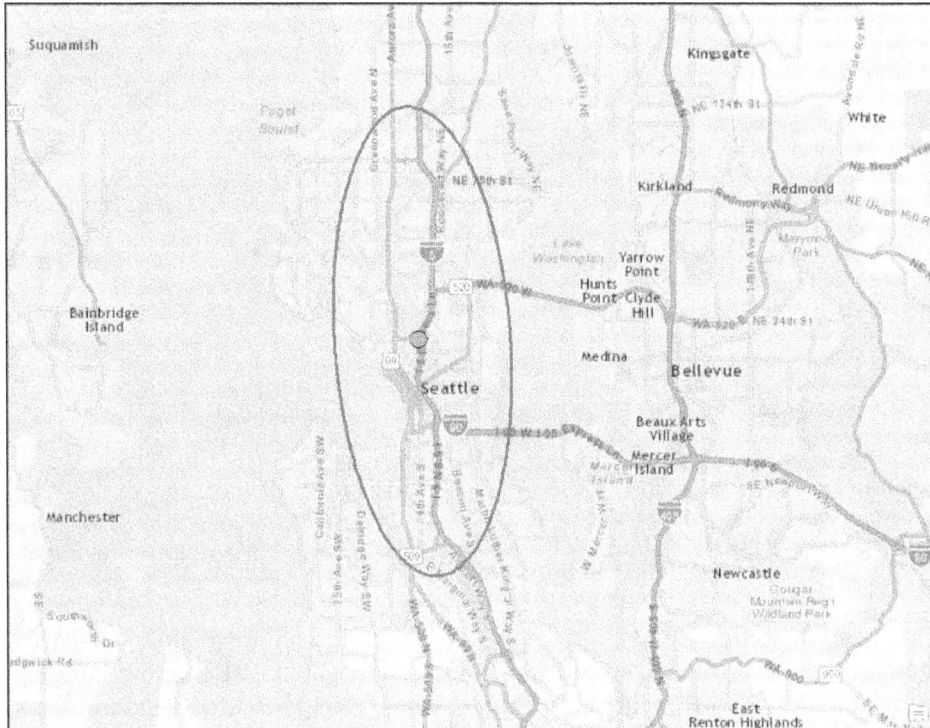

14. The final tool to be executed in this section is the **Optimized Hot Spot Analysis** tool found in the **Mapping Clusters** toolbox. Using weighted features, this tool will create a map of statistically significant hot and cold spots using the **Getis-Ord Gi*** statistic.

15. Find the tool and double-click on it to display the parameters.

16. For **Input Feature Class**, select `Seattle_BG_Burglary_2015`.

17. Define `Seattle_BG_Burglary_2015_Hotspot` as **Output Feature Class**.

18. Select `Join_Count` as the **Weight** field.

19. Click on **OK** to execute the tool. The progress dialog will indicate statistical processing information during processing, as shown in the following screenshot:

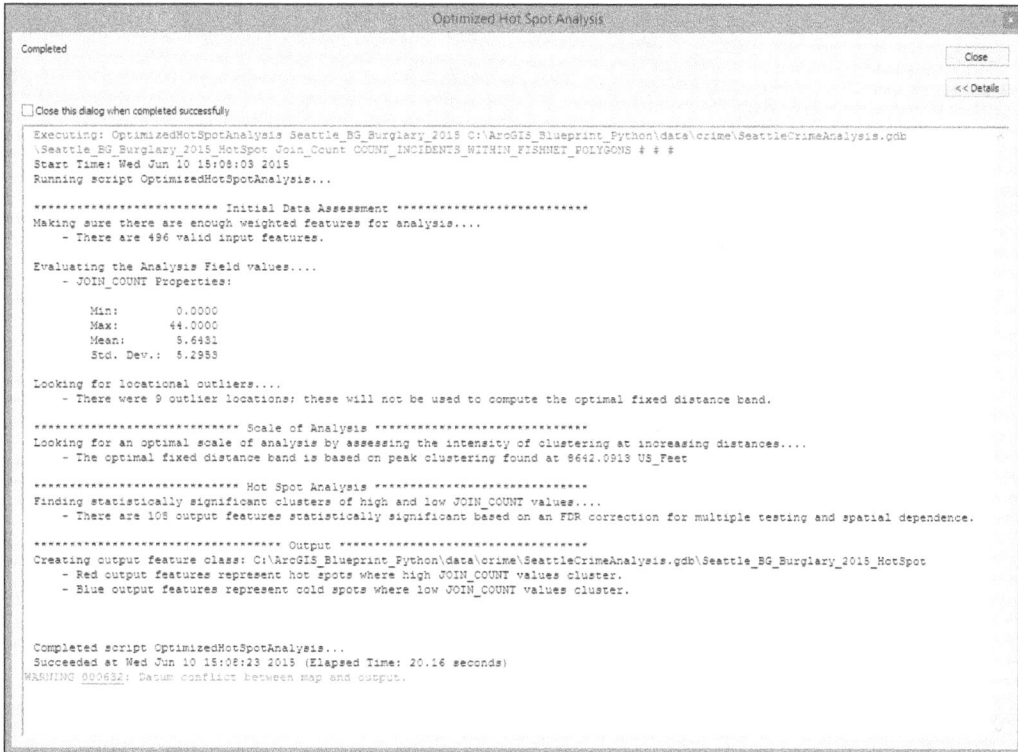

20. The output feature class will contain features that are found to be hotspots, cold spots, or not significant. Hot and cold spots will be generated at 90%, 95%, and 99% confidence levels. Hotspots are indicated by a red shade, as shown in the following screenshot. This corresponds to the output from the other tools that were executed in this section.

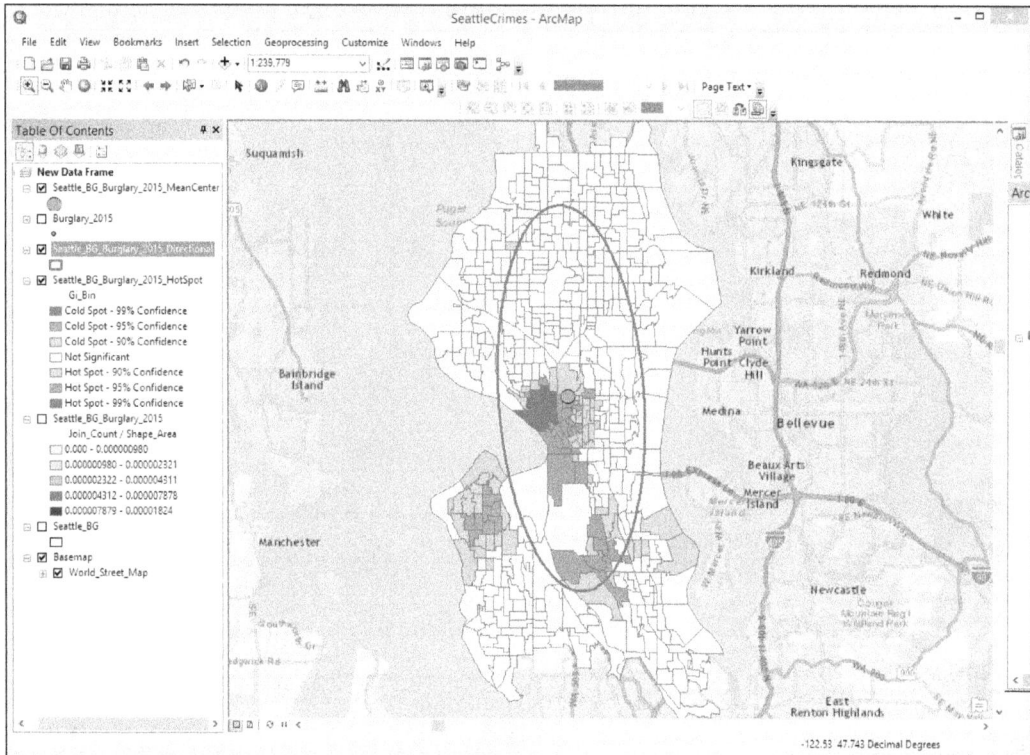

21. Save the map document file.

Summary

This chapter, which covered crime analysis with ArcGIS and Python, concentrated on the development of geoprocessing tools that can be used to import crimes and aggregate them to polygon boundary layers. In addition, a tool to automate the process of creating, exporting, and printing maps was implemented. The next chapter will concentrate on the development of tools that can be used for reporting and analysis using **ArcPy** and **Plotly**.

In the next chapter, we'll build on our efforts in this chapter by creating **Plotly** charts and graphs that can be added to the **ArcMap** layout view.

5
Analyzing Crime Patterns with ArcGIS Desktop, ArcPy, and Plotly (Part 2)

In this, the second of two chapters that cover the creation of crime analysis tools in ArcGIS, we will concentrate primarily on the development of charts and graphs using a combination of ArcGIS Desktop and `ArcPy` along with `Plotly` at `https://plot.ly`. `Plotly` is an online analytics and data visualization tool for graphs, analytics, and statistics. It includes a Python library that can be integrated with GIS data to supplement maps and analysis generated with ArcGIS Desktop.

The **Crime Analysis** toolbox created in the last chapter will be the focus as we add several new tools. The first tool, **Create Neighborhood Bar Chart**, will create a bar chart of crimes by Seattle neighborhood. Next, the **Create Line Plot** tool will graph the number of crimes over time on a line plot. This tool can be used to find seasonal patterns of crime. Finally, we'll enhance both tools to write their output to the **ArcMap** layout view and update the **Create Map** tool to export the product.

In this chapter, we will cover the following topics:

- Creating ArcGIS Desktop Python toolboxes
- Using `Plotly` to create bar charts and line plots
- Automating the process of creating, exporting, and printing maps using `ArcPy` mapping

Design

This is the second of two chapters that cover the development of tools for crime analysis, so we'll be building on what we developed in the first chapter. In the `CrimeAnalysis.pyt` toolbox, two new tools will be created: `NeighborhoodBarChart` and `LinePlot`. Both tools will use the Plotly Python library to create graphs based on crime data stored in a file geodatabase. The `NeighborhoodBarChart` tool will create a bar chart depicting the crime data by neighborhood, while the `LinePlot` tool will create a line plot graph showing the crime data over time. The charts will be exported to PNG format image files and displayed in an **ArcMap** layout. Finally, the **Create Map** tool will be updated to export the maps and charts stored in the layout view. The following diagram shows the whole process:

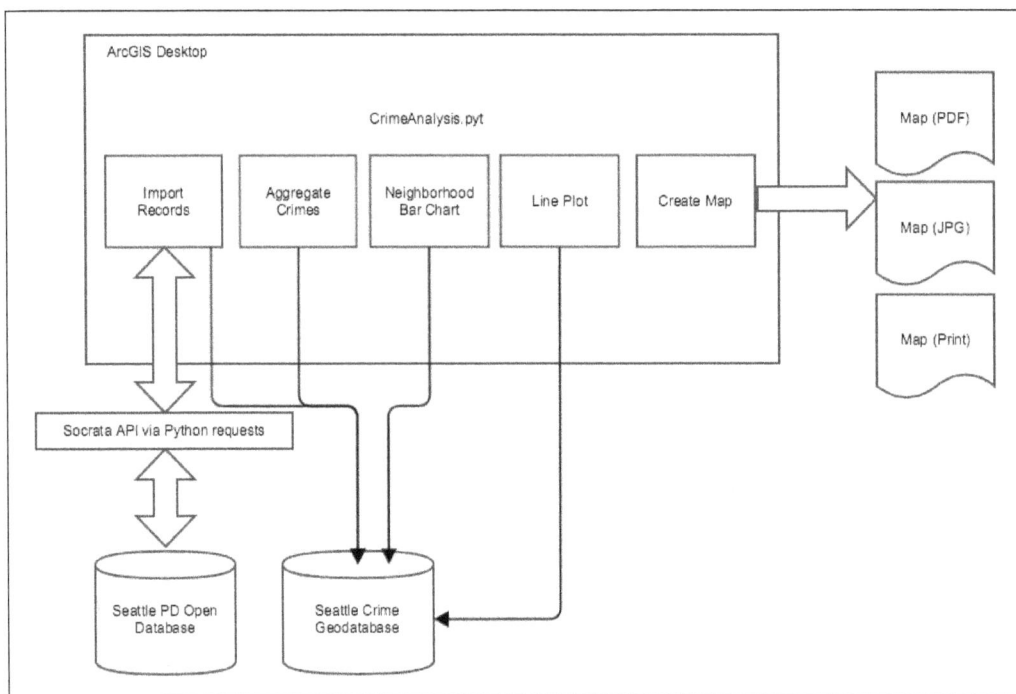

Let's get started and build the application.

Getting to know Plotly

Plotly (`plot.ly`) will be used to build the graphs and charts that are part of the tools built in this chapter. It's important to understand some fundamental concepts of how Plotly works before getting started.

Plotly is a cloud service specializing in data visualization and statistical analysis. Using Plotly's web-based interface, it is possible to upload data in various formats, including Microsoft Excel and Access, CSV, TSV, Matlab, and spreadsheets from Google Drive. Once imported, the web interface can then be used to create various types of visualization, including bar and pie charts, line graphs, scatter plots, area charts, histograms, box plots, heat maps (not GIS-based heat maps), and others. These visualizations can then be shared with others. The following screenshot depicts the web interface for Plotly:

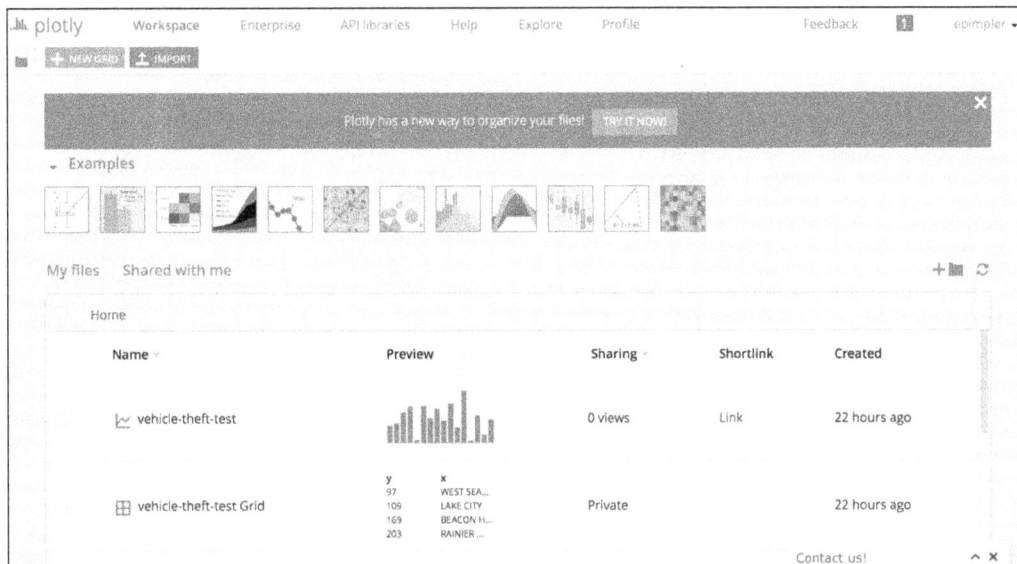

In addition to the web-based interface, there are several programming languages that can be used to dynamically create visualizations without first uploading data. These include Python, Matlab, R, Arduino, Julia, Perl, and a basic REST API. Python, combined with `ArcPy` and `Plotly`, provides a lot of data visualization flexibility.

Other features of Plotly include visualization interactivity, which means that end users can hover the mouse over graphs to obtain attribute information, along with zooming and panning the visualization. Also, there is a lot of flexibility in the layout formatting, including the ability to customize fonts, colors, annotation, and other graphing options.

Creating the Neighborhood Bar Chart tool

The **Neighborhood Bar Chart** tool will visualize the number of crimes for each major neighborhood in Seattle in the form of a bar chart. This tool will be added to the existing **Crime Analysis** toolbox created in the previous chapter. It will use a combination of ArcPy and Plotly to create the graph.

> The Plotly Python library will need to be installed before any of the tools in this chapter can be created. Use pip to install the Plotly Python library using the following command. If you haven't installed pip yet, refer to *Chapter 1, Extracting Real-Time Wildfire Data from ArcGIS Server with the ArcGIS REST API*, for detailed instructions on installing pip. Here's the command:
>
> `pip install plotly`

The **Create Neighborhood Bar Chart** tool will accept several parameters, including an input feature class and fields that provide a reference to a polygon layer containing neighborhood boundaries with the aggregated crime data, chart title, and output file name and location. The following steps will guide you to create a **Neighborhood Bar Chart** tool:

1. The use of the Plotly Python library requires an account along with an API key. Go to https://plot.ly/ and click on the **SIGN UP** button to create a new account, as shown in the following screenshot:

2. After creating an account, you can select your **Username** from the Plotly interface and then select **Settings** to display the API settings that will be used when creating charts and graphs. My credentials have been displayed as follows. Note, though, that the actual API key number for my account has been changed:

3. In your Python development environment, you will want to set up the credentials file on your computer. The credentials file is used to supply the account authentication information required to create charts. Rather than embedding this information directly into a script, a credentials file can be created and referenced each time. From the Python shell window of your development environment, enter the following two lines of code. This step only needs to be done once. The credentials file is then referenced each time a script accesses the Plotly library:

```
import plotly.tools as tls
tls.set_credentials_file(username="your username", api_key="your api key")
```

4. Open **ArcMap** with a blank map document file and add the **Streets** base map. Zoom in to the Seattle, WA, area. **Save** the file as `C:\ArcGIS_Blueprint_Python\ch5\SeattleCrimes.mxd`.

5. Locate the `CrimeAnalysis.pyt` toolbox created in the previous chapter.

6. Open the code for the toolbox in your Python development environment.

7. Copy and paste one of the existing tools to the bottom of the `CrimeAnalysis.pyt` file. Remove the existing content from the `execute()` and `getParameterInfo()` methods for the new class.

8. Rename the class `CreateNeighborhoodBarChart`. Update the `self.label` and `self.description` properties, as shown in the following code:

```
class CreateNeighborhoodBarChart(object):
    def __init__(self):
        """Define the tool (tool name is the name of the
class)."""
        self.label = "Create Neighborhood Bar Chart"
        self.description = "Creates a bar chart of crime data"
        self.canRunInBackground = False
```

9. Add import statements for the plotly modules:

```
import arcpy
import requests
import json
import datetime
import os
import time
import plotly.plotly as py
from plotly.graph_objs import *
```

10. Now it's time to add the `Input` and `Output` parameters for the tool. Add the following lines of code to the `getParameterInfo()` method for the `CreateNeighborhoodBarChart` class:

```
def getParameterInfo(self):
        """Define parameter definitions"""

        crimeFC = arcpy.Parameter(displayName = "Input Crimes to
Graph",
                name="crimeFC",
                datatype="GPFeatureLayer",
                parameterType="Required",
                direction="Input")
        crimeFC.filter.list = ['Polygon']

        crimeField = arcpy.Parameter(displayName = "Input Field to
Graph",
                name="crimeField",
                datatype="String",
                parameterType="Required",
                direction="Input")

        neighborhoodField = arcpy.Parameter(displayName =
"Neighborhood Field to Group By",
                name="neighborhoodField",
                datatype="String",
                parameterType="Required",
                direction="Input")

        ## chart title
        chartTitle = arcpy.Parameter(
            displayName="Chart Title",
            name="chartTitle",
            datatype="String",
            multiValue="False",
            parameterType="Required",
            direction="Input")

        ## chart title
        fileLocation = arcpy.Parameter(
            displayName="Save Chart",
            name="fileLocation",
            datatype="DEFile",
```

```
                         multiValue="False",
                         parameterType="Required",
                         direction="Output")

            params = [crimeFC, crimeField, neighborhoodField,
     chartTitle, fileLocation]
               return params
```

11. The first three parameters are related to the neighborhood feature class that will be used to create the chart. The `crimeFC` parameter references the feature class, while the `crimeField` parameter defines the field used to chart the data along the *y* axis (the `Join_Count` field if the **Aggregate Crimes** tool is used), and the `neighborhoodField` parameter defines the field containing the neighborhood names used for the *x* axis.

    The final two parameters include a textbox that will capture the title of the chart and a parameter used to define the output location and filename for the chart that will be exported. Finally, all the parameters will be added to the `params` list.

12. Find the `UpdateParameters()` method inside the `CreateNeighborhoodBaChart` class. This method is executed any time one of the parameters in the input dialog changes. In this particular case, there are two input parameters (`crimeField` and `neighborhoodField`) that need to be updated with the fields in the feature class selected by the user for the `crimeFC` input parameter. In other words, we want to populate the list of fields based on the feature class the user selects for **Input Crimes to Graph**. Add the following code block to accomplish this:

```
def updateParameters(self, parameters):
    """Modify the values and properties of parameters before
internal validation is performed. This method is called whenever a
parameter has been changed."""
    if parameters[0].value:
      desc = arcpy.Describe(parameters[0].value)
      fields = desc.fields
      listNumeric = []
      listString = []
      for f in fields:
```

```
        if f.type in ['Double', 'Integer', 'Single']:
            listNumeric.append(f.name)
        elif f.type == 'String':
            listString.append(f.name)
    parameters[1].filter.list = listNumeric
    parameters[2].filter.list = listString
return
```

This code block executes the `arcpy.Describe()` function against the feature class selected by the user and obtains a list of fields in the feature class. The `crimeField` parameter needs to contain numeric attribute fields, while `neighborhoodField` should contain attribute fields with a text data type. The field types are tested and then placed in a `list` variable corresponding to a type. Finally, the lists are assigned as filters to each of the parameters.

13. Add the tool to the `self.tools[]` list inside the `Toolbox` class as shown in the following code:

```
self.tools = [ImportRecords, AggregateCrimes, CreateMap,
CreateNeighborhoodBarChart]
```

14. Find the `execute()` method inside the `CreateNeighborhoodBarChart` class. This method will contain the functionality of the tool.

15. Add the following lines of code to capture the input variables submitted to the tool:

```
def execute(self, parameters, messages):
        """The source code of the tool."""
        crimeFC = parameters[0].valueAsText
        crimeField = parameters[1].valueAsText
        neighborhoodField = parameters[2].valueAsText
        chartTitle = parameters[3].valueAsText
        fileLocation = parameters[4].valueAsText
```

16. Add the `try`/`except` exception-handling structures just below the code you just added. The rest of the code in this section should go inside the `try` block:

```
try:
except Exception as e:
    arcpy.AddMessage(e.message)
```

17. The first code block added to this tool will require a little explanation before coding. This tool is to be used with a polygon feature class that contains neighborhood boundaries. Each `neighborhood` boundaries has attribute fields that define the neighborhood name (`S_HOOD`), along with a secondary attribute field that defines a larger neighborhood (`L_HOOD`). This can be seen in the following screenshot, where the **Ballard** neighborhood is seen as a group of several small neighborhoods:

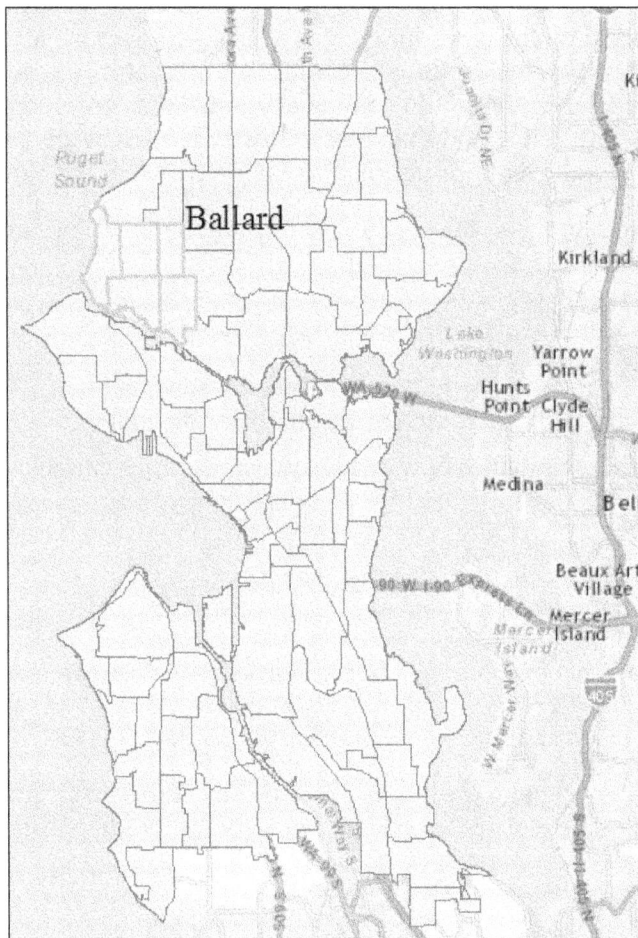

Ultimately, the chart created by this tool will group records based on the neighborhood with a larger geographic extent, as shown in the following figure:

Vehicle Thefts in 2015

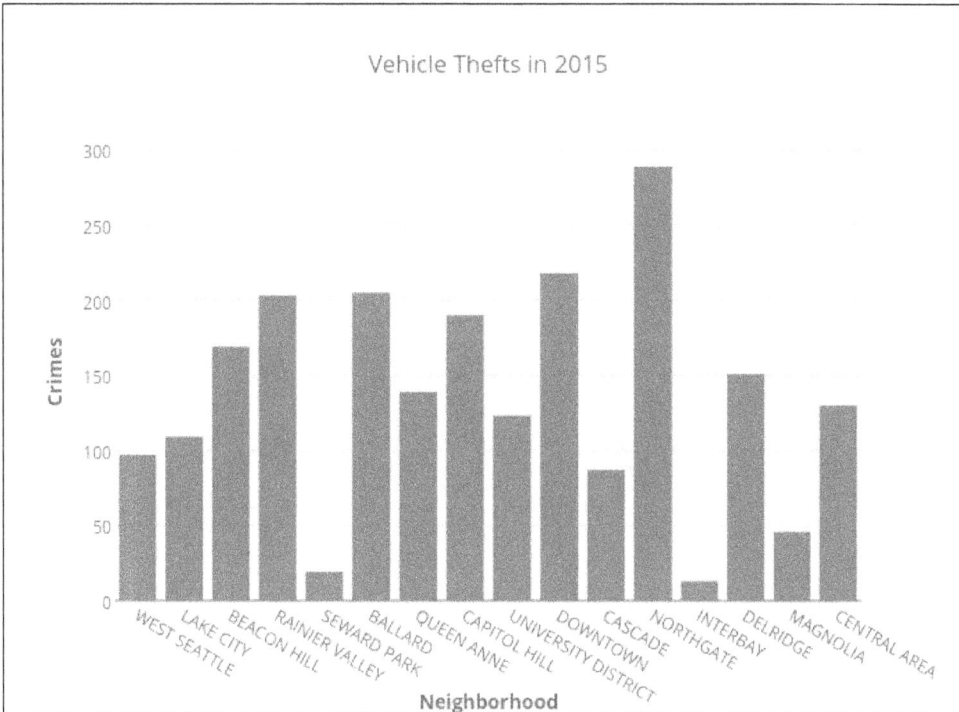

The tool will, therefore, loop through all the records in the neighborhood feature class and sum the number of crimes for each neighborhood.

18. With this background, it is time to code the block that will populate a Python dictionary containing a unique set of neighborhood names (`keys`) along with the total number of crimes in each neighborhood (`values`). Each neighborhood polygon in the `feature` class is assigned to a larger neighborhood defined in the `L_HOOD` attribute field. Add the following code to the `try` statement:

```
try:
    #aggregate by the larger neighborhood area L_HOOD
    dictHoods = {}
    with arcpy.da.SearchCursor(crimeFC, (crimeField,
neighborhoodField)) as cursor:
        for row in cursor:
          if row[1] in dictHoods:
                currentCount = dictHoods[row[1]]
                dictHoods[row[1]] = currentCount + row[0]
          else:
                dictHoods[row[1]] = row[0]
```

19. The `Plotly Bar` object that creates the bar chart will accept two list variables as parameters: one for the x axis and the other for the y axis. In this next code block, we'll pull out the `keys` and `values` from the dictionary created in the last step and divide it into two list variables. Some neighborhoods were not assigned to a larger neighborhood; thus, their values have been coded as No Broader Term. We'll exclude these values for this chart. Add the code block highlighted in the following code:

```
try:
    #aggregate by the larger neighborhood area L_HOOD
    dictHoods = {}
    with arcpy.da.SearchCursor(crimeFC, (crimeField,
neighborhoodField)) as cursor:
        for row in cursor:
            if row[1] in dictHoods:
                currentCount = dictHoods[row[1]]
                dictHoods[row[1]] = currentCount + row[0]
            else:
                dictHoods[row[1]] = row[0]
    x = []
    y = []
    for hood in dictHoods:
        #exclude if not part of a larger neighborhood group
        if hood != "NO BROADER TERM":
            x.append(hood)
            y.append(dictHoods[hood])
```

20. Assign the lists to a Plotly `data` object:

```
x = []
y = []
for hood in dictHoods:
    #exclude if not part of a larger neighborhood group
    if hood != "NO BROADER TERM":
        x.append(hood)
        y.append(dictHoods[hood])

#assign the data
data = Data([Bar(x=x,y=y)])
```

21. Create the Plotly `layout` object for the graph. This block of code should go just below the last line you added:

```
#layout of the graph
layout = Layout(
    title = chartTitle,
    xaxis = XAxis(
```

```
                title='<b>Neighborhood</b>'
            ),
        yaxis = YAxis(
                title='<b>Crimes</b>'
            )
    )
```

22. Create the `Figure` object that contains the `data` and `layout` objects. This line of code should be placed just below the last line you added:

    ```
    fig = Figure(data=data,layout=layout)
    ```

23. Save the graph:

    ```
    fileToSave = fileLocation  + ".png"
    py.image.save_as(fig, filename=fileToSave)

    arcpy.AddMessage("Created and saved chart to: " + fileLocation)
    ```

24. You can check your work by examining the `C:\ArcGIS_Blueprint_ Python\solutions\ch5\CreateBarChart.py` solution file. Refer to the `CreateBarChart` class.

25. Save the file and exit your Python development environment.

26. Now it's time to test the tool. If required, open **ArcMap** with the `SeattleCrimes.mxd` file and the `Crime Analyst` toolbox.

27. Run the `Import Records` tool to create a new `feature` class called `VehicleThefts_2014` with the **Begin Crime Date** 1/1/2014 and the **End Crime Date** 12/31/2014. The `Crime Type` should be VEHICLE THEFT. You can see the parameters in the following screenshot. This will import all vehicle theft crimes for Seattle in the year 2014.

28. Click on **OK** to run the tool. This should import approximately 3,300 records, and the feature class will be added to the **ArcMap** table of contents.

29. These records can be aggregated to neighborhood boundaries for Seattle by running the **Aggregate Crimes** tool in the **Crime Analysis** toolbox. Double-click on the tool and fill out the parameters, as shown in the following screenshot:

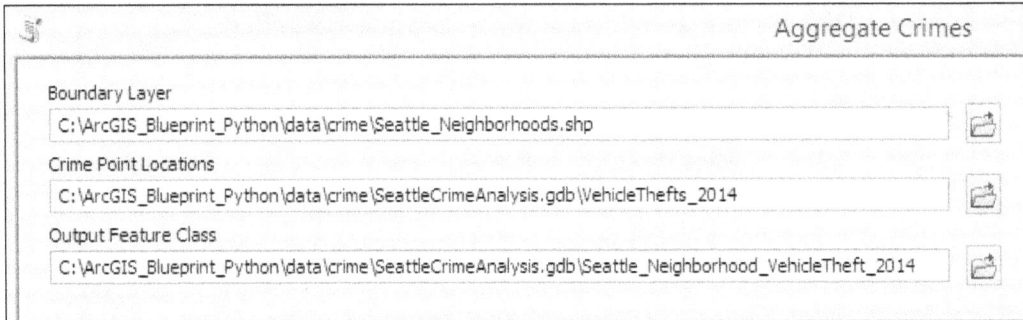

Aggregate Crimes

Boundary Layer
C:\ArcGIS_Blueprint_Python\data\crime\Seattle_Neighborhoods.shp

Crime Point Locations
C:\ArcGIS_Blueprint_Python\data\crime\SeattleCrimeAnalysis.gdb\VehicleThefts_2014

Output Feature Class
C:\ArcGIS_Blueprint_Python\data\crime\SeattleCrimeAnalysis.gdb\Seattle_Neighborhood_VehicleTheft_2014

30. Click on **OK** to run the tool. Now, the **Create Neighborhood Bar Chart** tool can be run against the Seattle_Neighborhood_VehicleTheft_2014 feature class. The feature class will be added to the **ArcMap** table of contents.

31. Double-click on the **Create Neighborhood Bar Chart** tool in the **Crime Analysis** toolbox and fill out the parameters, as shown in the following screenshot:

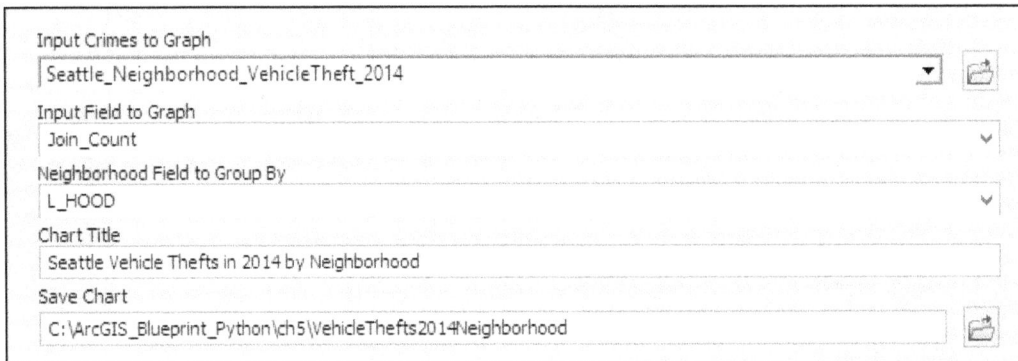

Input Crimes to Graph
Seattle_Neighborhood_VehicleTheft_2014

Input Field to Graph
Join_Count

Neighborhood Field to Group By
L_HOOD

Chart Title
Seattle Vehicle Thefts in 2014 by Neighborhood

Save Chart
C:\ArcGIS_Blueprint_Python\ch5\VehicleThefts2014Neighborhood

32. Click on **OK** to run the tool. This should create an image file called `VehicleThefts2014Neighborhood.png`. This is depicted in the following screenshot:

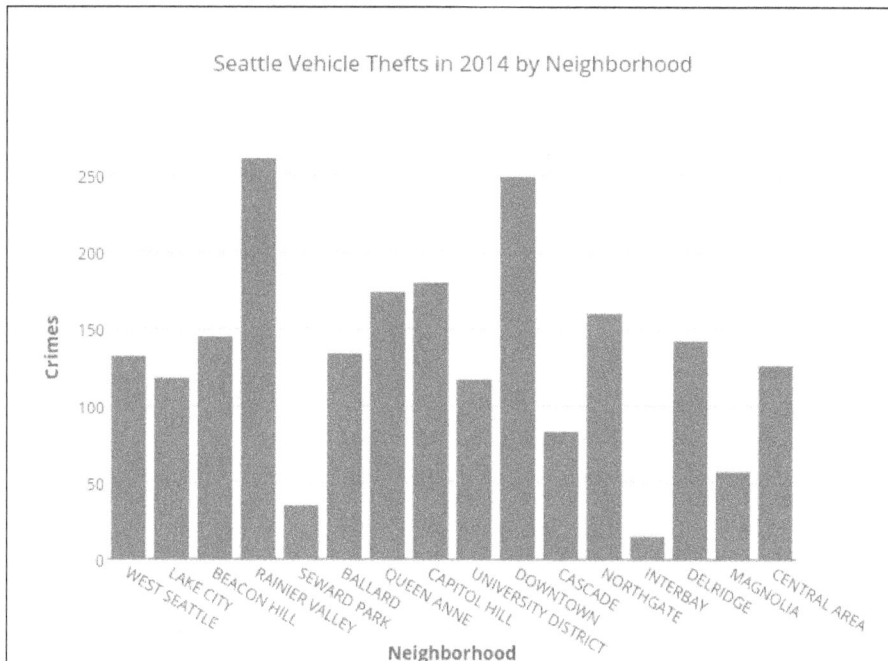

Creating the Create Line Plot tool

The **Create Line Plot** tool will create a line plot that depicts the number of crimes over time. This tool will be added to the **Crime Analysis** toolbox created in the previous chapter. It will use a combination of ArcPy and Plotly to create the graph.

This tool will use an existing point feature class layer imported with the **Import Records** tool. It will aggregate the total number of crimes by the month and plot the results to a line plot graph. Input parameters will include the feature class and field to be charted, along with a chart title, location, and filename for the output chart:

1. If required, open `C:\ArcGIS_Blueprint_Python\ch5\SeattleCrimes.mxd` in **ArcMap**.

2. Locate the `CrimeAnalysis.pyt` toolbox that was created in the last chapter.

3. Open the code for the toolbox in your Python development environment.

4. Copy and paste one of the existing tools at the bottom of the
    `CrimeAnalysis.pyt` file.

5. Rename the class `CreateLinePlot`.

6. Remove the code inside the `getParameterInfo()`, `getUpdateParameters()`,
    and `execute()` methods for the new `CreateLinePlot` class.

7. Update the `self.label` and `self.description` properties of the
    `CreateLinePlot` class, as shown in the following code:

```
class CreateLinePlot(object):
    def __init__(self):
        """Define the tool (tool name is the name of the
class)."""
        self.label = "Create Line Plot"
        self.description = "Creates a line plot
 of crime data"
        self.canRunInBackground = False
```

8. Add the following code block to the `getParameterInfo()` method to define
    the parameters to capture the input feature class and the field to be graphed
    along with the chart title, output location, and filename:

```
def getParameterInfo(self):
        """Define parameter definitions"""

        crimeFC = arcpy.Parameter(displayName = "Input Crimes to
Graph",
            name="crimeFC",
            datatype="GPFeatureLayer",
            parameterType="Required",
            direction="Input")
        crimeFC.filter.list = ['Point']

        crimeField = arcpy.Parameter(displayName = "Input Field
(Month)",
            name="crimeField",
            datatype="String",
            parameterType="Required",
            direction="Input")

        ## chart title
        chartTitle = arcpy.Parameter(
```

```
            displayName="Chart Title",
            name="chartTitle",
            datatype="String",
            multiValue="False",
            parameterType="Required",
            direction="Input")

    ## file location
    fileLocation = arcpy.Parameter(
            displayName="Save Chart",
            name="fileLocation",
            datatype="DEFile",
            multiValue="False",
            parameterType="Required",
            direction="Output")

    params = [crimeFC, crimeField, chartTitle, fileLocation]
    return params
```

9. In the `Toolbox` class, add the `CreateLinePlot` tool to the `self.tools` list, as shown in the following code:

```
self.tools = [ImportRecords, AggregateCrimes, CreateMap,
CreateNeighborhoodBarChart, CreateLinePlot]
```

10. Find the `UpdateParameters()` method and add the following code block, which will update the values in the `crimeField` parameter based on the `feature` class selected by the layer for the `crimeFC` parameter:

```
def updateMessages(self, parameters):
    """Modify the messages created by internal validation for each
tool
    parameter.  This method is called after internal
validation."""
        if parameters[0].value:
            desc = arcpy.Describe(parameters[0].value)
            fields = desc.fields
            listFields = []
            for f in fields:
                if f.type == 'String':
                    listFields.append(f.name)
            parameters[1].filter.list = listFields
        return
```

11. Find the `execute()` method and add the following lines of code to accept the parameter information submitted by the user:

```
def execute(self, parameters, messages):
        """The source code of the tool."""
        crimeFC = parameters[0].valueAsText
        crimeField = parameters[1].valueAsText
        chartTitle = parameters[2].valueAsText
        fileLocation = parameters[3].valueAsText
```

12. Next, set the workspace variable, as shown in the following code:

```
def execute(self, parameters, messages):
        """The source code of the tool."""
        crimeFC = parameters[0].valueAsText
        crimeField = parameters[1].valueAsText
        chartTitle = parameters[2].valueAsText
        fileLocation = parameters[3].valueAsText

        arcpy.env.workspace = "C:/ArcGIS_Blueprint_Python/ch5"
```

13. Add the `try`/`except` exception-handling structures right below the code you just added. The rest of the code in this section should go inside the `try` block:

```
try:
except Exception as e:
    arcpy.AddMessage(e.message)
```

14. The next block of code will use an `arcpy.da.SearchCursor` object to loop though all the records in the `crime` feature class selected by the user. As it loops through the records, it will obtain a unique list of values in the **Input Field** selected by the user (this should be the **MONTH** field). A second loop will loop through each month, create a feature layer containing the records that match the month, count the number of records, and add them to a list containing the total number of crimes. Add the following code block:

```
try:
    #aggregate the crimes by month
    listCrimes = []
    listMonths = []
    with arcpy.da.SearchCursor(crimeFC, (crimeField)) as cursor:
        for row in cursor:
```

```
        listMonths.append(row[0])
    listMonths = list(sorted(set(listMonths), key=int))
    for m in listMonths:
        arcpy.MakeFeatureLayer_management(crimeFC, "EachMonth.
lyr", crimeField + " = " + "\'" +  m + "\'")
        result = arcpy.GetCount_management("EachMonth.lyr")
        numCrimes = int(result.getOutput(0))
        listCrimes.append(int(numCrimes))
```

15. Create a `Scatter` object, passing in the list of months and the list of crimes:

```
for m in listMonths:
    arcpy.MakeFeatureLayer_management(crimeFC, "EachMonth.lyr",
crimeField + " = " + "\'" +  m + "\'")
    result = arcpy.GetCount_management("EachMonth.lyr")
    numCrimes = int(result.getOutput(0))
    listCrimes.append(int(numCrimes))


trace = Scatter(x=listMonths, y=listCrimes)
```

16. Create a Data object by passing in the trace variable:

```
trace = Scatter(x=listMonths, y=listCrimes)
data = Data([trace])
```

17. Create the layout for the graph:

```
trace = Scatter(x=listMonths, y=listCrimes)
data = Data([trace])

#layout of the graphfor m in listMonths:
layout = Layout(
    title = chartTitle,
    xaxis = XAxis(
        title='<b>Month</b>'
    ),
    yaxis = YAxis(
        title='<b>Crimes</b>'
    )
)
```

18. Create a `Figure` object that will contain the data as well as the layout. This line should be placed just below the code block you added in the previous step:

```
fig = Figure(data=data,layout=layout)
```

19. Save the chart to a PNG image file:

```
fig = Figure(data=data,layout=layout)

fileToSave = fileLocation  + ".png"
py.image.save_as(fig, filename=fileToSave)

arcpy.AddMessage("Created and saved chart to: " + fileLocation)
```

20. Save the file and close your development environment.

21. Now it's time to test the tool. If required, open **ArcMap** and the `SeattleCrimes.mxd` file in the `C:\ArcGIS_Blueprint_Python\ch 5` folder and the **Crime Analyst** toolbox.

22. Double-click on the **Create Line Plot** tool. Add the following parameters and click on **OK** to execute the tool. The **Input Crimes to Graph** feature class is a point feature class that contains the imported vehicle thefts from 2014 (**VehicleThefts_2014**). This feature class was created in the last section of the chapter, when we imported records for the **Neighborhood Bar Chart** tool. **MONTH** should be chosen as the **Input Field (Month)** to be mapped. The tool will sum all vehicle thefts by month:

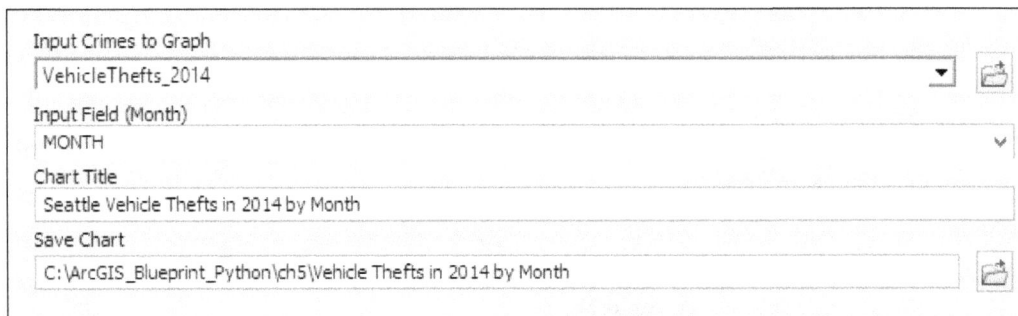

23. The output line chart should appear as shown in the following screenshot:

Seattle Vehicle Thefts in 2014 by Month

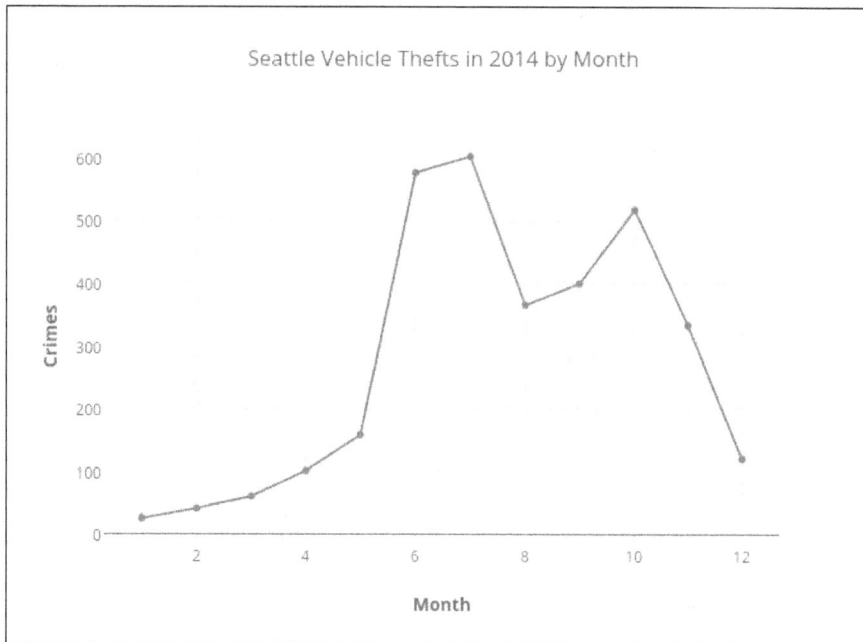

Creating the output

The final section of this chapter will focus on the creation of a visualization product that combines the maps and charts that can be created with the tools built in the previous two chapters. The **ArcMap** layout view will be used in combination with coding in order to automate the process of creating our final visualization product. The following steps will guide you through the creation of the final visualization product:

1. If required, open `C:\ArcGIS_Blueprint_Python\ch5\SeattleCrimes.mxd` in **ArcMap**.

2. The final product will have three data frames: **Charts** and **Graphs**, **Maps**, and **Spatial Statistics Maps**. Rename the default **Layers** data frame **Charts** and **Graphs** and leave the `VehicleThefts_2014`, `Seattle_Neighborhood_VehicleTheft_2014`, and `World_Street_Map` layers in the data frame.

3. Create the **Maps** and **Spatial Statistics Maps** data frames by navigating to **Insert | Data Frame** from the main **ArcMap** menu. Rename them accordingly.

4. Copy and paste the `World_Street_Map` and `Seattle_Neighborhood_VehicleTheft_2014` layers to the new **Maps** data frame that you created. Do this for the **Spatial Statistics Maps** data frame as well.

5. Activate the **Spatial Statistics Maps** data frame.

6. Open `ArcToolbox` and find the **Mean Center** tool in the **Spatial Statistics** toolbox. It is in the **Measuring Geographic Distributions** toolset.

7. Double-click on the **Mean Center** tool and define `Seattle_Neighborhood_VehicleTheft_2014` as **Input Feature Class** and **Output Feature Class** of `Seattle_Neighborhood_VehicleTheft_2014_MeanCenter` inside the `SeattleCrimeAnaysis` geodatabase. Select `Join_Count` as **Weight Field**. Click on **OK** to execute the tool.

8. In the **Measuring Geographic Distributions** toolset, double-click on the **Directional Distribution** tool and define `Seattle_Neighborhood_VehicleTheft_2014` as `Input Feature Class` and `Output Feature Class` of `Seattle_Neighborhood_VehicleTheft_2014_Directional` inside the `SeattleCrimeAnaysis` geodatabase. Leave one standard deviation as **Ellipse Size** and select `Join_Count` as **Weight Field**. Click on **OK** to execute the tool.

9. In the **Mapping Clusters** toolset, click on the **Hot Spot Analysis** tool. Select `Seattle_Neighborhoold_VehicleTheft_2014` as `Input Feature Class` and `Join_Count` as **Input Field**. Define **Output Feature Class** of `Seattle_Neighborhood_VehicleTheft_2014_Hotspot`. Select **Zone of Indifference** as **Conceptualization of Spatial Relationships** and leave the rest of the parameters to the defaults. Click on **OK** to run the tool. Your view should now appear as shown in the following screenshot:

10. Switch to the **Layout** view and select the `LandscapeModernInset` template by clicking on the **Change Layout** button in **Layout Toolbar**.

11. Arrange the **Layout** view so that the main map is pulled from the **Maps** data frame while the secondary map is retrieved from the **Spatial Statistics Maps** data frame. This can be seen in the following screenshot:

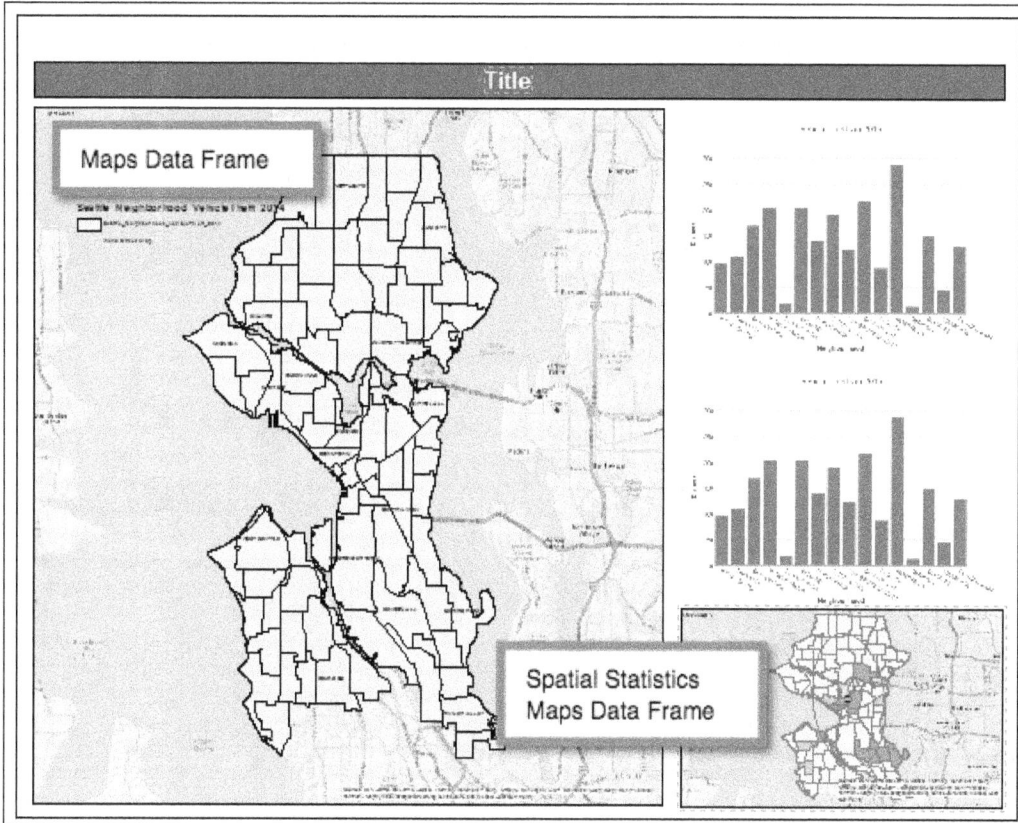

12. Add two **Picture** elements by navigating to **Insert | Picture** and arranging them as shown in the following screenshot. You can use the C:\ArcGIS_ Blueprint_Python\ch5\ChartPlaceholder.png file for both. This will serve as a temporary placeholder for the charts we create with the tools:

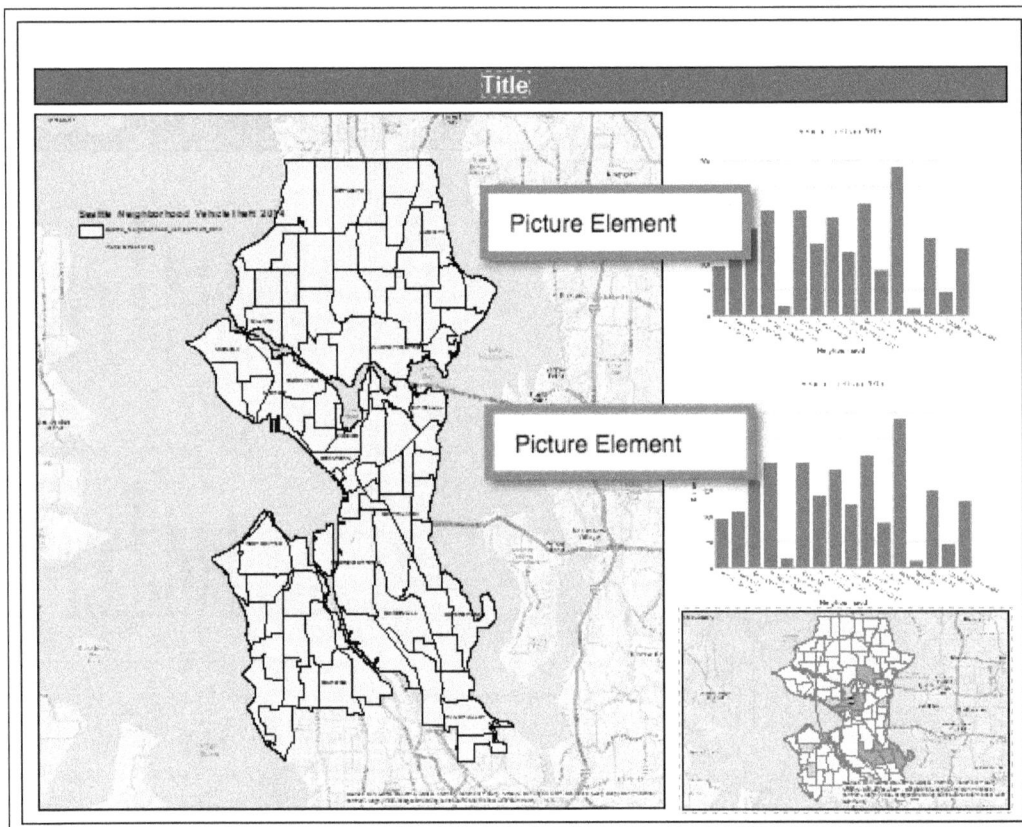

13. Right-click on the upper **Picture** element and select **Properties**. Under the **Size** and **Position** tab, give it the **Element Name** BarChart, as shown in the following screenshot:

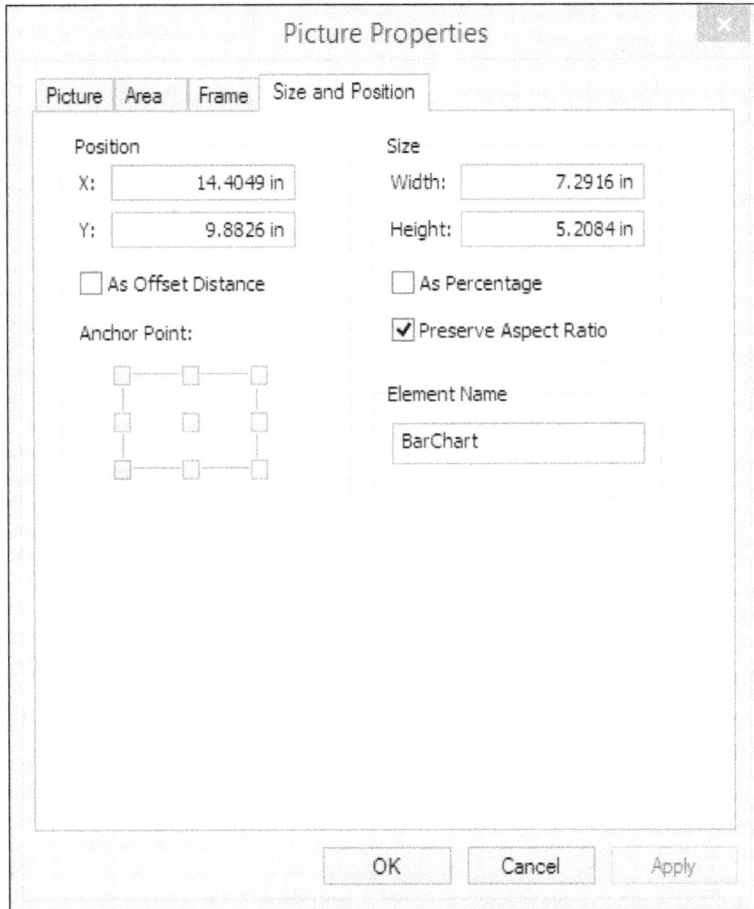

14. Right-click on the lower **Picture** element and select **Properties**. Under the **Size** and **Position** tab, give it the **Element Name** Line Chart.

15. Right-click on the **Title** element and select **Properties**. Under the **Size** and **Position** tab, give it the **Element Name** CrimeTitle.

16. Save the map document file.

17. In the next few steps, we'll update the code of our tools to automatically place the generated charts into the layout view. Open the Python development environment with the loaded CrimeAnalysis.pyt code.

18. Locate the `CreateNeighborhoodBarChart` class and the `execute()` method inside this class.

19. At the bottom of the `execute()` method, add the following code block. This will insert the bar chart directly into the layout view:

```
fig = Figure(data=data,layout=layout)
fileToSave = fileLocation + ".png"
py.image.save_as(fig, filename=fileToSave)

mxd = arcpy.mapping.MapDocument("CURRENT")
for elm in arcpy.mapping.ListLayoutElements(mxd, "PICTURE_
ELEMENT"):
    if elm.name == "BarChart":
        elm.sourceImage = fileToSave
mxd.save()

    arcpy.AddMessage("Created and saved chart to: " +
fileLocation)
except Exception as e:
    arcpy.AddMessage(e.message)
```

20. Locate the `CreateLinePlot` class and the `execute()` method inside this class.

21. At the bottom of the `execute()` method, add the following code block. This will insert the line chart directly into the layout view:

```
fig = Figure(data=data,layout=layout)
fileToSave = fileLocation + ".png"
py.image.save_as(fig, filename=fileToSave)
mxd = arcpy.mapping.MapDocument("CURRENT")
for elm in arcpy.mapping.ListLayoutElements(mxd, "PICTURE_
ELEMENT"):
    if elm.name == "LineChart":
        elm.sourceImage = fileToSave
mxd.save()
arcpy.AddMessage("Created and saved chart to: " + fileLocation)
```

22. Locate the `CreateMap` class and the `execute()` method inside this class.

23. Update the line of code that retrieves the data frame, which is shown as follows:

```
df = arcpy.mapping.ListDataFrames(mxd, "Maps")[0]
```

24. Save the code.

25. Now, it's time to test. Activate the **Charts** and **Graphs** data frame and run the **Create Neighborhood Bar Chart** tool with the parameters specified in the following screenshot. The execution of the tool should result in the graph being created as a .png image file and being placed into the **Layout** view.

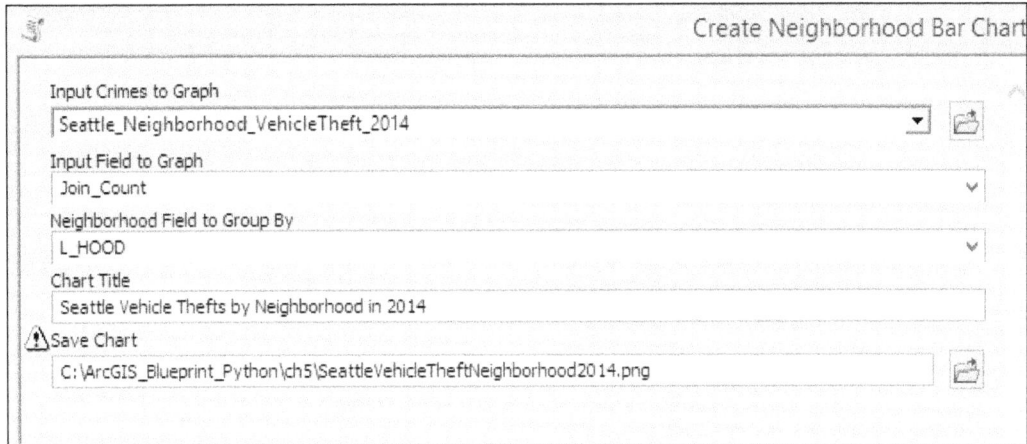

26. Run the **Create Line Plot** tool with the parameters specified in the following screenshot. The execution of the tool should result in the graph being created as a .png image file and being placed into the **Layout** view.

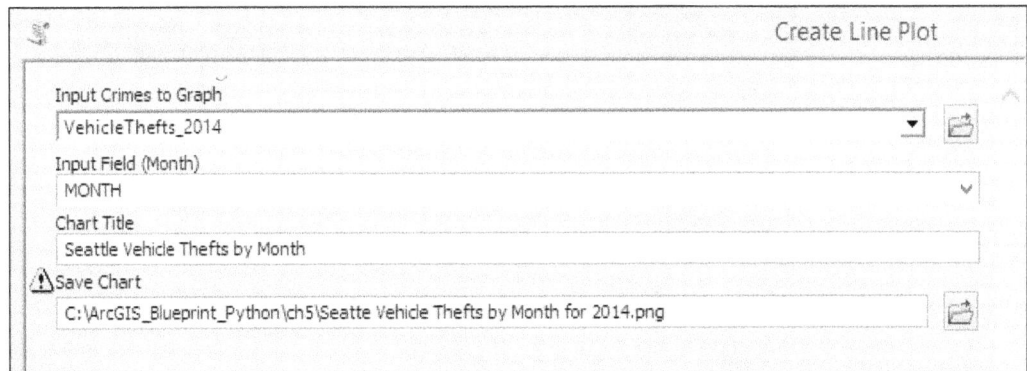

27. Activate the **Maps** data frame and run the **Create Map** tool. Fill in the parameters as follows:

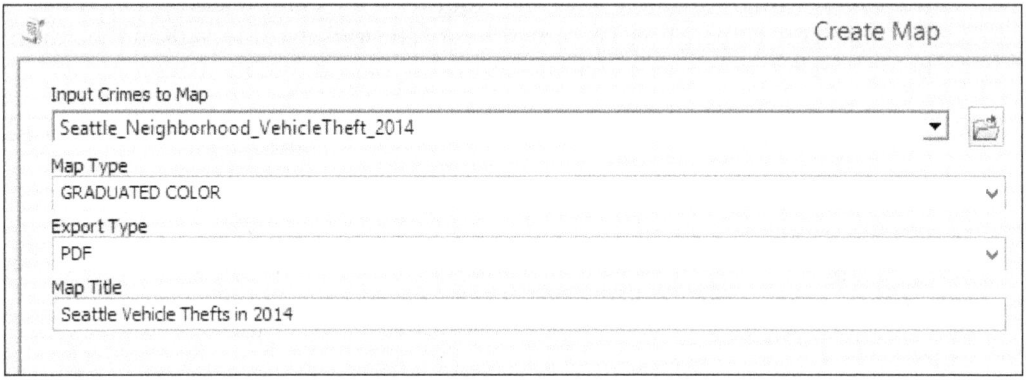

28. A new PDF file called `Seattle Vehicle Theft in 2014.pdf` should now be in your `C:\ArcGIS_Blueprint_Python\ch5` folder. It should appear as shown in the following screenshot:

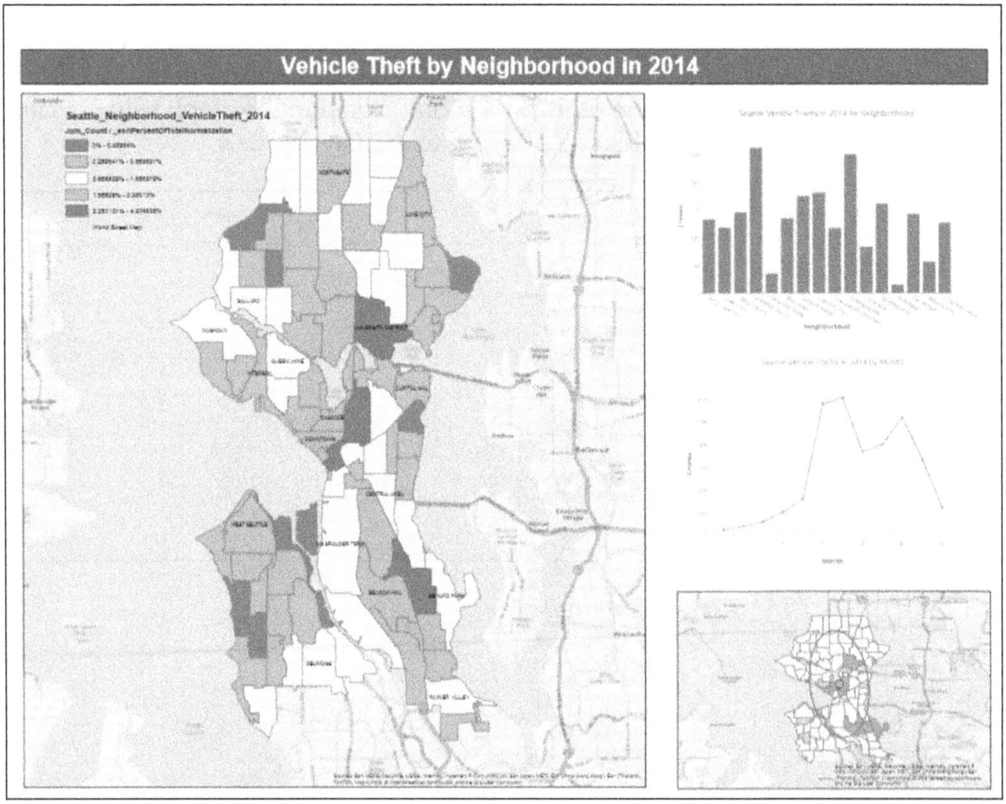

Summary

In this chapter, two tools were added to the **Crime Analysis** toolbox to provide crime data visualization products that supplement and enhance the mapping products created in the previous chapter. Using the crime data imported and aggregated from the open data Seattle police department, the Plotly graphing and charting Python library was used to create bar charts and line plots. The Plotly library can be used to create a wide variety of data visualization products in addition to the simple graphs created in this chapter. Take some time to explore the variety of additional data visualization products available through the Plotly library.

In the next chapter, we'll explore the use of wxPython to create advanced, engaging user interfaces for ArcGIS Desktop.

6
Viewing and Querying Parcel Data

One of the primary limitations of working with Python in ArcGIS Desktop has been the lack of tools for the development of Graphical User Interfaces (GUI). Previous chapters in this book covered the use of custom ArcGIS script tools as well as Python add-ins to capture user input, but both of these options are limited. The core Python library includes *Tkinter* for user interface development, but it doesn't provide a modern look and can be difficult to work with. This chapter will cover the use of wxPython to build advanced user interfaces for ArcGIS Desktop.

wxPython is a GUI toolkit for Python that enables the creation of advanced user interfaces. This library is a Python extension module (native code), that wraps the wxWidgets cross-platform GUI library written in C++. This chapter uses the wxPython library in combination with an ArcGIS Desktop add-in written in Python to capture the user input to query and view parcel data.

In this chapter, we will cover the following topics:

- GUI development with wxPython
- Creating an ArcGIS Desktop add-in with Python
- Using the ArcPy data access module to search for parcel data
- Using the webbrowser module to open a web browser

Design

Conceptually, building the ArcGIS application in this chapter is pretty simple. However, it is going to require a lot of code. We're going to build an application that will query and view parcel data for Kendall County, TX. The graphical user interface (GUI) for the application will be built using wxPython. The interface will be created in a Python script file called Interface.py. wxPython is an excellent choice for GUI development and will improve your ability to develop user interfaces that have only been possible with a combination of **ArcObjects** and .NET in the past. To display the user interface, we will build an ArcGIS Desktop add-in that will consist of an extension along with a button in a toolbar. When the user clicks on the button, it will trigger the display of the user interface built in wxPython. The interface will provide multiple options to query the parcel feature class. Records returned by a query will be displayed in a grid-type structure. Users can then select a record from the grid to have the application zoom in to the parcel and open the web-based **Kendall County Appraisal District** search results for the selected property. The following diagram shows the structure and functions of **Parcel Feature Class**:

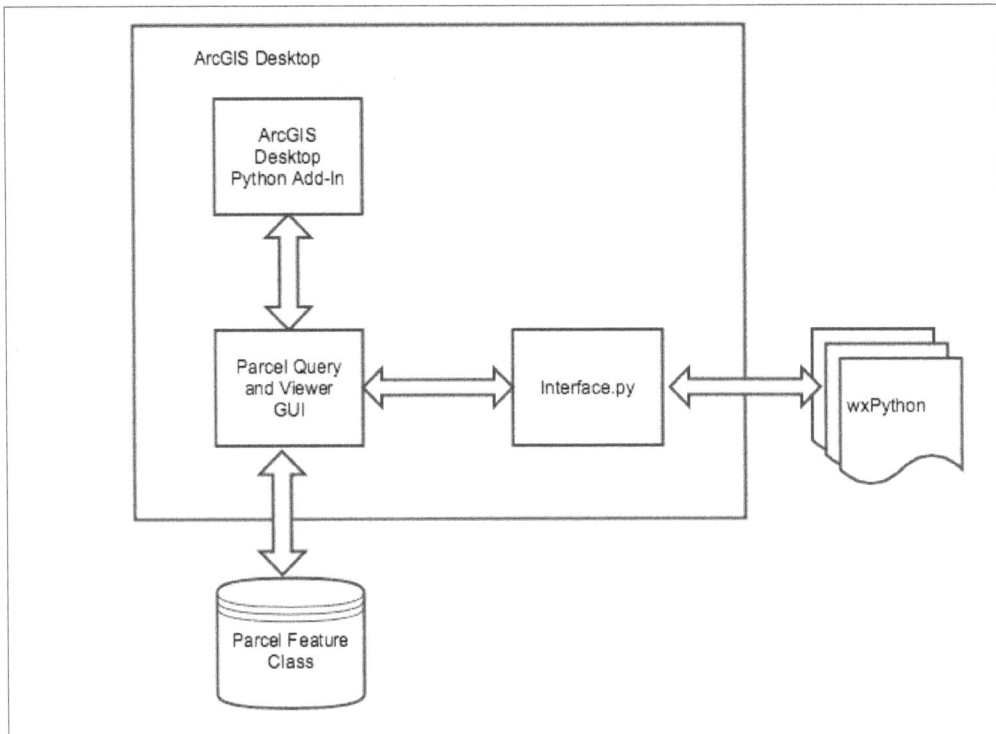

Let's get started and build the application.

Creating a user interface with wxPython

In this step, you will learn how to use wxPython to create a user interface that captures information that will be used to perform various types of queries against a parcel layer. After completing this section, you will have a user interface that allows the end user to search for parcels by the owner name, address, unique identifier, and an advanced search that allows for a combination of search terms. The final user interface will appear as shown in the following screenshot:

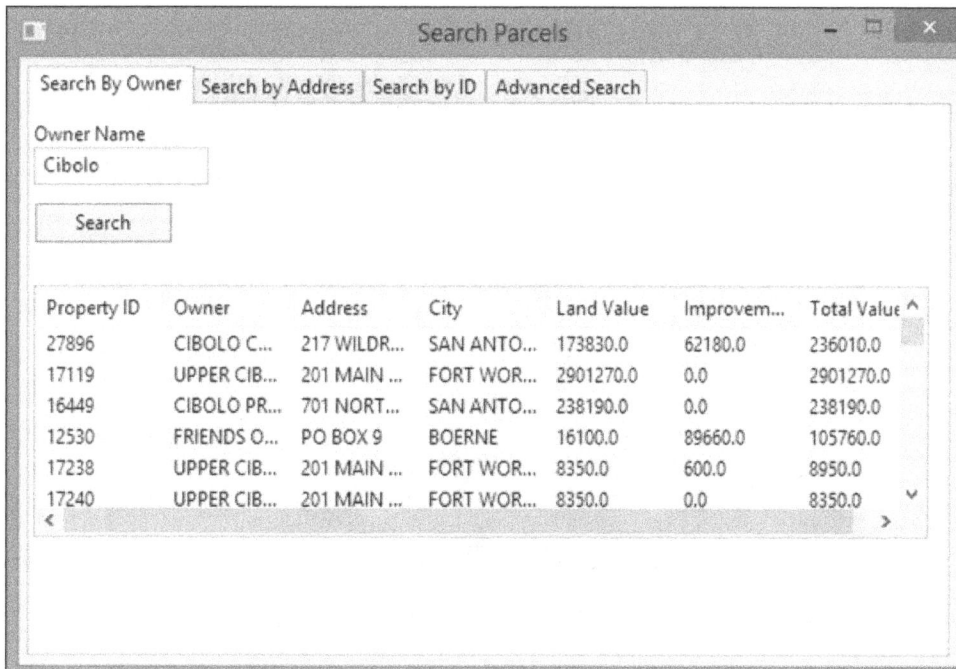

The interface consists of four tabs: **Search by Owner**, **Search by Address**, **Search by ID**, and **Advanced Search**. The preceding screenshot displays the **Search by Owner** interface that is used to search by the owner name or a portion of the owner name. All four tabs also include a grid control that displays the results of the search. When a record is selected from the grid, ArcGIS Desktop will zoom in to the extent of that feature and also open the **Kendall County Appraisal District** parcel search website for the selected record, as shown in the following screenshot:

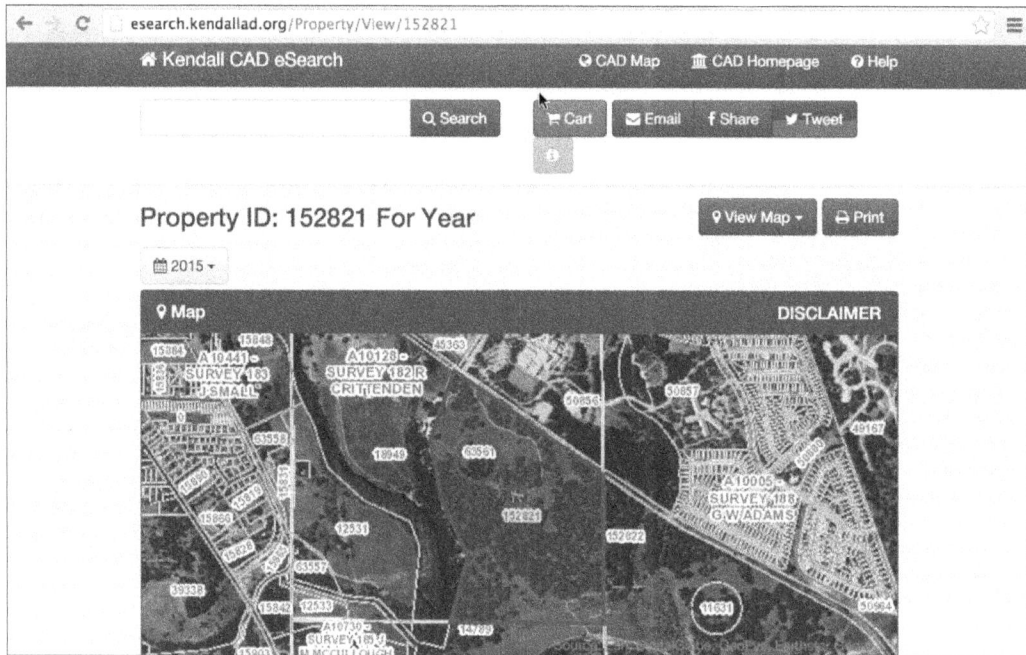

The **Search by ID** interface is similar to the **Search by Owner** interface with the exception that it searches by a unique property identifier rather than an owner name. Otherwise, it works the same as the **Search by Owner** interface.

The **Search by Address** tab includes user interface controls required to capture a street name (or a portion of a street name), and/or a subdivision. The subdivision control is a ComboBox filled with predefined values that can be selected by the user.

Finally, the **Advanced Search** tab contains user interface controls required to capture any combination of an owner name, street name, subdivision, and the minimum and maximum property values. This can be seen in the following screenshot:

The wxPython library will need to be installed before continuing with this section.

If you haven't already installed wxPython, you can download and install the library using an installer found at the mentioned URL. Versions 10.1 and higher of ArcGIS Desktop use Python 2.7, so you will want to install wxPython 3.0 for this version. The URL is http://www.wxpython.org/download.php#msw.

The following steps will guide you to create the user interface with wxPython:

1. Open your Python development environment and create a new script. Save it as C:\ArcGIS_Blueprint_Python\ch6\Interface.py.

2. Import the wx, arcpy, pythonaddins, and webbrowser modules:

```
import wx
import arcpy
import pythonaddins
import webbrowser
```

3. Create a class called `MainFrame` that accepts `wx.Frame` as the only parameter and defines the initialization method:

```
class MainFrame(wx.Frame):
    def __init__(self):
```

4. Inside the `__init__` function, define the initialization for `wx.Frame`. A **Frame** widget in wxPython acts as a top-level widget that essentially defines the window for the user interface. The parameters passed into the initialization function will define the title for **Frame** along with the size and style:

```
class MainFrame(wx.Frame):
    def __init__(self):
        wx.Frame.__init__(self, None, title="Search Parcels",
            size=(600,400),style=wx.DEFAULT_FRAME_STYLE &
            ~wx.MAXIMIZE_BOX ^ wx.RESIZE_BORDER)
```

5. Create a **Panel** inside **Frame**. The **Panel** will serve as a container for the widgets in the user interface:

```
class MainFrame(wx.Frame):
    def __init__(self):
        wx.Frame.__init__(self, None, title="Search Parcels",
            size=(600,400),style=wx.DEFAULT_FRAME_STYLE &
            ~wx.MAXIMIZE_BOX ^ wx.RESIZE_BORDER)

        self.Bind(wx.EVT_CLOSE, self.OnClose)
        # Here we create a panel and a notebook on the panel
        panel = wx.Panel(self)
```

6. We'll return to the `MainFrame` class shortly, but now you need to create a second class called `NotebookParcel`, as shown in the following code. This class accepts a single `wx.Notebook` object in the constructor. A Notebook widget in wxPython acts like a tabbed widget that has multiple panes:

```
class NotebookParcel(wx.Notebook):
    def __init__(self, parent):
        wx.Notebook.__init__(self, parent, id=wx.ID_ANY, style=wx.
BK_DEFAULT)
```

7. Now, return to the `MainFrame` class and create a new object instance of `NotebookParcel`:

```
class MainFrame(wx.Frame):
    def __init__(self):
        wx.Frame.__init__(self, None, title="Search Parcels",s
ize=(600,400),style=wx.DEFAULT_FRAME_STYLE & ~wx.MAXIMIZE_BOX ^
wx.RESIZE_BORDER)
```

```
# Here we create a panel and a notebook on the panel
panel = wx.Panel(self)

notebook = NotebookParcel(panel)
```

8. Create a vertical `BoxSizer` object and add the new instance of `NotebookParcel`. The `BoxSizer` object is used for layout management in the placement of widgets on the user interface. Rather than using absolute positioning when adding widgets to the interface, we use `Sizer` objects. These `Sizer` objects give a lot of flexibility in that the widgets inside the `Sizer` resize and reposition themselves if the user expands or contracts the container window:

```
class MainFrame(wx.Frame):
    def __init__(self):
        wx.Frame.__init__(self, None, title="Search Parcels",s
ize=(600,400),style=wx.DEFAULT_FRAME_STYLE & ~wx.MAXIMIZE_BOX ^
wx.RESIZE_BORDER)

        self.Bind(wx.EVT_CLOSE, self.OnClose)
        # Here we create a panel and a notebook on the panel
        panel = wx.Panel(self)

        arcpy.env.workspace = r"C:\ArcGIS_Blueprint_Python\data\
Kendall"

        notebook = NotebookParcel(panel)
        sizer = wx.BoxSizer(wx.VERTICAL)
        sizer.Add(notebook, 1, wx.ALL|wx.EXPAND, 5)

        panel.SetSizer(sizer)
        self.Layout()
```

9. Over the next few steps, we'll create several classes, each of which will define one of the tabs on the user interface. Let's start with the `PageOwner` class, which will define the tab used to capture the input to perform a parcel search by the owner. Add the following code, and then we'll discuss how it works:

```
class PageOwner(wx.Panel):
    def __init__(self,parent):
        wx.Panel.__init__(self, parent)
        sizer = wx.BoxSizer(wx.VERTICAL)

        staticTxtOwner = wx.StaticText(self, -1, "Owner Name")
        self.txtOwner = wx.TextCtrl(self, wx.ID_ANY, "")

        btnOK = wx.Button(self, 1, 'Search')
        btnOK.Bind(wx.EVT_BUTTON, self.OnOk, id=1)
```

```
        self.lc = wx.ListCtrl(self, -1, style=wx.LC_REPORT)
        self.lc.Bind(wx.EVT_LIST_ITEM_SELECTED, self.OnClick,
self.lc)
        self.lc.InsertColumn(0, 'Property ID')
        self.lc.InsertColumn(1, 'Owner')
        self.lc.InsertColumn(2, 'Address')
        self.lc.InsertColumn(3, 'City')
        self.lc.InsertColumn(4, 'Land Value')
        self.lc.InsertColumn(5, 'Improvement Value')
        self.lc.InsertColumn(6, 'Total Value')

        sizer.AddSpacer(10)
        sizer.Add(staticTxtOwner)
        sizer.Add(self.txtOwner)
        sizer.AddSpacer(10)
        sizer.Add(btnOK)
        sizer.AddSpacer(25)
        sizer.Add(self.lc)

        self.SetSizer(sizer)
```

The `PageOwner` class accepts a single parameter in the constructor. This parameter is an instance of `wx.Panel`, which acts as a container for other widgets. Inside the `__init__` function, we create the `panel` and `BoxSizer` objects for the layout management of **Panel**. This tab will contain four widgets: `StaticText`, `TextCtrl`, `Button`, and `ListCtrl`. A `StaticText` widget simply holds text that normally functions as descriptive information about another widget. The `TextCtrl` widget is used to capture an owner name or a portion of an owner name. Finally, the `ListCtrl` widget acts like a grid control to display the results of the query.

The two lines of the following code that you added create a `StaticText` object with the **Owner Name** text and a `TextCtrl` object that will capture an owner name from the user:

```
staticTxtOwner = wx.StaticText(self, -1, "Owner Name")
self.txtOwner = wx.TextCtrl(self, wx.ID_ANY, "")
```

The next two lines of code create the `Button` object and bind an event to this object. An event is an action that takes place within the application. In this case, the event is simply a button click with which we want to initiate the search. When the event takes place, we want the code to run in response to that event. We'll define the contents of the event handler in a later step:

```
btnOK = wx.Button(self, 1, 'Search')
btnOK.Bind(wx.EVT_BUTTON, self.OnOk, id=1)
```

The `ListCtrl` object is created with the code block you see in the following code. When created in **Report** mode, as is the case with our example, `ListCtrl` is made to appear like a grid. In addition to defining various columns to hold the results of the query, an event handler is also created to respond to a row that has been selected from the object. The event-handling code will be defined in a later step:

```
self.lc = wx.ListCtrl(self, -1, style=wx.LC_REPORT)
self.lc.Bind(wx.EVT_LIST_ITEM_SELECTED, self.OnClick, self.lc)
self.lc.InsertColumn(0, 'Property ID')
self.lc.InsertColumn(1, 'Owner')
self.lc.InsertColumn(2, 'Address')
self.lc.InsertColumn(3, 'City')
self.lc.InsertColumn(4, 'Land Value')
self.lc.InsertColumn(5, 'Improvement Value')
self.lc.InsertColumn(6, 'Total Value')
```

Finally, the widgets are added to `BoxSizer`, and `BoxSizer` is applied to **Panel**.

10. Add a new class called `PageAddress`. This class will define the content for the tab that is used to search for parcels by the address. Add the following code:

```
class PageAddress(wx.Panel):
    def __init__(self, parent):
        wx.Panel.__init__(self, parent)
        sizer = wx.BoxSizer(wx.VERTICAL)


        staticTxtStreetName = wx.StaticText(self, -1, "Street
Name")
        self.txtStreetName = wx.TextCtrl(self, wx.ID_ANY, "")

        staticTxtSubdivision = wx.StaticText(self, -1,
"Subdivision")
        subdivisions = ['Stone Creek Ranch', 'Cordillera',
'Waterstone', 'Coveney Ranch' ]
        self.comboBoxSub = wx.ComboBox(self, -1, size=(150, -1),
choices=subdivisions, style=wx.CB_READONLY)

        btnOK = wx.Button(self, 1, 'Search')
        btnOK.Bind(wx.EVT_BUTTON, self.OnOk, id=1)
```

```
        self.lc = wx.ListCtrl(self, -1, style=wx.LC_REPORT)
        self.lc.Bind(wx.EVT_LIST_ITEM_SELECTED, self.OnClick,
    self.lc)
        self.lc.InsertColumn(0, 'Property ID')
        self.lc.InsertColumn( 1, 'Owner')
        self.lc.InsertColumn(2, 'Address')
        self.lc.InsertColumn(3, 'City')
        self.lc.InsertColumn(4, 'Land Value')
        self.lc.InsertColumn(5, 'Improvement Value')
        self.lc.InsertColumn(6, 'Total Value')

        sizer.AddSpacer(10)
        sizer.Add(staticTxtStreetName)
        sizer.Add(self.txtStreetName)
        sizer.AddSpacer(10)
        sizer.Add(staticTxtSubdivision)
        sizer.Add(self.comboBoxSub)
        sizer.AddSpacer(10)
        sizer.Add(btnOK)
        sizer.AddSpacer(25)
        sizer.Add(self.lc)
        self.SetSizer(sizer)
```

There are quite a few similarities between the `PageAddress` class and the `PageOwner` class, so I won't go into as much detail in the code. In this class, we created `StaticText` objects for the street number and street name along with the subdivision. Two `TextCtrl` objects were defined to capture the street number and street name information from the user. A ComboBox object was also created and populated with a list of values. The following code illustrates how this was accomplished. There are many more subdivisions in Kendall County, TX, than the four that were defined in this list, but this will keep things simple for our example application. The rest of the code should be very familiar to you from the `PageOwner` class. It simply involves the creation of the `ListCtrl` widget and the addition of all the widgets to `BoxSizer`:

```
subdivisions = ['Stone Creek Ranch', 'Cordillera', 'Waterstone',
'Coveney Ranch' ]
comboBoxSub = wx.ComboBox(self, -1, size=(150, -1),
choices=subdivisions, style=wx.CB_READONLY)
```

11. Next, we'll create the `PageID` class that is essentially the same as the `PageOwner` class, so we won't spend much time discussing this code block. Add the following code:

```
class PageID(wx.Panel):
    def __init__(self, parent):
        wx.Panel.__init__(self, parent)
        sizer = wx.BoxSizer(wx.VERTICAL)
        staticTxtID = wx.StaticText(self, -1, "Unique Identifier")
        self.txtID = wx.TextCtrl(self, wx.ID_ANY, "")
        btnOK = wx.Button(self, 1, 'Search')
        btnOK.Bind(wx.EVT_BUTTON, self.OnOk, id=1)

        self.lc = wx.ListCtrl(self, -1, style=wx.LC_REPORT)
        self.lc.Bind(wx.EVT_LIST_ITEM_SELECTED, self.OnClick,
        self.lc)
        self.lc.InsertColumn(0, 'Property ID')
        self.lc.InsertColumn(1, 'Owner')
        self.lc.InsertColumn(2, 'Address')
        self.lc.InsertColumn(3, 'City')
        self.lc.InsertColumn(4, 'Land Value')
        self.lc.InsertColumn(5, 'Improvement Value')
        self.lc.InsertColumn(6, 'Total Value')


        sizer.AddSpacer(10)
        sizer.Add(staticTxtID)
        sizer.Add(self.txtID)
        sizer.AddSpacer(10)
        sizer.Add(btnOK)
        sizer.AddSpacer(25)
        sizer.Add(self.lc)
        self.SetSizer(sizer)
```

The only difference between the `PageID` and `PageOwner` class is that, in the `PageID` class, we are capturing a unique property identifier rather than an owner name.

12. The last class that we'll create is the `PageAdvanced` class. While there is more code in this class than the others, there shouldn't be any new widgets. Add the following code block:

```
class PageAdvanced(wx.Panel):
    def __init__(self, parent):
        wx.Panel.__init__(self, parent)
        sizer = wx.BoxSizer(wx.VERTICAL)
```

```
        sizer.AddSpacer(10)

        inputOneSizer   = wx.BoxSizer(wx.HORIZONTAL)
        inputTwoSizer   = wx.BoxSizer(wx.HORIZONTAL)
        inputThreeSizer = wx.BoxSizer(wx.HORIZONTAL)
        inputFourSizer = wx.BoxSizer(wx.HORIZONTAL)
        btnSizer        = wx.BoxSizer(wx.HORIZONTAL)

        staticTxtOwner = wx.StaticText(self, -1, "Owner Name")
        self.txtOwner = wx.TextCtrl(self, wx.ID_ANY, "")

        staticTxtStreetName = wx.StaticText(self, -1, "Street
Name")
        self.txtStreetName = wx.TextCtrl(self, wx.ID_ANY, "")

        inputOneSizer.Add(staticTxtOwner)
        inputOneSizer.Add(self.txtOwner)
        inputOneSizer.AddSpacer(10)
        inputOneSizer.Add(staticTxtStreetName)
        inputOneSizer.Add(self.txtStreetName)

        staticTxtSubdivision = wx.StaticText(self, -1,
"Subdivision")
        subdivisions = ['Stone Creek Ranch', 'Cordillera',
'Waterstone', 'Coveney Ranch' ]
        self.comboBoxSub = wx.ComboBox(self, -1, size=(150, -1),
choices=subdivisions, style=wx.CB_READONLY)

        inputTwoSizer.Add(staticTxtSubdivision)
        inputTwoSizer.Add(self.comboBoxSub)

        staticTxtMinVal = wx.StaticText(self, -1, "Minimum Value")
        self.txtMinVal = wx.TextCtrl(self, wx.ID_ANY, "")

        staticTxtMaxVal = wx.StaticText(self, -1, "Maximum Value")
        self.txtMaxVal = wx.TextCtrl(self, wx.ID_ANY, "")

        inputThreeSizer.Add(staticTxtMinVal)
        inputThreeSizer.Add(self.txtMinVal)
        inputThreeSizer.AddSpacer(10)
        inputThreeSizer.Add(staticTxtMaxVal)
        inputThreeSizer.Add(self.txtMaxVal)

        btnOK = wx.Button(self, 1, 'Search')
        btnOK.Bind(wx.EVT_BUTTON, self.OnOk, id=1)
```

```
        btnSizer.Add(btnOK)

        self.lc = wx.ListCtrl(self, -1, style=wx.LC_REPORT)
        self.lc.Bind(wx.EVT_LIST_ITEM_SELECTED, self.OnClick,
self.lc)
        self.lc.InsertColumn(0, 'Property ID')
        self.lc.InsertColumn(1, 'Owner')
        self.lc.InsertColumn(2, 'Address')
        self.lc.InsertColumn(3, 'City')
        self.lc.InsertColumn(4, 'Land Value')
        self.lc.InsertColumn(5, 'Improvement Value')
        self.lc.InsertColumn(6, 'Total Value')
        inputFourSizer.Add(self.lc)

        sizer.Add(inputOneSizer)
        sizer.AddSpacer(25)
        sizer.Add(inputTwoSizer)
        sizer.AddSpacer(25)
        sizer.Add(inputThreeSizer)
        sizer.AddSpacer(25)
        sizer.Add(btnSizer)
        sizer.AddSpacer(25)
        sizer.Add(inputFourSizer)

        self.SetSizer(sizer)
```

13. Return to the `NotebookParcel` class. Add the following code block. This code block creates new instances of the `PageOwner`, `PageAddress`, `PageID`, and `PageAdvanced` classes and adds them to `NotebookParcel`. Remember that `NotebookParcel` is a tabbed structure, and each of the Page classes is an individual tab in the container. The second parameter passed into the `AddPage()` function defines the text that should be displayed in the tab:

```
class NotebookParcel(wx.Notebook):
    def __init__(self, parent):
        wx.Notebook.__init__(self, parent, id=wx.ID_ANY, style=wx.
BK_DEFAULT)

        tabOwner = PageOwner(self)
        self.AddPage(tabOwner, "Search By Owner")

        tabAddress = PageAddress(self)
        self.AddPage(tabAddress, "Search by Address")

        tabID = PageID(self)
```

```
self.AddPage(tabID, "Search by ID")

tabAdvanced = PageAdvanced(self)
self.AddPage(tabAdvanced, "Advanced Search")
```

14. At this point, we have defined the basic look of the interface. Now, we need to return to each of the Page classes and add the code that responds to the events that we have bound. Find the PageOwner class, then the lines of code that you see here:

```
btnOK = wx.Button(self, 1, 'Search')
btnOK.Bind(wx.EVT_BUTTON, self.OnOk, id=1)

self.lc = wx.ListCtrl(self, -1, style=wx.LC_REPORT)
self.lc.Bind(wx.EVT_LIST_ITEM_SELECTED, self.OnClick, self.lc)
```

In the first highlighted line of code, we bind the EVT_BUTTON event to a method called OnOk. This basically means that, when the button is clicked on in the interface, a method called OnOK will be executed. The second highlighted line of code binds the EVT_LIST_ITEM_SELECTED event to a method called OnClick. This means that, when the user clicks on or selects a row in the ListCtrl widget, the OnClick method will be executed. The OnOk and OnClick methods both need to be written.

15. Add a new method called OnOK() to the PageOwner class. It should line up exactly with the __init__ method:

```
def OnOk(self, event):
```

16. Add the following code block to the OnOk method, and then we'll discuss what the code accomplishes:

```
def OnOk(self, event):
        if self.txtOwner.GetValue():
            owner = self.txtOwner.GetValue().upper()
            queryString =  "file_as_na LIKE \'%" + owner + "%\'"

            self.lc.DeleteAllItems()

            with arcpy.da.SearchCursor("Kendall_Parcels.shp",
("PROP_ID","file_as_na", "situs_num" , "situs_st_1", "situs_st_2",
"situs_city", "land_val", "imprv_val", "market"), queryString) as
cursor:
                    flag = False
                    for row in cursor:
                        flag = True
                        pos = self.lc.InsertStringItem(0, str(row[0]))
```

```
                  self.lc.SetStringItem(pos,1,row[1])
                  self.lc.SetStringItem(pos,2,row[2] + " " +
row[3] + " " + row[4])
                  self.lc.SetStringItem(pos,3,row[5])
                  self.lc.SetStringItem(pos,4,str(row[6]))
                  self.lc.SetStringItem(pos,5,str(row[7]))
                  self.lc.SetStringItem(pos,6,str(row[8]))

              if not flag:
                  pythonaddins.MessageBox("No records found",
"Query Error", 0)
          else:
              pythonaddins.MessageBox("Enter an owner name or
portion of a name", "Query Error", 0)
```

This code block will execute any time a user clicks on the **Search** button in the **Search by Owner** tab. The code first retrieves the input from the txtOwner widget and converts the string to uppercase. The data in the Kendall_Parcels shapefile for Kendall County is stored in uppercase for text data fields. In the second line of the following code, note that the Kendall_Parcels shapefile uses the LIKE clause along with wildcard (&) characters to define the WHERE clause that will be used to search for parcels:

```
owner = self.txtOwner.GetValue().upper()
queryString =  "file_as_na LIKE \'%" + owner + "%\'"
```

The DeleteAllItems() function is then called on the ListCtrl object to remove any existing records for past searches.

Next, a SearchCursor object is created against the Kendall_Parcels shapefile. In creating the SearchCursor object, the SearchCursor() function is called, with the Kendall_Parcels.shp shapefile passed in as the first parameter, and a tuple containing the list of fields to be returned is passed as the second parameter. A for loop is then used to loop through the contents of the cursor, and the ListCtrl widget is populated:

```
with arcpy.da.SearchCursor("Kendall_Parcels.shp", ("PROP_
ID","file_as_na", "situs_num" , "situs_st_1", "situs_st_2",
"situs_city", "land_val", "imprv_val", "market"), queryString) as
cursor:
              flag = False
              for row in cursor:
                  flag = True
                  pos = self.lc.InsertStringItem(0, str(row[0]))
                  self.lc.SetStringItem(pos,1,row[1])
                  self.lc.SetStringItem(pos,2,row[2] + " " +
row[3] + " " + row[4])
```

```
        self.lc.SetStringItem(pos,3,row[5])
        self.lc.SetStringItem(pos,4,str(row[6]))
        self.lc.SetStringItem(pos,5,str(row[7]))
        self.lc.SetStringItem(pos,6,str(row[8]))


    if not flag:
        pythonaddins.MessageBox("No records found",
"Query Error", 0)
```

17. Add a new method to the `PageOwner` class called `OnClick`. The `OnClick` method will execute any time a use clicks on a populated row in the `ListCtrl` widget. Add the following code block, and then we'll discuss how it works:

```
def OnClick(self,event):
        prop_id = event.GetText()
        try:
            mxd = arcpy.mapping.MapDocument("CURRENT")

            arcpy.MakeFeatureLayer_management("Kendall_Parcels.
shp","parcels_lyr")
            arcpy.SelectLayerByAttribute_management("parcels_lyr",
"NEW_SELECTION", "PROP_ID = " + prop_id)
            result = arcpy.GetCount_management("parcels_lyr")
            count = int(result.getOutput(0))

            df = arcpy.mapping.ListDataFrames(mxd)[0]
            layer = arcpy.mapping.ListLayers(mxd, "parcels_lyr",
df)[0]
            df.extent = layer.getSelectedExtent()

            webbrowser.open_new('http://esearch.kendallad.org/
Property/View/' + prop_id)

        except Exception as e:
            pythonaddins.MessageBox(e.message, "Query Error", 0)
```

The `prop_id = event.GetText()` line of code retrieves the property ID from the selected record in `ListCtrl`. Inside the `try` statement, an instance of the current map document is retrieved, and a feature layer is created with the lines of code you see here:

```
mxd = arcpy.mapping.MapDocument("CURRENT")
arcpy.MakeFeatureLayer_management("Kendall_Parcels.shp","parcels_
lyr")
```

The `SelectLayerByAttribute` tool is then executed against the feature layer with a `WHERE` clause set to the property `ID` retrieved from `ListCtrl`. This will select a feature from the `parcels_lyr` feature layer:

```
arcpy.SelectLayerByAttribute_management("parcels_lyr", "NEW_
SELECTION", "PROP_ID = " + prop_id)
```

If a matching record is found, zoom in to the extent of the selected feature:

```
df = arcpy.mapping.ListDataFrames(mxd)[0]
layer = arcpy.mapping.ListLayers(mxd, "parcels_lyr", df)[0]
df.extent = layer.getSelectedExtent()
```

Finally, open a web browser and the **Kendall County Appraisal District** web page that contains information about this particular property at `webbrowser.open_new('http://esearch.kendallad.org/Property/View/' + prop_id)`.

18. The `OnClick` and `OnOk` methods for the `PageID` class are very similar to the `PageOwner` class. The only difference is that we're now querying by the property `ID` instead of the owner name. Add the following code blocks to the `PageID` class to implement these event handlers:

```
def OnClick(self,event):
    prop_id = event.GetText()
    try:
        mxd = arcpy.mapping.MapDocument("CURRENT")

        arcpy.MakeFeatureLayer_management("Kendall_Parcels.
shp","parcels_lyr")
        arcpy.SelectLayerByAttribute_management("parcels_lyr",
"NEW_SELECTION", "PROP_ID = " + prop_id)
        result = arcpy.GetCount_management("parcels_lyr")
        count = int(result.getOutput(0))

        df = arcpy.mapping.ListDataFrames(mxd)[0]
        layer = arcpy.mapping.ListLayers(mxd, "parcels_lyr", df)
[0]
        df.extent = layer.getSelectedExtent()

        webbrowser.open_new('http://esearch.kendallad.org/
Property/View/' + prop_id)

    except Exception as e:
        pythonaddins.MessageBox(e.message, "Query Error", 0

def OnOk(self, event):
```

```
          if self.txtID.GetValue():
              id = self.txtID.GetValue()
              queryString =  "PROP_ID = " + id

              self.lc.DeleteAllItems()

              with arcpy.da.SearchCursor("Kendall_Parcels.shp", ("PROP_
ID","file_as_na", "situs_num" , "situs_st_1", "situs_st_2",
"situs_city", "land_val", "imprv_val", "market"), queryString) as
cursor:
                  flag = False
                  for row in cursor:
                      flag = True
                      pos = self.lc.InsertStringItem(0, str(row[0]))
                      self.lc.SetStringItem(pos,1,row[1])
                      self.lc.SetStringItem(pos,2,row[2] + " " + row[3]
+ " " + row[4])
                      self.lc.SetStringItem(pos,3,row[5])
                      self.lc.SetStringItem(pos,4,str(row[6]))
                      self.lc.SetStringItem(pos,5,str(row[7]))
                      self.lc.SetStringItem(pos,6,str(row[8]))

                  if not flag:
                      pythonaddins.MessageBox("No records found", "Query
Error", 0)
              else:
                  pythonaddins.MessageBox("Enter an ID", "Query Error", 0)
```

19. It's time to implement the event handlers for the PageAddress class. Find the PageAddress class in your code. First, we'll implement the OnOk() method. Add the following code block, and then we'll discuss how it works:

```
def OnOk(self, event):
        strStreetName = self.txtStreetName.GetValue().upper()
        strSubdivision = self.comboBoxSub.GetValue().upper()

        if strStreetName or strSubdivision:
            if strStreetName and not strSubdivision:
                queryString =  "situs_st_1 LIKE \'%" +
strStreetName + "%\'"
            elif strSubdivision and not strStreetName:
                queryString =  "DESC_ LIKE \'%" + strSubdivision +
"%\'"
            elif strSubdivision and strStreetName:
```

```
                queryString =  "DESC_ LIKE \'%" + strSubdivision +
"%\' and situs_st_1 LIKE \'%" + strStreetName + "%\'"



            self.lc.DeleteAllItems()

            with arcpy.da.SearchCursor("Kendall_Parcels.shp",
("PROP_ID","file_as_na", "situs_num" , "situs_st_1", "situs_st_2",
"situs_city", "land_val", "imprv_val", "market"), queryString) as
cursor:
                flag = False
                for row in cursor:
                    flag = True
                    pos = self.lc.InsertStringItem(0, str(row[0]))
                    self.lc.SetStringItem(pos,1,row[1])
                    self.lc.SetStringItem(pos,2,row[2] + " " +
row[3] + " " + row[4])
                    self.lc.SetStringItem(pos,3,row[5])
                    self.lc.SetStringItem(pos,4,str(row[6]))
                    self.lc.SetStringItem(pos,5,str(row[7]))
                    self.lc.SetStringItem(pos,6,str(row[8]))


                if not flag:
                    pythonaddins.MessageBox("No records found",
"Query Error", 0)
        else:
            pythonaddins.MessageBox("Enter a street name or
subdivision", "Query Error", 0)
```

The first two lines in the following code capture the values (if any) entered by the user:

```
strStreetName = self.txtStreetName.GetValue().upper()
strSubdivision = self.comboBoxSub.GetValue().upper()
```

The next code block, seen as follows, defines the SQL query that will be used when creating the `SearchCursor` object against the `Kendall_Parcels` shapefile:

```
if strStreetName or strSubdivision:
        if strStreetName and not strSubdivision:
            queryString =  "situs_st_1 LIKE \'%" +
strStreetName + "%\'"
        elif strSubdivision and not strStreetName:
            queryString =  "DESC_ LIKE \'%" + strSubdivision +
"%\'"
```

```
       elif strSubdivision and strStreetName:
              queryString =  "DESC_ LIKE \'%" + strSubdivision +
"%\' and situs_st_1 LIKE \'%" + strStreetName + "%\'"
```

The query will be created based on the input provided by the user. It is not required that both the street name and subdivision be defined. The user can provide the input for the street name or the subdivision, or they can provide both.

The last block of code should look familiar because it's essentially the same code block created in the OnOk() method for the PageOwner and PageID classes:

```
with arcpy.da.SearchCursor("Kendall_Parcels.shp", ("PROP_
ID","file_as_na", "situs_num" , "situs_st_1", "situs_st_2",
"situs_city", "land_val", "imprv_val", "market"), queryString) as
cursor:
              flag = False
              for row in cursor:
                  flag = True
                  pos = self.lc.InsertStringItem(0, str(row[0]))
                  self.lc.SetStringItem(pos,1,row[1])
                  self.lc.SetStringItem(pos,2,row[2] + " " +
row[3] + " " + row[4])
                  self.lc.SetStringItem(pos,3,row[5])
                  self.lc.SetStringItem(pos,4,str(row[6]))
                  self.lc.SetStringItem(pos,5,str(row[7]))
                  self.lc.SetStringItem(pos,6,str(row[8]))

              if not flag:
                  pythonaddins.MessageBox("No records found",
"Query Error", 0)
```

20. Finally, add the OnClick() method to PageAddress. The code should be exactly the same as the OnClick() event for the PageOwner and PageID classes. Refer to the previous steps, if you are unsure about the code.

21. Finally, we'll add the OnOk() and OnClick() event handlers to the PageAdvanced class. Find the class now and add the OnOk() code block seen here:

```
def OnOk(self, event):

        queryString = ""
        flagOwner = False
        if self.txtOwner.GetValue():
            flagOwner = True
```

```
        owner = self.txtOwner.GetValue().upper()
        queryString =  "file_as_na LIKE \'%" + owner + "%\'"

    flagStreet = False
    if self.txtStreetName.GetValue():
        flagStreet = True
        strStreetName = self.txtStreetName.GetValue().upper()
        if flagOwner:
            queryString = queryString + " AND " + "situs_st_1
LIKE \'%" + strStreetName + "%\'"
        else:
            queryString = "situs_st_1 LIKE \'%" +
strStreetName + "%\'"

    if self.comboBoxSub.GetValue():
        strSubdivision = self.comboBoxSub.GetValue().upper()
        if flagOwner or flagStreet:
            queryString = queryString + " AND " + "DESC_ LIKE
\'%" + strSubdivision + "%\'"
        else:
            queryString = "DESC_ LIKE \'%" + strSubdivision +
"%\'"

    if self.txtMinVal.GetValue():
        numMinVal = long(self.txtMinVal.GetValue())
        queryString = queryString + " AND market > " +
str(numMinVal)

    if self.txtMaxVal.GetValue():
        numMaxVal = long(self.txtMaxVal.GetValue())
        queryString = queryString + " AND market < " +
str(numMaxVal)

    if not queryString:
        pythonaddins.MessageBox("Enter one or more search
parameters", "Query Error", 0)
    else:
        self.lc.DeleteAllItems()

        with arcpy.da.SearchCursor("Kendall_Parcels.shp",
("PROP_ID","file_as_na", "situs_num" , "situs_st_1", "situs_st_2",
"situs_city", "land_val", "imprv_val", "market"), queryString) as
cursor:
            flag = False
            for row in cursor:
```

```
                          flag = True
                          pos = self.lc.InsertStringItem(0, str(row[0]))
                          self.lc.SetStringItem(pos,1,row[1])
                          self.lc.SetStringItem(pos,2,row[2] + " " +
row[3] + " " + row[4])
                          self.lc.SetStringItem(pos,3,row[5])
                          self.lc.SetStringItem(pos,4,str(row[6]))
                          self.lc.SetStringItem(pos,5,str(row[7]))
                          self.lc.SetStringItem(pos,6,str(row[8]))


                 if not flag:
                          pythonaddins.MessageBox("No records found",
"Query Error", 0)
```

This is the largest of the OnOk() methods, primarily due to the complexity
of the query. The following block of code builds the query based on the
input provided or not provided in the input widgets. The code starts with
an empty string assigned to the queryString variable. Next, if the user
has provided input in the txtOwner widget, it adds a SQL statement to the
queryString variable. The code continues to build the SQL statement in this
fashion by examining the content of each user interface widget and adding a
SQL if the user has provided an input value:

```
        queryString = ""
        flagOwner = False
        if self.txtOwner.GetValue():
            flagOwner = True
            owner = self.txtOwner.GetValue().upper()
            queryString =  "file_as_na LIKE \'%" + owner + "%\'"

        flagStreet = False
        if self.txtStreetName.GetValue():
            flagStreet = True
            strStreetName = self.txtStreetName.GetValue().upper()
            if flagOwner:
                queryString = queryString + " AND " + "situs_st_1
LIKE \'%" + strStreetName + "%\'"
            else:
                queryString = "situs_st_1 LIKE \'%" +
strStreetName + "%\'"

        if self.comboBoxSub.GetValue():
            strSubdivision = self.comboBoxSub.GetValue().upper()
            if flagOwner or flagStreet:
```

```
                    queryString = queryString + " AND " + "DESC_ LIKE
\'%" + strSubdivision + "%\'"
            else:
                    queryString = "DESC_ LIKE \'%" + strSubdivision +
"%\'"

        if self.txtMinVal.GetValue():
            numMinVal = long(self.txtMinVal.GetValue())
            queryString = queryString + " AND market > " +
str(numMinVal)

        if self.txtMaxVal.GetValue():
            numMaxVal = long(self.txtMaxVal.GetValue())
            queryString = queryString + " AND market < " +
str(numMaxVal)
```

22. The rest of the code in the `OnOk()` method is exactly the same as the other classes.

23. Add an `OnClick()` event to the `PageAdvanced` class, which is the same as the other classes. Refer to the previous steps, if you need assistance.

24. Save `Interface.py`. We'll return to it later to add a few lines of code.

Creating the ArcGIS Python add-in

With the user interface complete for **Parcel Viewer**, we need a way to display the dialog box for the user input. The easiest way to accomplish this in ArcGIS Desktop is to tie the interface to an **ArcGIS Python Add-In Wizard**. In this step, we'll create a **Button** add-in that will display the user interface when clicked on:

1. In *Chapter 3, Automating the Production of Map Books with Data-Driven Pages and ArcPy*, you learned how to use the **ArcGIS Python Add-In Wizard** to create add-ins. If you need to refresh your memory on this topic, refer to the *Exporting the map series with ArcPy Mapping* and *Automating the Production of Map Books with Data-Driven Pages and the ArcPy* section in the same chapter.

2. Create a folder in `C:\ArcGIS_Blueprint_Python\ch6` called `ParcelViewer`. This will be the container folder for the Python add-in.

3. Start the **ArcGIS Python Add-In Wizard** and add the settings shown under the Project Settings tab in the following screenshot:

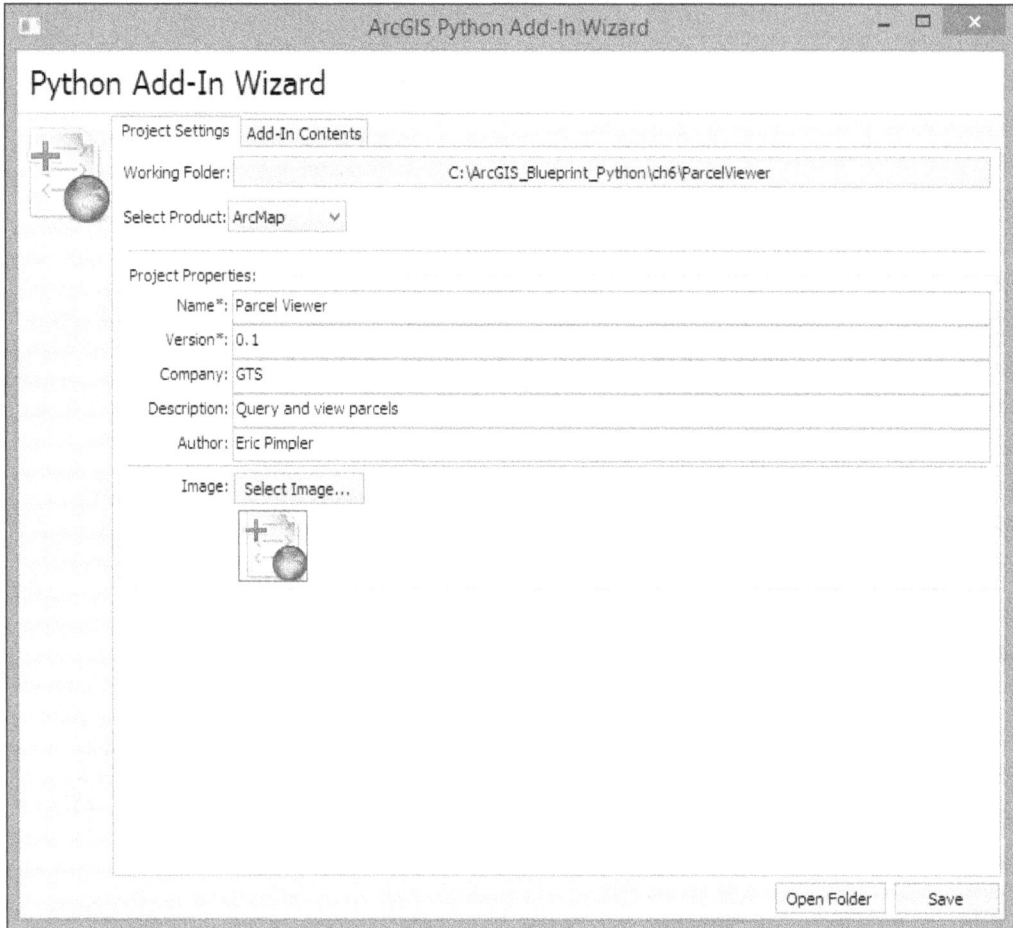

4. Click on the **Add-In Contents** tab and create a new **Extension** with the properties seen in the following screenshot. Make sure you click on the startup method:

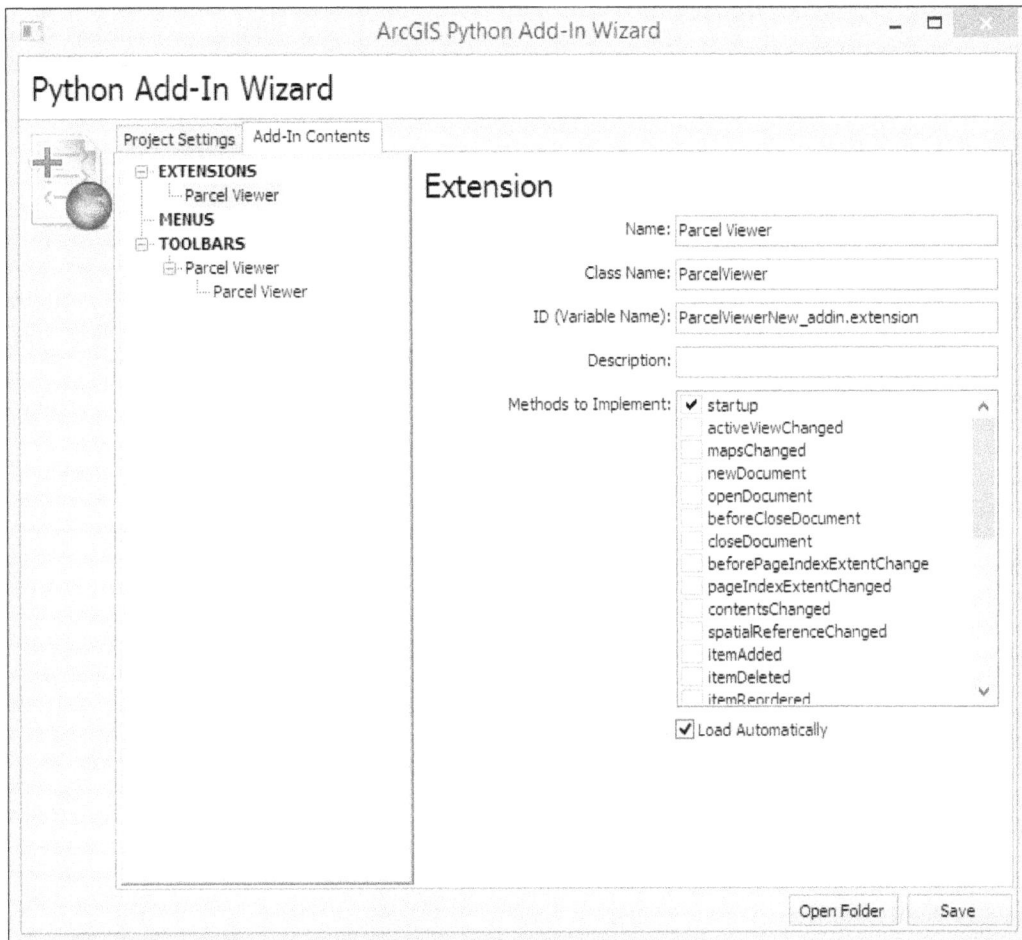

5. Click on the **Save** button.

6. Create a new **Toolbar** with the properties shown in the following screenshot:

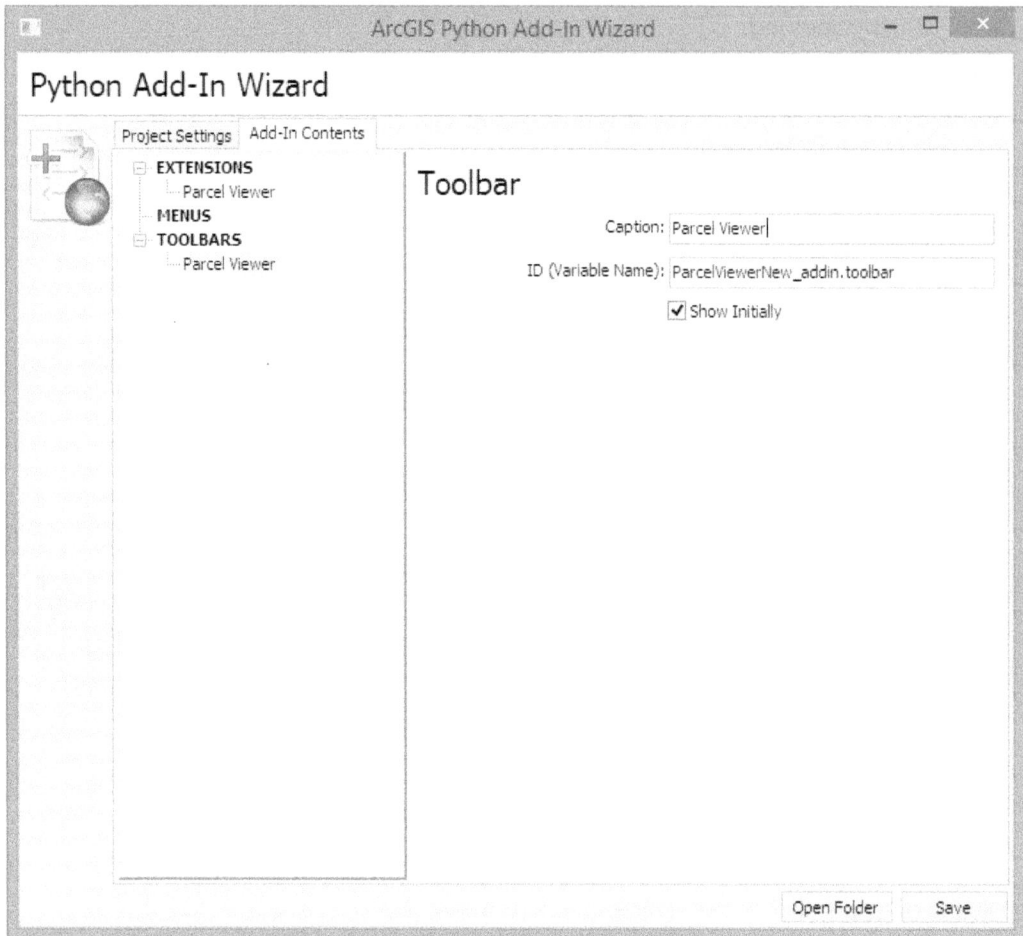

7. Right-click on the **Parcel Viewer** toolbar and create a new **Button** add-in with the properties shown in the following screenshot:

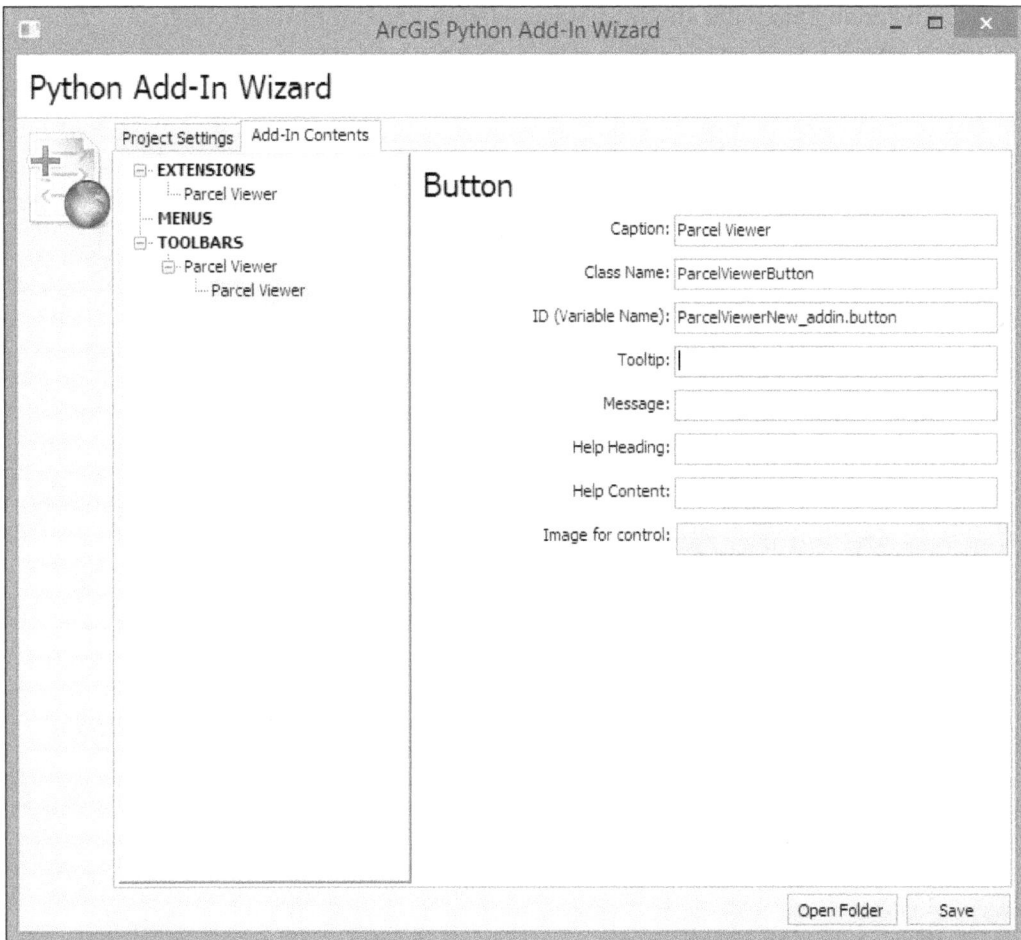

8. Click on the **Save** button.

9. Click on the **Open Folder** button to display the contents of the **Add-In Contents**.

10. Inside the `Install` folder is a file called `ParcelViewer_addin.py`. Open this file in your Python development environment. We're going to make some additions and changes that will support the display of our `wxPython` user interface.

11. The code in this file should appear as seen in the following code:

```
import arcpy
import pythonaddins

class ParcelViewer(object):
    """Implementation for ParcelViewerNew_addin.extension2
(Extension)"""
    def __init__(self):
        # For performance considerations, please remove all unused
methods in this class.
        self.enabled = True
    def startup(self):
        pass


class ParcelViewerButton(object):
    """Implementation for ParcelViewerNew_addin.button (Button)"""
    def __init__(self):
        self.enabled = True
        self.checked = False
    def onClick(self):
        pass
```

12. Add the following import statements. The sys.path.append() method is related to displaying the extension on startup:

```
import os
import sys
sys.path.append(os.path.dirname(__file__))
import arcpy
import pythonaddins
```

13. Make the following changes in the __init__ method of the ParcelViewer class:

```
class ParcelViewer(object):
    """Implementation for ParcelViewer_addin.extension
(Extension)"""
    def __init__(self):
        # For performance considerations, please remove all unused
methods in this class.
        self._wxApp = None
        self._enabled = None
    def startup(self):
        try:
            from wx import PySimpleApp
            self._wxApp = PySimpleApp()
```

```
        self._wxApp.MainLoop()
    except:
        pythonaddins.MessageBox("Error starting Parcel Viewer
extension.", "Extension Error", 0)
```

The first two lines of code in the __init__ function, seen as follows, set a couple of global variables that we'll use in the startup() method:

```
self._wxApp = None
self._enabled = None
```

The startup() method is called when the extension is first loaded. Inside the startup() method are three statements that are important to loading the wxPython user interface for the application:

```
from wx import PySimpleApp
self._wxApp = PySimpleApp()
self._wxApp.MainLoop()
```

The first statement imports the PySimpleApp class. The second statement calls the constructor for PySimpleApp() and assigns the object to the self._wxApp variable. PySimpleApp is the main application class for a wxPython application.

14. Next, add the following code. This code enables or disables the button associated with the add-in when the extension is turned on or off:

```
@property
def enabled(self):
    """Enable or disable the  button when the extension is
turned on or off."""
    if self._enabled == False:
        wxpybutton.enabled = False
    else:
        wxpybutton.enabled = True
    return self._enabled


@enabled.setter
def enabled(self, value):
    """Set the enabled property of this extension when
the extension is turned on or off in the Extension Dialog of
ArcMap."""
    self._enabled = value
```

15. Now, it's time to turn our attention to the `ParcelViewerButton` class. Add the following code block:

```
class ParcelViewerButton(object):
    """Implementation for ParcelViewer_addin.button (Button)"""
    _dlg = None

    @property
    def dlg(self):
        """Return the MainFrame dialog."""
        if self._dlg is None:
            from Interface import MainFrame
            self._dlg = MainFrame()
        return self._dlg

    def __init__(self):
        self.enabled = True
        self.checked = False
```

This code block creates a global variable called `_dlg` and creates a property called `dlg` that will be used to display the user interface of the application when the **Add-In** button is clicked on. If the `_dlg` variable has not been set, the property will import the `MainFrame` class created in the `Interface.py` file; call the constructor for this class, and assign it to the `_dlg` variable. Essentially, this is what triggers the display of the user interface.

16. Alter the `onClick()` method in the `ParcelViewerButton` class, as shown in the following code:

```
def onClick(self):
        try:
            self.dlg.Show(True)
        except Exception as e:
            pythonaddins.MessageBox(e.message, "Error", 0)
```

17. The `self.dlg.Show(True)` method sets the `dlg` property and calls the `Show()` method on the `MainFrame` class.

18. Save the `ParcelViewer_addin.py` file.

19. If required, open `Interface.py` in your Python editor.

20. Find the `MainFrame` class and add the following code:

```
class MainFrame(wx.Frame):
    def __init__(self):
```

```
        wx.Frame.__init__(self, None, title="Search Parcels",s
ize=(600,400),style=wx.DEFAULT_FRAME_STYLE & ~wx.MAXIMIZE_BOX ^
wx.RESIZE_BORDER)

        self.Bind(wx.EVT_CLOSE, self.OnClose)
        # Here we create a panel and a notebook on the panel
        panel = wx.Panel(self)

        arcpy.env.workspace = r"C:\ArcGIS_Blueprint_Python\data\
Kendall"

        notebook = NotebookParcel(panel)
        sizer = wx.BoxSizer(wx.VERTICAL)
        sizer.Add(notebook, 1, wx.ALL|wx.EXPAND, 5)

        panel.SetSizer(sizer)
        self.Layout()
    def OnClose(self, event):
        """Close the frame. Do not use destroy."""
        self.Show(False)
```

21. The highlighted code creates a new event handler for the EVT_CLOSE event. This event is triggered when the application is closed. The handler for this event is the OnClose() method. This method simply closes the user interface. Also, there is a line of code that sets the workspace environment variable to the folder that contains the Kendall_Parcels shapefile used in the application.

22. Save Interface.py, and you can also close the editor.

23. In Windows Explorer, copy the Interface.py file found in the C:\ArcGIS_Blueprint_Python\ch6 folder to the C:\ArcGIS_Blueprint_Python\ch6\ParcelViewer\Install folder, which contains the ParcelViewer_addin.py file.

24. You can check your work against the solution file by going to C:\ArcGIS_Blueprint_Python\solutions\ch6 and examining the Interface.py and ParcelViewer_addin.py files.

25. In the C:\ArcGIS_Blueprint_Python\ch6\ParcelViewer folder, double-click on the makeaddin.py file to create the ParcelViewer.esriaddin file. This file will be created in the same directory.

26. Double-click on ParcelViewer.esriaddin to install the add-in.

27. Now, it's time to test the application. Open **ArcMap** and load the
ParcelViewer.mxd file found in the C:\ArcGIS_Blueprint_Python\ch6
folder. You should see a single layer called Kendall_Parcels. If required,
navigate to **Customize | Toolbars | Parcel Viewer** to display the add-in,
as shown in the following screenshot:

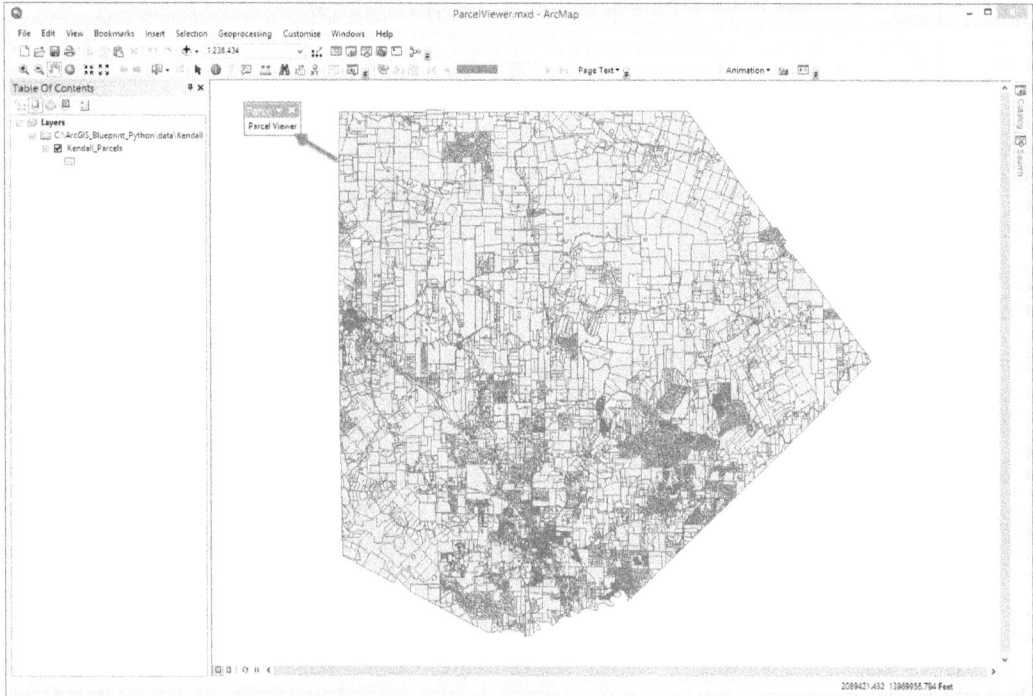

28. Click on **Parcel Viewer** to display the user interface. The interface will look like the following screenshot:

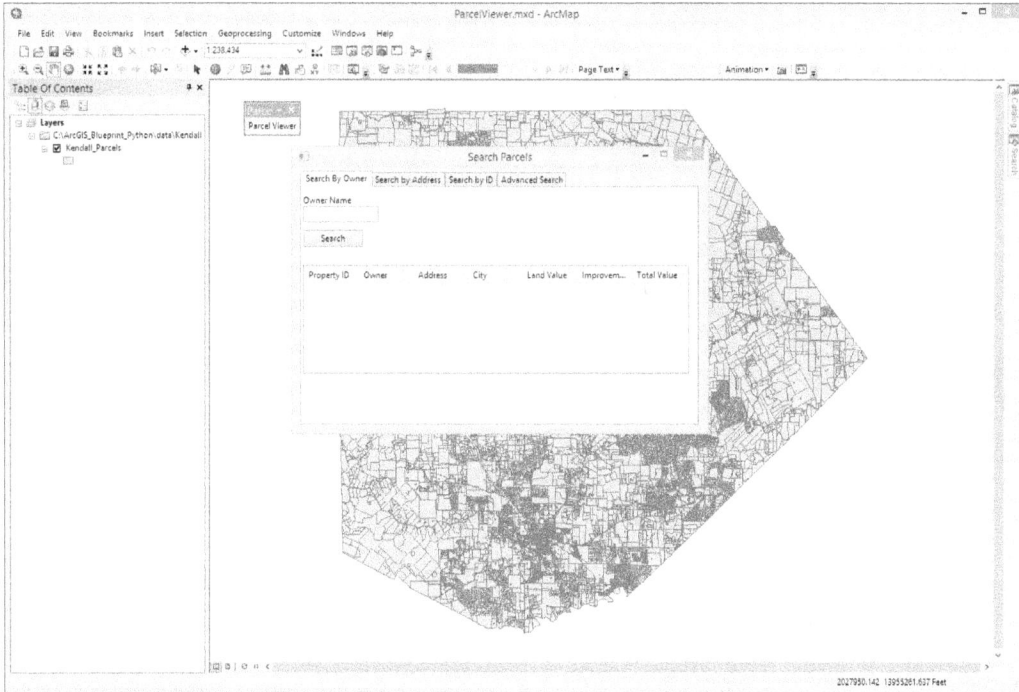

29. In the **Search by Owner** tab, enter **Owner Name of Cibolo** and click on the **Search** button. You should see the result shown in the following screenshot:

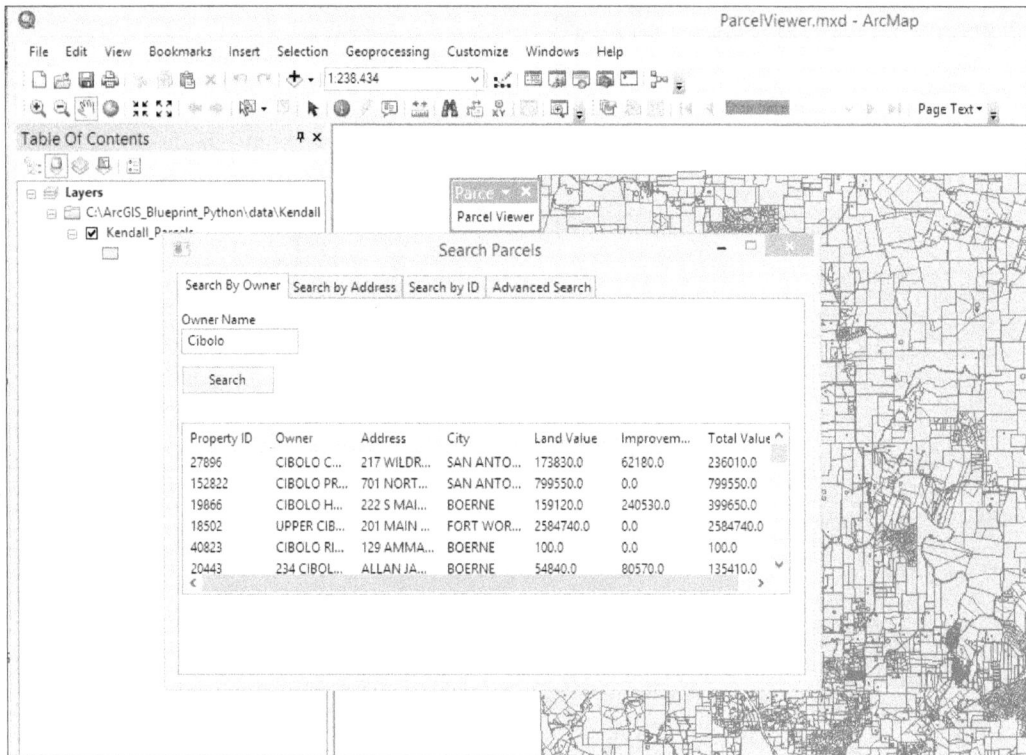

30. Next, click on one of the returned records to see the information returned in a web browser with the map zoomed to the parcel, as shown in the following screenshot:

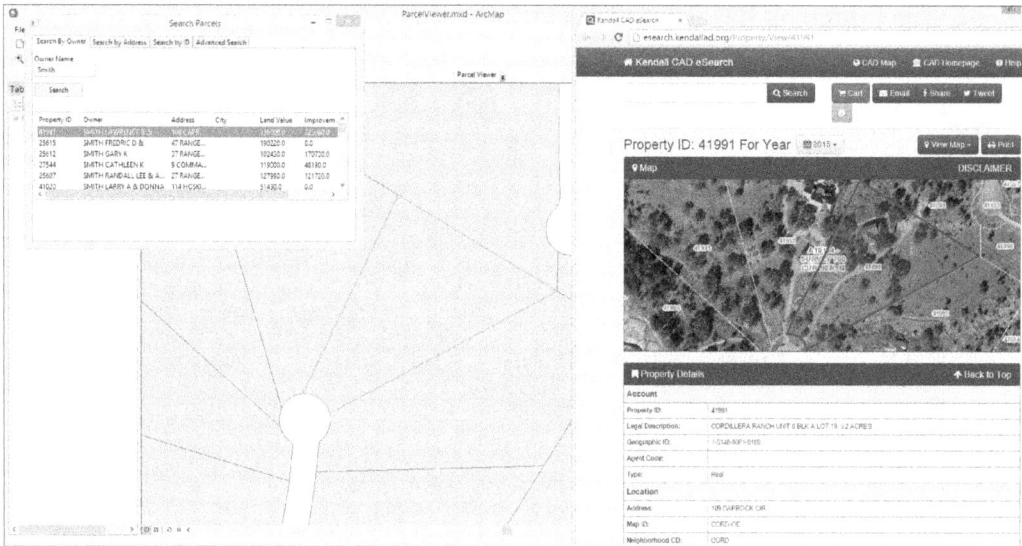

31. The same basic functionality will be present in the others tabs. The following are some screenshots to help you test the application across each of the tabs. **Search by Address** is seen in the following screenshot:

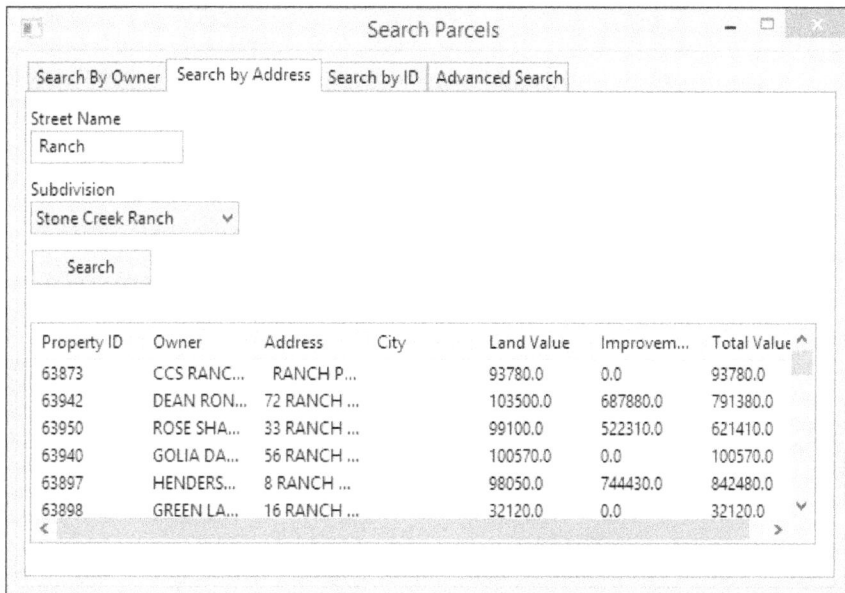

32. The **Search by ID** tab is seen in the next screenshot:

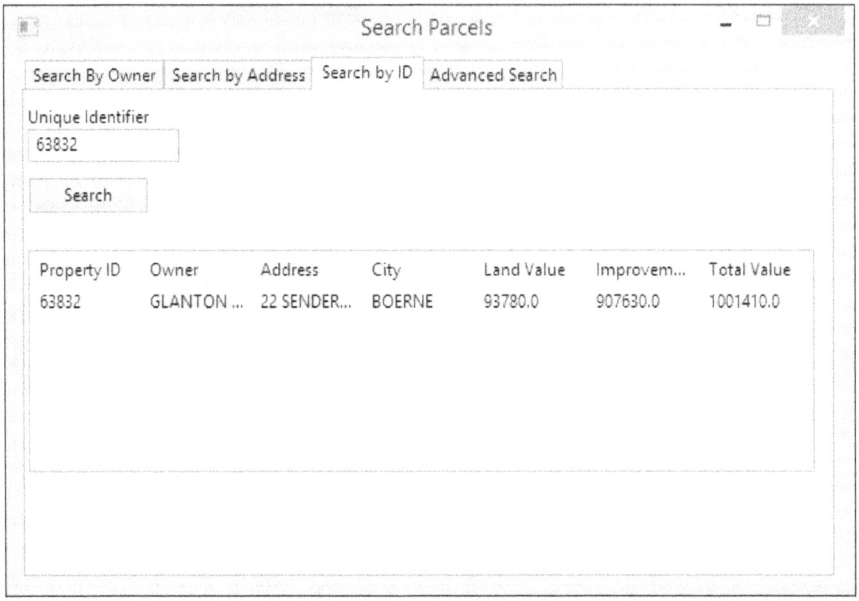

33. Finally, the **Advanced Search** functionality is seen in this screenshot:

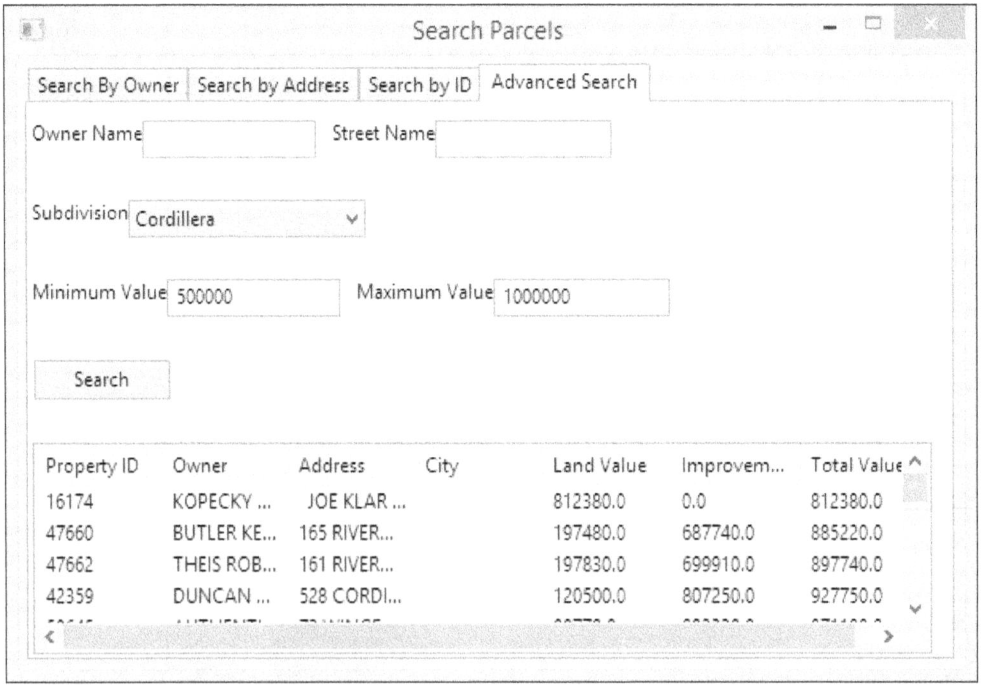

Summary

In this chapter, you learned how to use the wxPython library to build a graphical user interface (GUI) and attach it to an ArcGIS Desktop add-in. Now that you know the basics of the wxPython library, you should be able to build complex user interfaces that can handle any sort of input needs. Without a library such as wxPython, your options to create user interfaces for Python applications are severely limited. Libraries such as wxPython go a long way toward replacing the user interface functionality traditionally limited to the use of **ArcObjects** with .NET.

In the next chapter, you will learn how to create demographic reports using the ArcGIS REST API, ArcPy, and GeoEnrichment Service for retail site selection.

7
Using Python with the ArcGIS REST API and the GeoEnrichment Service for Retail Site Selection

The ArcGIS REST API provides access to a wide variety of web services, including ready-to-use ArcGIS Online services hosted by Esri, and services you or other organizations have published. Hosted ArcGIS Online services include basemaps, geocoding and place search, directions and routing, demographic and lifestyle attributes, spatial analysis, and elevation analysis. In this chapter, the GeoEnrichment service that provides demographic and lifestyle information will be queried with the ArcGIS REST API through the Python requests module in support of a site-selection application.

The application built in this chapter will support the site-selection process for a new coffee store in Denver, CO. Specifically, the application will include a tool to identify census block groups that meet general age and income-related variables and that are outside the trade area of competing stores. A second tool will allow the analyst to identify specific locations within these selected census block groups as sites for potential new coffee stores. Finally, the last tool we'll build in this chapter will attach lifestyle expenditure information to each of the potential stores by querying the GeoEnrichment service.

In this chapter, we will cover the following topics:

- Accessing the GeoEnrichment service for demographic and lifestyle information
- Using the ArcGIS REST API with the Python requests module
- Creating a custom tool that allows the end user to interactively define new point locations
- Attaching demographic information to features using the `ArcPy` data access module

Design

The design of this application will include the creation of three custom tools inside an **ArcGIS Desktop Python** toolbox. The first tool, **Census Block Group Selection**, will query an existing census block group layer to find suitable areas that meet the income and population characteristics defined by the tool. It will also remove any census block groups from consideration if they are within the boundaries of the trade area of an existing, competing coffee shop. The second tool, **Potential Stores**, will be an interactive tool that will allow the end user to define point locations for potential new coffee stores within the boundaries identified with the **Census Block Group Selection** tool. The final tool, **Enrich Potential Stores**, will take the stores defined by the **Potential Stores** tool, pass them to the GeoEnrichment service, and take the lifestyle information returned by the service and write it back to the potential store location layer. The following figure shows how these tools work with ArcGIS Desktop:

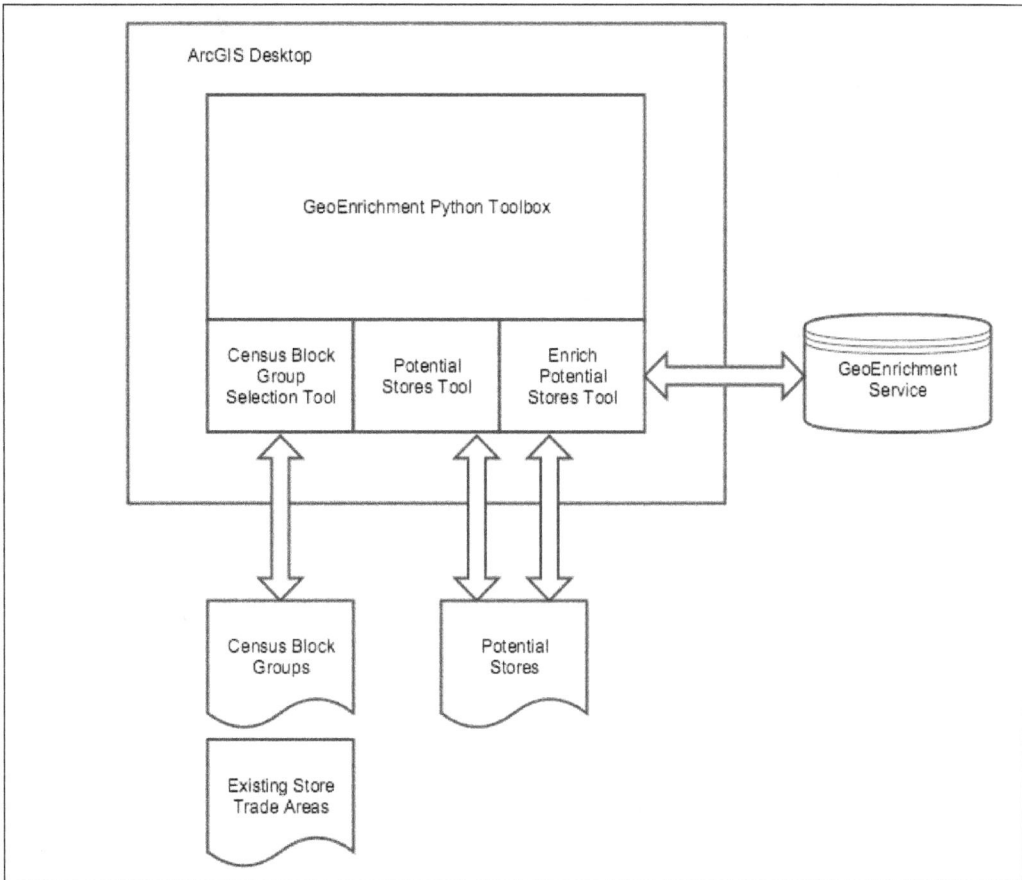

Let's get started and build the application.

Creating the Census Block Group selection tool

In this step, you'll create the **GeoEnrichment Python** toolbar and **Census Block Group Initial Selection** tool. The GeoEnrichment toolbar will serve as the container for all the three tools that will be built in this chapter. The tool that will be created in this section is the **Census Block Group Initial Selection** tool. This tool is designed to create an initial selection set of census block groups that meet the average household income and the percentage of the population between the ages of 20-50, which is the criteria defined by the user.

The user interface for the tool will appear as shown in the following screenshot:

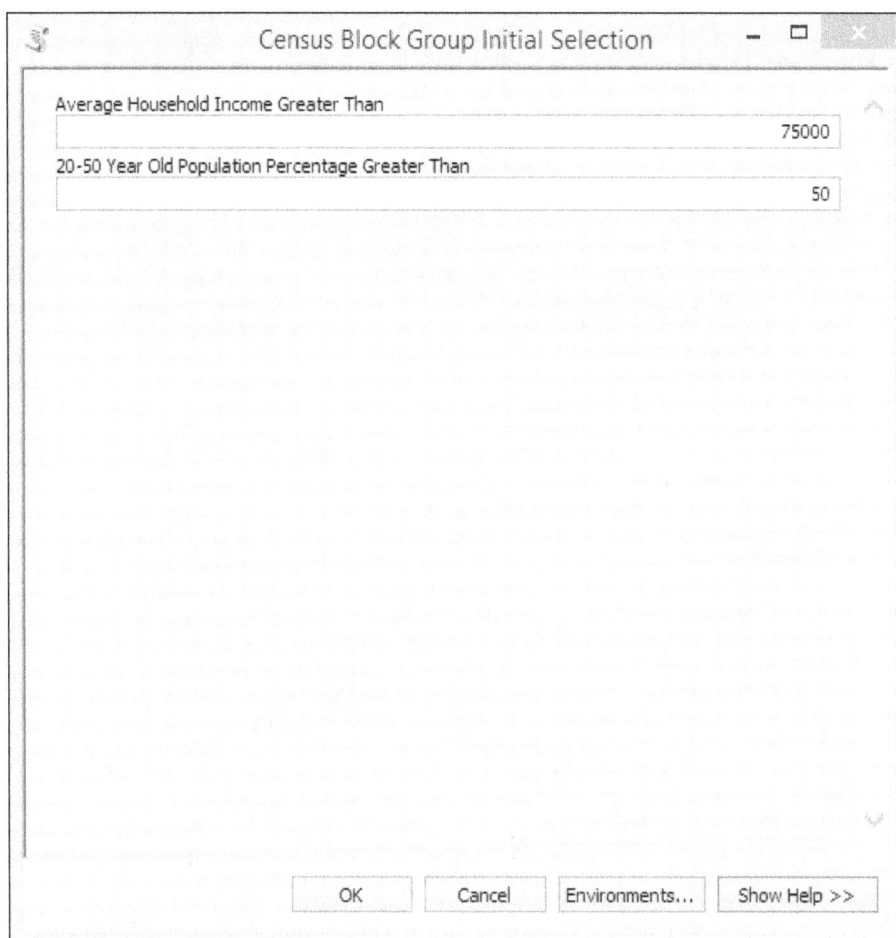

After creating the initial selection set of census block groups that match these criteria, the tool will then remove any census block groups that intersect the half-mile buffer zone of an existing coffee shop. Finally, the tool will copy the remaining census block group features to an output feature class. These census block groups form the initial geographic boundaries that have been determined to meet our criteria for a new coffee establishment.

Follow these steps to create the **Census Block Group Initial Selection** tool:

1. Open **ArcMap** and display the **ArcCatalog** pane. In the **Toolboxes** folder, under **My Toolboxes**, create a new **Python Toolbox** and call it `GeoEnrichment.pyt`.

2. Open the `GeoEnrichment.mxd` file in **ArcMap** and spend some time examining the data. In addition to a `basemap`, the map document includes a layer called **American Community Survey (ACS)** that contains demographic and socio-economic data by the census block group. Also included are layers for competing coffee shops in the Denver area, along with a layer containing half-mile buffers for each store, street, and city boundary as well as a layer that will contain the results of this tool, called `CensusBlockGroupResults`.

3. Open the code for the toolbox in your Python development environment.

4. Rename the `Tool` class `CensusBlockGroupSelection`. Also, update the `self.label` and `self.description` properties, as shown in the following code:

    ```
    class CensusBlockGroupSelection(object):
        def __init__(self):
            """Define the tool (tool name is the name of the
    class)."""
            self.label = "Census Block Group Initial Selection"
            self.description = "Performs the initial selection of
    candidates"
    ```

5. Add the tool to the `self.tools` list in the `Toolbox` class:

    ```
    self.tools = [CensusBlockGroupSelection]
    ```

6. Add `import` statements at the top of your script for the `requests`, `arcpy`, `arcpy.mapping`, and `json` modules. You should already have the `requests` module installed from a previous chapter, but if not, you'll want to install this module before continuing:

    ```
    import arcpy
    import arcpy.mapping as mapping
    import requests
    import json
    ```

7. Next, add the input parameters for the tool. This tool will include two parameters. Both parameters will be numeric values. The socio-economics of high-end retail coffee establishments indicate that we need to target individuals with high incomes and people between the ages of 20-50. The first parameter will capture a value indicating the average household income, and the second will capture the percentage of the population that should be between the ages we are targeting. Define the default values for each using the `Parameter.value` property. In the `getParameterInfo()` method, add the following code:

```python
def getParameterInfo(self):
    param0 = arcpy.Parameter(displayName = "Average Household
Income Greater Than", \
                    name="avgHHInc", \
                    datatype="GPLong", \
                    parameterType="Required",\
                    direction="Input")
    param0.value = 75000

    param1 = arcpy.Parameter(displayName = "20-50 Year Old
Population Percentage Greater Than", \
                    name="percPop", \
                    datatype="GPLong",\
                    parameterType="Required",\
                    direction="Input")
    param1.value = 50

    params = [param0, param1]
    return params
```

8. Next, find the `execute()` method and capture the input parameters that will be passed into the tool, as shown in the following code:

```python
def execute(self, parameters, messages):
        """The source code of the tool."""
        avgHHInc = parameters[0].valueAsText
        percPop = parameters[1].valueAsText
```

9. Set the following environment variable, which will set the current workspace and allow the script to overwrite an existing file:

```python
def execute(self, parameters, messages):
        """The source code of the tool."""
        avgHHInc = parameters[0].valueAsText
```

```
      percPop = parameters[1].valueAsText

      arcpy.env.overwriteOutput = True

      arcpy.env.workspace = r"C:\ArcGIS_Blueprint_Python\data\
Denver"
```

10. Create the `try`/`except` block:

```
def execute(self, parameters, messages):
      """The source code of the tool."""
      avgHHInc = parameters[0].valueAsText
      percPop = parameters[1].valueAsText

      arcpy.env.overwriteOutput = True

      arcpy.env.workspace = r"C:\ArcGIS_Blueprint_Python\data\
Denver"

      try:


      except Exception as e:
          arcpy.AddMessage(e.message)
```

11. Find the ACS layer that contains the census block group boundaries containing the demographic and socio-economic data:

```
try:
    mxd = mapping.MapDocument("CURRENT")
    for lyr in mapping.ListLayers(mxd):
        if lyr.name == "ACS":
except Exception as e:
    arcpy.AddMessage(e.message)
```

12. Select all census block groups where the population in an age range of 20-50 is greater than the value indicated by the user. Save the selection set to a `FeatureLayer` object called `acs_layer`:

```
mxd = mapping.MapDocument("CURRENT")
for lyr in mapping.ListLayers(mxd):
    if lyr.name == "ACS":
        arcpy.MakeFeatureLayer_management(lyr,"acs_lyr")

        #select features where 20-50 population percentage is
greater than
        arcpy.SelectLayerByAttribute_management("acs_lyr", "NEW_
SELECTION", "Pop20_50_P > " + percPop)
```

13. Create a subset selection (a selection from the existing selection), where the average household income is greater than the value indicated by the user:

```
mxd = mapping.MapDocument("CURRENT")
for lyr in mapping.ListLayers(mxd):
    if lyr.name == "ACS":
        arcpy.MakeFeatureLayer_management(lyr,"acs_lyr")

        #select features where 20-50 population percentage is
greater than
        arcpy.SelectLayerByAttribute_management("acs_lyr", "NEW_
SELECTION", "Pop20_50_P > " + percPop)

        #select features where avg household income is greater
than
        arcpy.SelectLayerByAttribute_management("acs_lyr",
"SUBSET_SELECTION", "AVG_HH_INC > " + avgHHInc)
```

14. Finally, remove any selected census block groups that intersect a half-mile buffer of an existing coffee shop and copy the results to an output shapefile:

```
mxd = mapping.MapDocument("CURRENT")
for lyr in mapping.ListLayers(mxd):
    if lyr.name == "ACS":
        arcpy.MakeFeatureLayer_management(lyr,"acs_lyr")

        #select features where 20-50 population percentage is
greater than
        arcpy.SelectLayerByAttribute_management("acs_lyr", "NEW_
SELECTION", "Pop20_50_P > " + percPop)

        #select features where avg household income is greater
than
        arcpy.SelectLayerByAttribute_management("acs_lyr",
"SUBSET_SELECTION", "AVG_HH_INC > " + avgHHInc)

        #deselect features within the buffer
        arcpy.SelectLayerByLocation_management("acs_lyr",
"INTERSECT", "DenverCoffeeStoreBuffer", selection_type="REMOVE_
FROM_SELECTION")

        arcpy.CopyFeatures_management("acs_lyr",
"CensusBlockGroupResults.shp")
```

15. Refresh the active view, as shown in the highlighted code here. You can also review the entire `try`/`except` block that you've created for accuracy:

```
try:
    mxd = mapping.MapDocument("CURRENT")
    for lyr in mapping.ListLayers(mxd):
        if lyr.name == "ACS":
            arcpy.MakeFeatureLayer_management(lyr,"acs_lyr")


            #select features where 20-50 population percentage is
greater than
            arcpy.MakeFeatureLayer_management(lyr,"acs_lyr")


            #select features where 20-50 population percentage is
greater than
            arcpy.SelectLayerByAttribute_management("acs_lyr",
"NEW_SELECTION", "Pop20_50_P > " + percPop)


            #select features where avg household income is greater
than
            arcpy.SelectLayerByAttribute_management("acs_lyr",
"SUBSET_SELECTION", "AVG_HH_INC > " + avgHHInc)


            #deselect features within the buffer
            arcpy.SelectLayerByLocation_management("acs_lyr",
"INTERSECT", "DenverCoffeeStoreBuffer", selection_type="REMOVE_
FROM_SELECTION")


            arcpy.CopyFeatures_management("acs_lyr",
"CensusBlockGroupResults.shp")


    arcpy.RefreshActiveView()



except Exception as e:
    arcpy.AddMessage(e.message)
```

16. You can check your work by examining the `C:\ArcGIS_Blueprint_Python\solutions\ch7\CensusBlockGroupInitialSelection.py` solution file. Refer to the `getParameterInfo()` and `execute()` methods.

17. Save the file and exit your Python development environment.

18. In the **Catalog** view of **ArcMap**, double-click on the tool to test your code.

19. Set the **Average Household Income Greater Than** parameter to 65000 and the **20-50 Year Old Population Percentage Greater Than** parameter to 45, as shown in the following screenshot:

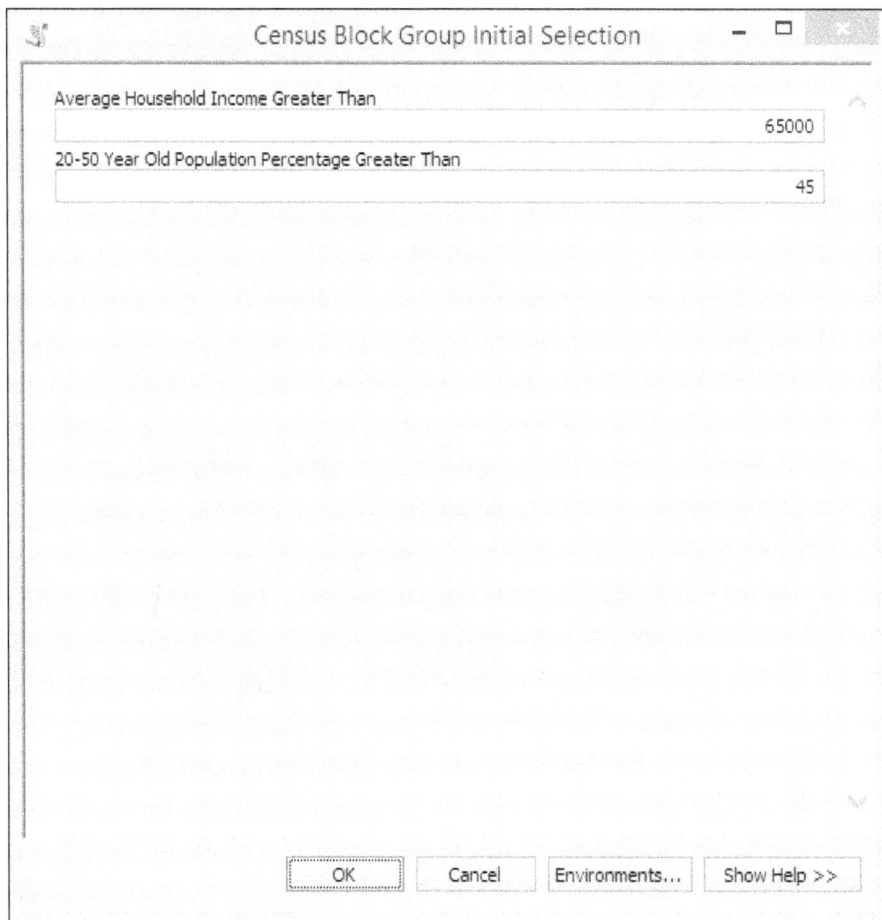

20. Click on **OK** to execute the tool. Upon completion, the
    CensusBlockGroupResults layer will be updated as shown in the following
    screenshot. This layer is symbolized with a transparent fill and a thick, red
    outline. These are the census block groups that match the initial criteria.
    The tool that we create in the next section will be used to define individual
    locations within one or more of these identified census block groups:

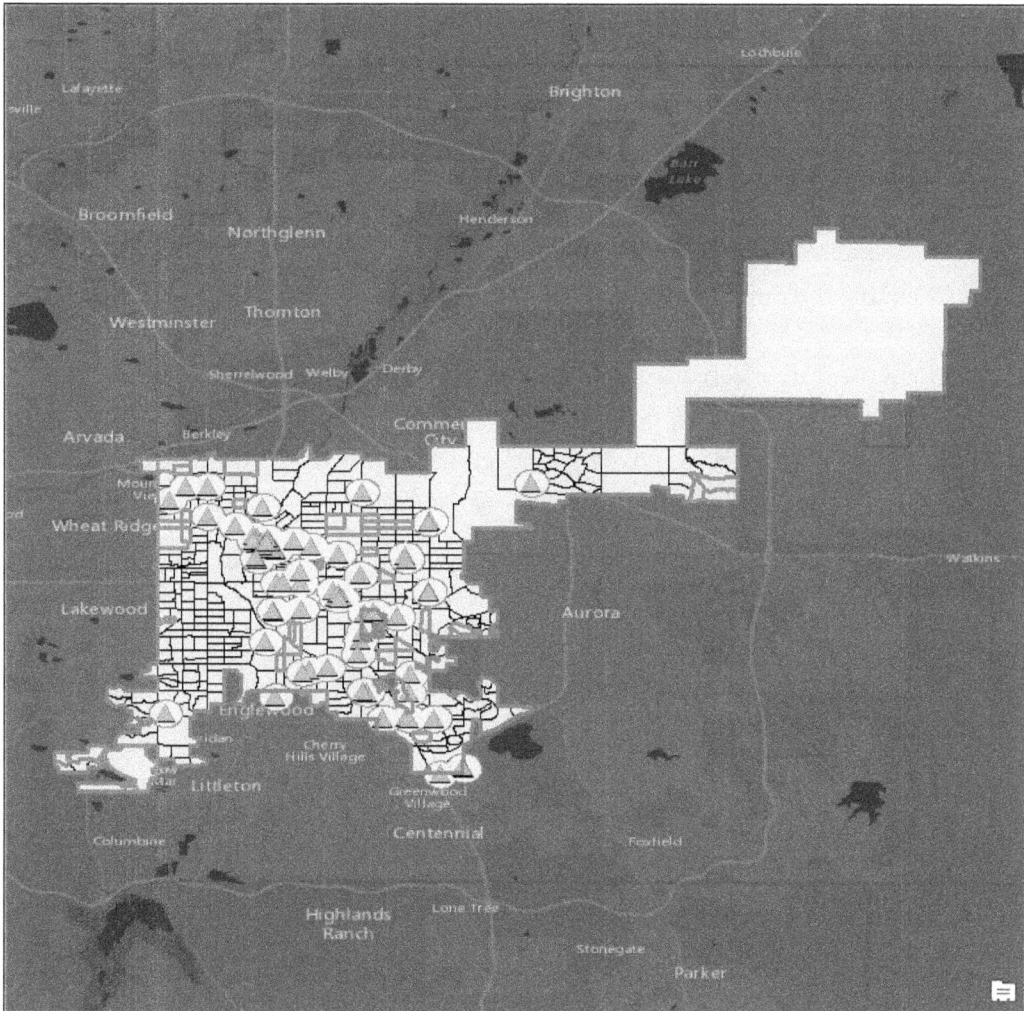

Creating the Define Potential Stores tool

Now that a tool that defines some initial areas that would be suitable for the development of a new coffee shop has been created, we want to turn our attention to refining individual locations for further analysis. The next step will be to create the **Define Potential Stores** tool. This tool will allow the end user to create new point locations representing potential coffee store locations. Socio-economic information will then be attached to each of these point locations along with the final tool that will be created in the last section of this chapter:

1. Open the Python development environment for the `GeoEnrichment.pyt` toolbar.

2. Duplicate the code that you have already created for the `CensusBlockGroupSelection` class by copying and pasting this class into the same `GeoEnrichment.pyt` file.

3. Rename the duplicated `CensusBlockGroupSelection` class `PotentialStores`.

4. Remove the code inside the `getParameterInfo()` and `execute()` methods for the new `PotentialStores` class.

5. Alter the `self.label` and `self.description` properties in the `__init__` method, as shown here:

```
def __init__(self):
        """Define the tool (tool name is the name of the
class)."""
        self.label = "Define Potential Stores"
        self.description = "Define Potential Stores"
        self.canRunInBackground = False
```

6. Set the workspace environment variable:

```
def __init__(self):
        """Define the tool (tool name is the name of the
class)."""
        self.label = "Define Potential Stores"
        self.description = "Define Potential Stores"

        arcpy.env.workspace = r"C:\ArcGIS_Blueprint_Python\data\
Denver\NewStoreLocations.gdb"
```

7. Find the `getParameterInfo()` method and add the following parameter. The `datatype` parameter should be set to `GPFeatureRecordSetLayer`. This is a data type we haven't worked with so far. The `GPFeatureRecordSetLayer` object will enable the end user to interactively add features to a `feature` class. This tool will use the `PotentialStores` layer that is defined as the default value as the layer to be used for the default **symbology**:

```
def getParameterInfo(self):
    param0 = arcpy.Parameter(displayName = "Create Potential
Stores", \
                        name="potentialStores", \
                        datatype="GPFeatureRecordSetLayer", \
                        parameterType="Required", \
                        direction="Input")
    param0.value = "PotentialStores"
```

8. Add a second parameter that captures an output feature class where the new points will be stored, and add both parameters to the list of parameters that will be returned:

```
def getParameterInfo(self):
    param0 = arcpy.Parameter(displayName = "Create Potential
Stores", \
                        name="potentialStores", \
                        datatype="GPFeatureRecordSetLayer", \
                        parameterType="Required", \
                        direction="Input")
    param0.value = "PotentialStores"

    param1 = arcpy.Parameter(displayName = "Output Feature Class
Name", \
                        name="name", \
                        datatype="GPString", \
                        parameterType="Required", \
                        direction="Input")

    params = [param0, param1]
    return params
```

9. Find the `execute()` method and add the following two lines of code that capture the user input parameters:

```
def execute(self, parameters, messages):
        """The source code of the tool."""
        in_featureset = parameters[0].valueAsText
        name = parameters[1].valueAsText
```

10. Create the `try`/`except` blocks:

```
def execute(self, parameters, messages):
        """The source code of the tool."""
        in_featureset = parameters[0].valueAsText
        name = parameters[1].valueAsText


        try:


        except Exception as e:
            arcpy.AddMessage(e.message)
```

11. Inside the `try` block, create a new feature class based on the input supplied by the user. It should be a point feature class, and it should inherit the schema from the `PotentialStores` feature class. Also, define a new `FeatureSet` object and save the content of the input feature class to this `FeatureSet`. The `FeatureSet` object is a lightweight representation of a feature class that contains data as well as a schema. This object is used when interactively capturing input features from a geoprocessing tool:

```
try:
    feature_class = arcpy.CreateFeatureclass_management(arcpy.env.
 workspace, name, "POINT", "PotentialStores")
    feature_set = arcpy.FeatureSet(in_featureset)
    feature_set.save(feature_class)
except Exception as e:
    arcpy.AddMessage(e.message)
```

12. Next, create a new `FeatureLayer` object from this `FeatureSet` method and add it to the `data` frame:

```
try:

    feature_class = arcpy.CreateFeatureclass_management(arcpy.env.
workspace, name, "POINT", "PotentialStores")

    feature_set = arcpy.FeatureSet(in_featureset)
    feature_set.save(feature_class)

    mxd = mapping.MapDocument('current')
    df = mapping.ListDataFrames(mxd)[0]

    arcpy.MakeFeatureLayer_management(feature_set, name)
    addLayer = mapping.Layer(name)
    mapping.AddLayer(df, addLayer, "TOP")

except Exception as e:
    arcpy.AddMessage(e.message)
```

13. Finally, create a reference to a predefined layer file that contains the symbology to be applied to the newly created feature class, and apply this symbology using the `UpdateLayer()` function:

```
try:

    feature_class = arcpy.CreateFeatureclass_management(arcpy.env.
workspace, name, "POINT", "PotentialStores")

    feature_set = arcpy.FeatureSet(in_featureset)
    feature_set.save(feature_class)

    mxd = mapping.MapDocument('current')
    df = mapping.ListDataFrames(mxd)[0]

    arcpy.MakeFeatureLayer_management(feature_set, name)
    addLayer = mapping.Layer(name)
    mapping.AddLayer(df, addLayer, "TOP")

    srcLayer = mapping.Layer(r"C:\ArcGIS_Blueprint_Python\data\
Denver\PotentialStores.lyr")
    mapping.UpdateLayer(df, mapping.ListLayers(mxd, name, df)[0],
srcLayer, True)

except Exception as e:
    arcpy.AddMessage(e.message)
```

14. You can check your work by examining the `C:\ArcGIS_Blueprint_Python\solutions\ch7\PotentialStores.py` solution file. Refer to the `getParameterInfo()` and `execute()` methods.

15. Save the file and exit your Python development environment.

16. In **ArcMap**, zoom in to a cluster of census block groups that were identified as good candidates. These will be the census block groups outlined in red from running the **Census Block Group Initial Selection** tool, as shown in the following screenshot:

17. Turn on the **Denver Streets** layer.

18. In the **Catalog** view of **ArcMap**, double-click on the **Define Potential Stores** tool to test your code. Make sure the **Attributes** window is displayed as well, as seen in the following screenshot. This will allow you to add names and descriptions to each of the potential sites:

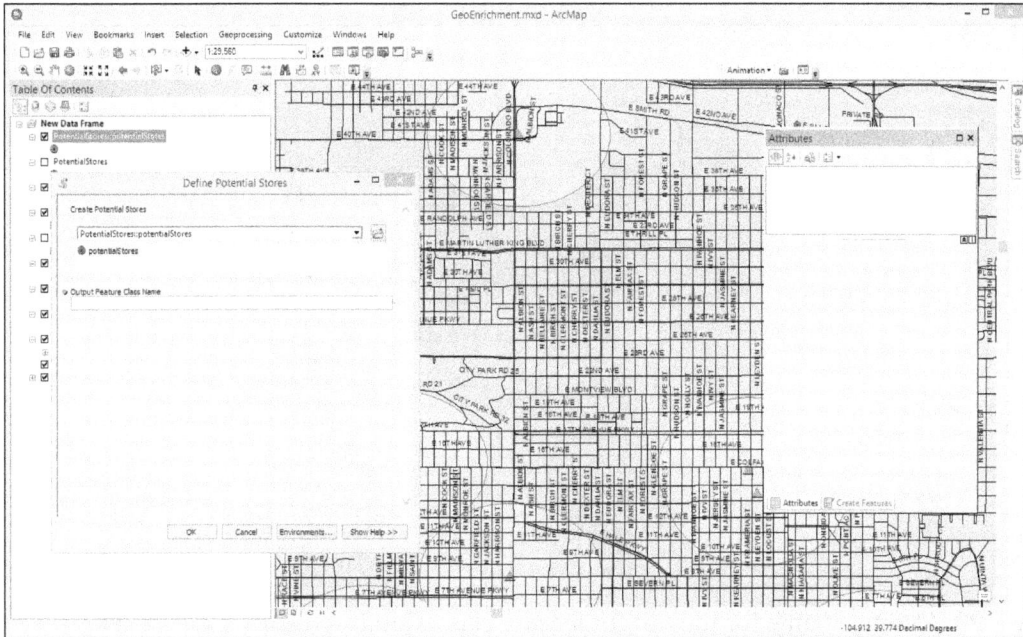

19. On the map, add three or four points by clicking on individual street locations, as shown in the following screenshot. You'll also want to add a name and, optionally, a description for each point using the **Attributes** window. However, don't add any attributes for the `AvgStar6` and `TotBfSpend` attributes. These attributes will be populated with the final tool that we create in this chapter:

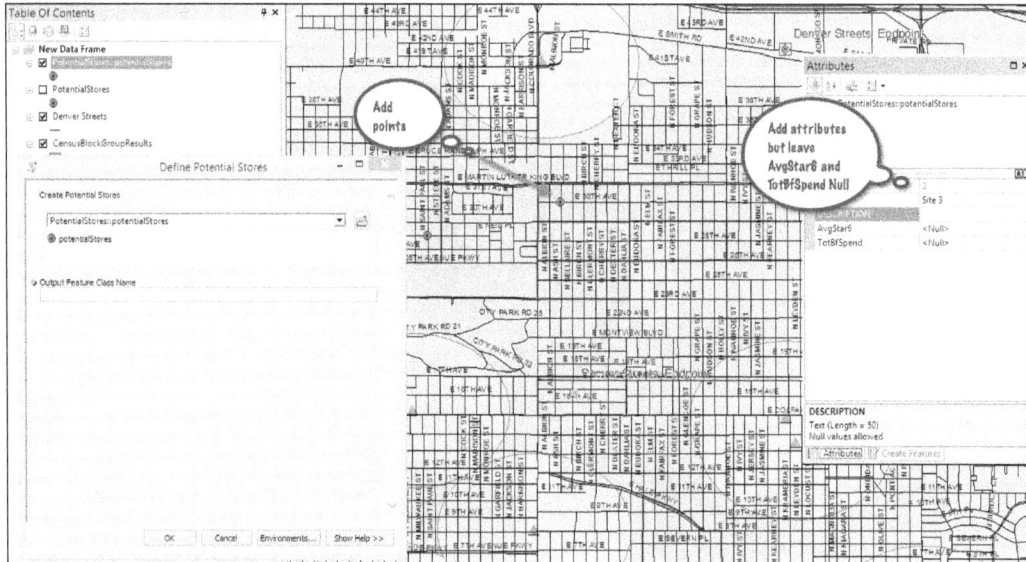

20. Define a name for the output feature class. Call it `PotentialCoffeeStores`. It will be saved to the `NewStoreLocations.gdb` geodatabase in the `C:\ArcGIS_Blueprint_Python\data\Denver` folder.

21. Click on **OK** to execute the tool.

Creating the Enrich Potential Stores tool

The final tool that we'll create in this chapter is the **Enrich Potential Stores** tool. It will assign socio-economic attributes pulled from the ArcGIS Online GeoEnrichment service to the features created with the **Define Potential Stores** tool. Using the Python `requests` module, it will pass a list of features and a drive time value to the GeoEnrichment service. The GeoEnrichment service will create a drive time polygon around each potential store and calculate socio-economic variables for the total *Starbucks* expenditures and the total fast food breakfast expenditures in that area. This information will then be written to the potential stores' feature class. The following steps will help you to create the **Enrich Potential Stores** tool:

1. Before coding this tool, some background information on the GeoEnrichment service and the ArcGIS REST API needs to be introduced. In your web browser, go to `http://resources.arcgis.com/en/help/arcgis-rest-api/index.html` to visit the main page for the ArcGIS REST API.

2. Select **Services** by navigating to **Esri | Demographic and Lifestyle attributes | The GeoEnrichment service | Accessing the service**.

3. To access the GeoEnrichment service, you will need to provide authentication credentials. This can be done by prompting the user for login information or by storing credentials with your application. We'll keep it simple in this exercise and simply pass in a token as part of the URL query string. Keep in mind that this is not a best practice. In this case, we're simply using this method to simplify things so that we can focus on other topics.

4. Authentication requires that you have an ArcGIS Online subscription through either an **Organization** plan or a **Developer** plan. Also, the use of the GeoEnrichment service requires the use of credits through your ArcGIS Online subscription.

5. Information on the URL request query string to generate a token is provided when accessing the service page. Again, to keep things simple, I'll just have you submit the request for a token by manually submitting the request. In a web browser, add the following query string to the address bar and press the *Enter* key. You'll need to insert your ArcGIS Online username and password information as well as a website for the referrer. If you don't know this information, contact your ArcGIS Online administrator at `https://www.arcgis.com/sharing/generateToken?username=yourUserName&password=yourPassword&referer=http://myserver/mywebapp&expiration=15&f=json`.

6. The return should appear similar to what is shown in the following code. Save the token information and keep in mind that you'll need to regenerate this token periodically:

```
{
        "token": "Zc07Ivtpoo-AWjVj4u-Is5NiwNQRXHs_2uI17IkTkLxFk5Fc
BnBr5jiYwko2cyMU",
        "expires": 1354427210436,
        "ssl": false
}
```

Spend some time reading the documentation for the GeoEnrichment service so that you'll have a better understanding of how this service works and what parameters can be passed to the service.

7. You can also get information on the available data variables for the service by going to **Services** from **Esri | Demographic and Lifestyle attributes | Get variables** and then selecting the data browser link on the help page. Click on any of the links in the data browser to start drilling down into the available variables, as shown in the following screenshot:

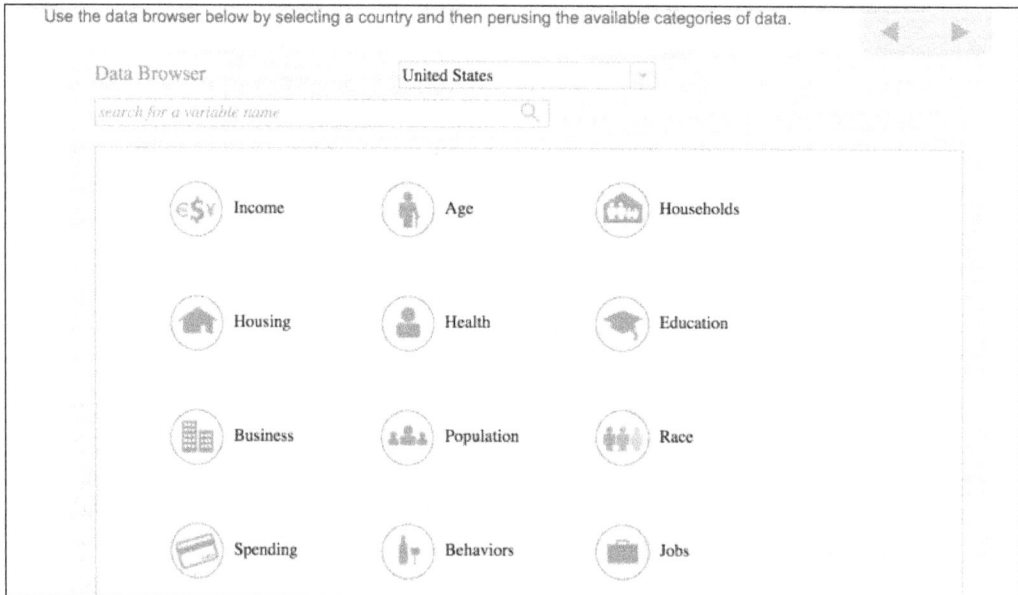

8. Open the Python development environment for the `GeoEnrichment.pyt` toolbar.

9. Duplicate the code that you have already created for the `CensusBlockGroupSelection` class by copying and pasting this class into the same `GeoEnrichment.pyt` file.

10. Rename the duplicated `CensusBlockGroupSelection` class `EnrichPotentialStores`.

11. Remove the code inside the `getParameterInfo()` and `execute()` methods for the new `EnrichPotentialStores` class.

12. Alter the `self.label` and `self.description` properties in the `__init__` method, as shown here:

```
def __init__(self):
        """Define the tool (tool name is the name of the
class)."""
        self.label = "Enrich Potential Stores"
        self.description = "Enrich Potential Stores"
        self.canRunInBackground = False
```

13. Find the `getParameterInfo()` method and add the two parameters shown here. The parameters will capture a `feature` class to enrich (this will be the point `feature` class you created with the last tool), and a drive time distance. Set a default value of 5 minutes for the drive time distance:

```
def getParameterInfo(self):
    param0 = arcpy.Parameter(displayName = "Layer to Enrich", \
                    name="layerToEnrich", \
                    datatype="DEFeatureClass", \
                    parameterType="Required", \
                    direction="Input")

    param1 = arcpy.Parameter(displayName = "Drive Time", \
                    name="driveTime", \
                    datatype="GPString", \
                    parameterType="Required", \
                    direction="Input")
    param1.value = 5

    params = [param0, param1]
    return params
```

14. Find the `execute()` method. This method will submit the list of features identified as potential stores to the ArcGIS Online GeoEnrichment service along with a drive time distance. The GeoEnrichment service will retrieve socio-economic information related to coffee establishments within the defined drive time for each store. This information will be returned to the script and written back to the potential store locations using the **ArcPy** data access module.

15. Retrieve the input parameters into new variables:

```
def execute(self, parameters, messages):
        """The source code of the tool."""
        in_features = parameters[0].valueAsText
        driveTime = parameters[1].valueAsText
```

16. The URL query string that will be submitted to the GeoEnrichment service will ultimately be quite long, so it will be easier to break this string into several parts, represented by individual variables. Add the following variables. The `url` variable will hold the first part of the URL string that points to the service, while the `studyAreaURL` variable will hold information related to the points that will be submitted for enrichment:

```
url = "http://geoenrich.arcgis.com/arcgis/rest/services/World/
geoenrichmentserver/GeoEnrichment/enrich?studyAreas=["

studyareaURL = ""
```

17. Add a `try`/`except` block.

```
def execute(self, parameters, messages):
    """The source code of the tool."""
    in_features = parameters[0].valueAsText
    driveTime = parameters[1].valueAsText

    url = "http://geoenrich.arcgis.com/arcgis/rest/services/World/
geoenrichmentserver/GeoEnrichment/enrich?studyAreas=["

    studyareaURL = ""

    try:

    except Exception as e:
        arcpy.AddMessage(e.message)
```

18. Inside the `try` block, add the following block of code. This code block creates a `SearchCursor` object that will loop through the input feature class containing the points to be enriched and will extract the coordinates of the point in addition to the attributes of the **Name** field. This information will be appended to the `studyareaURL` variable:

```
def execute(self, parameters, messages):
        """The source code of the tool."""
        in_features = parameters[0].valueAsText
        driveTime = parameters[1].valueAsText

        url = "http://geoenrich.arcgis.com/arcgis/rest/services/
World/geoenrichmentserver/GeoEnrichment/enrich?studyAreas=["

        studyareaURL = ""

        try:

            with arcpy.da.SearchCursor(in_features, ("SHAPE@XY",
"NAME")) as cursor:
                for row in cursor:
                    listCoords = row[0]
                    id = row[1]
```

```
            studyareaURL = studyareaURL +
"{\"geometry\":{\"x\":" + str(listCoords[0]) + ",\"y\":" +
str(listCoords[1]) + "},\"attributes\":{\"myID\":\"" + id +
"\"}},"

            studyareaURL = studyareaURL[:-1] + "]"


        except Exception as e:
            arcpy.AddMessage(e.message)
```

19. Create the three variables highlighted in the following code. The first variable, `studyAreaOptions`, defines the part of the URL query string that controls the size of the drive time buffer. Next, the `analysisVariables` variable defines the socio-economic variables to be pulled from the GeoEnrichment service.

    To keep things simple, the code will pull only two variables: the amount spent in the last 6 months at Starbucks and the amount spent on eating out for breakfast (the most common time to buy coffee). A more detailed analysis would most likely pull many other variables, but we'll keep things simple here in order to cut down on the amount of credits needed through ArcGIS Online to fulfill the request. The variable for the amount spent at Starbucks in the last 6 months is represented by `restaurants.MP29083_a_B`, and the variable that represents the total amount spent in the previous year eating out for breakfast is represented by `food.X1147_X`. You can find out more information on these and other demographic and lifestyle variables by visiting the data browser in the ArcGIS REST API help that we reviewed earlier.

20. Finally, the `urlSuffix` variable contains the output format and token. You will need to insert the token generated in a previous step. Finally, combine all the URL query strings:

```
try:

    with arcpy.da.SearchCursor(in_features,("SHAPE@XY", "NAME"))
as cursor:
        for row in cursor:
            listCoords = row[0]
            id = row[1]
```

```
                        studyareaURL = studyareaURL +
"{\"geometry\":{\"x\":" + str(listCoords[0]) + ",\"y\":" +
str(listCoords[1]) + "},\"attributes\":{\"myID\":\"" + id +
"\"}},"

                    studyareaURL = studyareaURL[:-1] + "]"

                studyareaOptions = "&studyAreasOptions={\"areaType\":\"D
riveTimeBuffer\",\"bufferUnits\":\"esriDriveTimeUnitsMinutes\",\"b
ufferRadii\":[" + driveTime + "]}"
                analysisVariables = "&analysisVariables=[\"restaurants.
MP29083a_B\",\"food.X1147_X\"]"
                urlSuffix = "&f=json&token=<your token here>"
                url = url + studyareaURL + studyareaOptions +
analysisVariables + urlSuffix
```

21. Submit the URL query string using the Python requests module and print out the response that is returned in the JSON format:

```
studyareaOptions = "&studyAreasOptions={\"areaType\":\"DriveTimeB
uffer\",\"bufferUnits\":\"esriDriveTimeUnitsMinutes\",\"bufferRad
ii\":[" + driveTime + "]}"
analysisVariables = "&analysisVariables=[\"restaurants.
MP29083a_B\",\"food.X1147_X\"]"
urlSuffix = "&f=json&token=<your token here"

url = url + studyareaURL + studyareaOptions + analysisVariables +
urlSuffix
r = requests.post(url)
arcpy.AddMessage(r.text)
```

22. Save your work and close the Python development environment.

23. In the **Catalog** view of **ArcMap**, double-click on the **Enrich Potential Stores** tool to test your code. Select the `PotentialCoffeeStores` feature class created in the previous step as **Layer to Enrich**, and leave the default **Drive Time** of 5 minutes. Click on **OK** to execute the tool. An example of the output can be seen in the following screenshot. Your output may differ slightly. If you get an error message or something that doesn't closely resemble the following output, you'll want to check the characters in the URL query strings. It's easy to make a typo with all these characters. You may also want to check the solution file against your file at `C:\ArcGIS_Blueprint_Python\solutions\ch7\EnrichPotentialStores.py`:

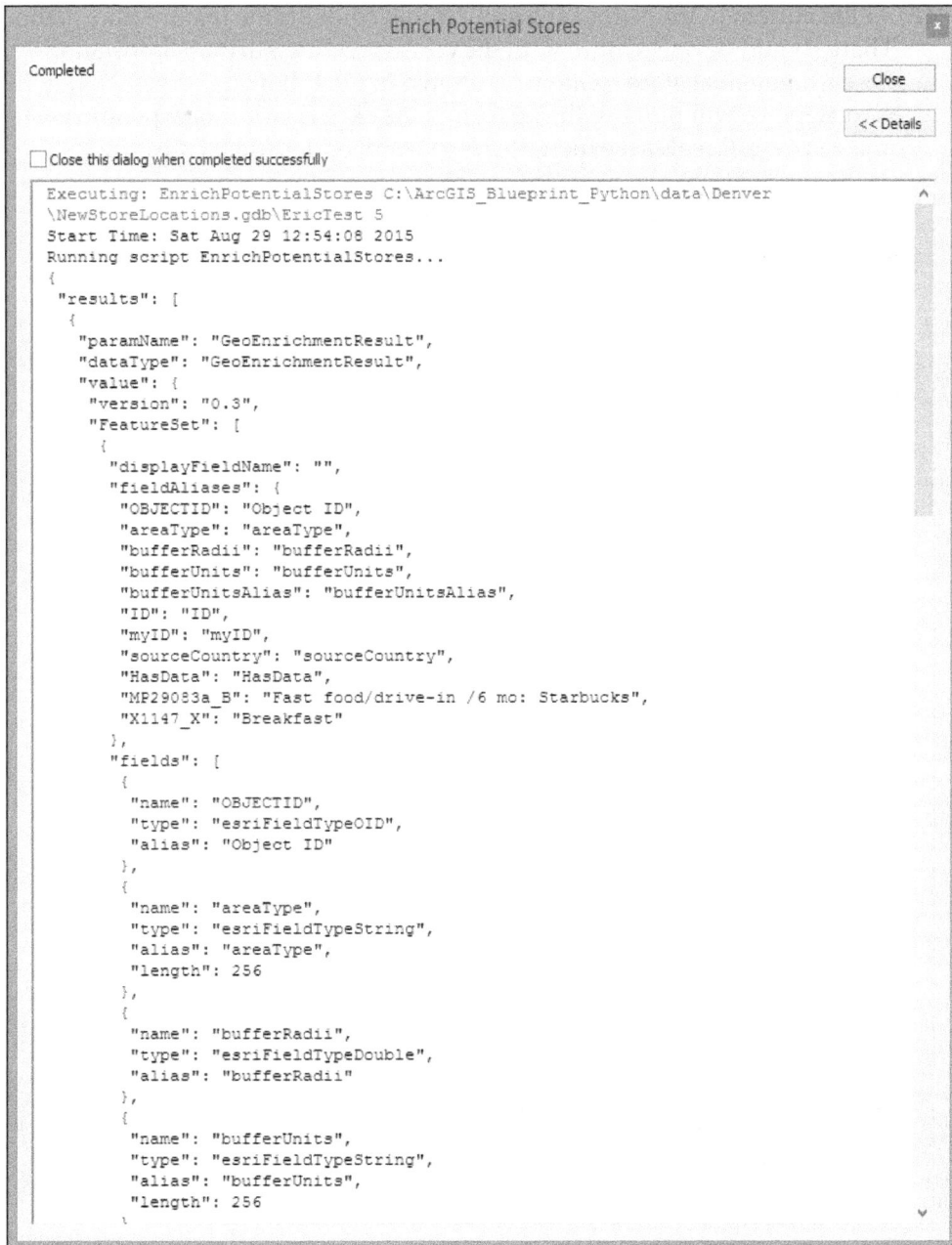

```
Executing: EnrichPotentialStores C:\ArcGIS_Blueprint_Python\data\Denver
\NewStoreLocations.gdb\EricTest 5
Start Time: Sat Aug 29 12:54:08 2015
Running script EnrichPotentialStores...
{
  "results": [
    {
      "paramName": "GeoEnrichmentResult",
      "dataType": "GeoEnrichmentResult",
      "value": {
        "version": "0.3",
        "FeatureSet": [
          {
            "displayFieldName": "",
            "fieldAliases": {
              "OBJECTID": "Object ID",
              "areaType": "areaType",
              "bufferRadii": "bufferRadii",
              "bufferUnits": "bufferUnits",
              "bufferUnitsAlias": "bufferUnitsAlias",
              "ID": "ID",
              "myID": "myID",
              "sourceCountry": "sourceCountry",
              "HasData": "HasData",
              "MP29083a_B": "Fast food/drive-in /6 mo: Starbucks",
              "X1147_X": "Breakfast"
            },
            "fields": [
              {
                "name": "OBJECTID",
                "type": "esriFieldTypeOID",
                "alias": "Object ID"
              },
              {
                "name": "areaType",
                "type": "esriFieldTypeString",
                "alias": "areaType",
                "length": 256
              },
              {
                "name": "bufferRadii",
                "type": "esriFieldTypeDouble",
                "alias": "bufferRadii"
              },
              {
                "name": "bufferUnits",
                "type": "esriFieldTypeString",
                "alias": "bufferUnits",
                "length": 256
```

24. At the bottom of the output is a section that contains the feature information. There should be one feature with the corresponding attribute information for each potential store location, as shown in the following screenshot. In the next step, we will pull out the myID, MP29083a_B (Starbucks expenditures), and X1147_X (breakfast expenditures at fast food locations) attributes.

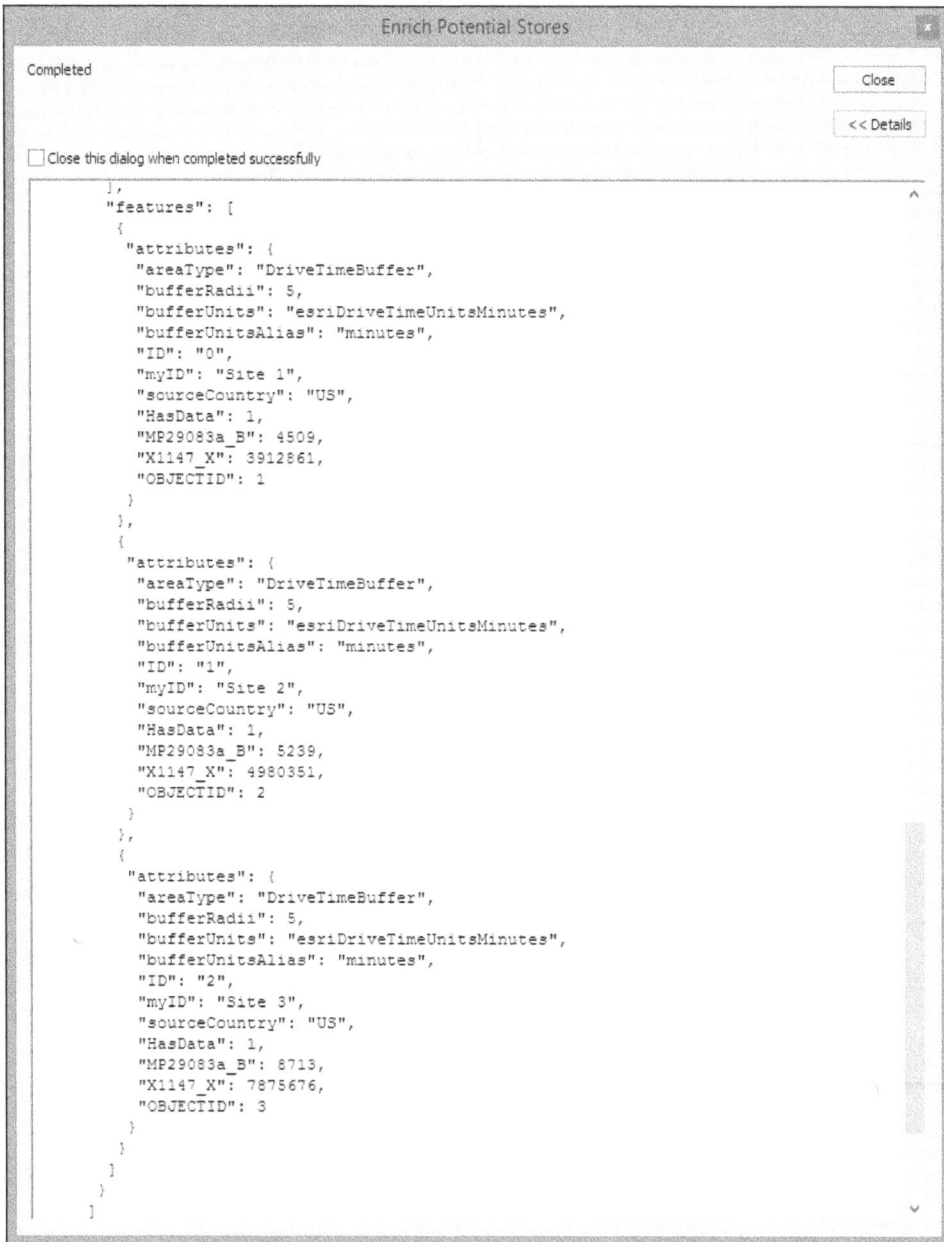

```
                                    Enrich Potential Stores                          [x]

Completed                                                              [  Close   ]

                                                                       [ << Details ]
☐ Close this dialog when completed successfully

                 ],                                                           ^
             "features": [
               {
                "attributes": {
                 "areaType": "DriveTimeBuffer",
                 "bufferRadii": 5,
                 "bufferUnits": "esriDriveTimeUnitsMinutes",
                 "bufferUnitsAlias": "minutes",
                 "ID": "0",
                 "myID": "Site 1",
                 "sourceCountry": "US",
                 "HasData": 1,
                 "MP29083a_B": 4509,
                 "X1147_X": 3912861,
                 "OBJECTID": 1
                }
               },
               {
                "attributes": {
                 "areaType": "DriveTimeBuffer",
                 "bufferRadii": 5,
                 "bufferUnits": "esriDriveTimeUnitsMinutes",
                 "bufferUnitsAlias": "minutes",
                 "ID": "1",
                 "myID": "Site 2",
                 "sourceCountry": "US",
                 "HasData": 1,
                 "MP29083a_B": 5239,
                 "X1147_X": 4980351,
                 "OBJECTID": 2
                }
               },
               {
                "attributes": {
                 "areaType": "DriveTimeBuffer",
                 "bufferRadii": 5,
                 "bufferUnits": "esriDriveTimeUnitsMinutes",
                 "bufferUnitsAlias": "minutes",
                 "ID": "2",
                 "myID": "Site 3",
                 "sourceCountry": "US",
                 "HasData": 1,
                 "MP29083a_B": 8713,
                 "X1147_X": 7875676,
                 "OBJECTID": 3
                }
               }
             ]
            }                                                            v
           ]
```

25. Reopen the Python development environment for this toolbox.

26. Convert the response from the JSON format to a Python dictionary and comment out the line of code you just added:

```
# arcpy.AddMessage(r.text)
decoded = json.loads(r.text)
```

27. The Python dictionary created from the JSON response returned by the GeoEnrichment service is quite complex. The next block of code will drill down into the Python dictionary and pull out the socio-economic information discussed earlier as well as the unique identifier associated with each record. Add the code highlighted here:

```
try:

    with arcpy.da.SearchCursor(in_features,("SHAPE@XY", "NAME"))
as cursor:
        for row in cursor:
            listCoords = row[0]
            id = row[1]
            studyareaURL = studyareaURL + "{\"geometry\":{\"x\":"
+ str(listCoords[0]) + ",\"y\":" + str(listCoords[1]) +
"},\"attributes\":{\"myID\":\"" + id + "\"}},"
            studyareaURL = studyareaURL[:-1] + "]"

        studyareaOptions = "&studyAreasOptions={\"areaType\":\"DriveTi
meBuffer\",\"bufferUnits\":\"esriDriveTimeUnitsMinutes\",\"bufferR
adii\":[" + driveTime + "]}"
        analysisVariables = "&analysisVariables=[\"restaurants.
MP29083a_B\",\"food.X1147_X\"]"
        urlSuffix = "&f=json&token=<your token here>

        url = url + studyareaURL + studyareaOptions +
analysisVariables + urlSuffix

        r = requests.post(url)
            decoded = json.loads(r.text)

        cntr = 1
        for rslt in decoded['results']:
            lstFeatures = rslt['value']['FeatureSet'][0]
['features']
            for ftr in lstFeatures:
                sixStar = (ftr['attributes']['MP29083a_B'])
                yrOutBreakfast = (ftr['attributes']['X1147_X'])
                siteID = (ftr['attributes']['myID'])
```

28. Finally, create an `UpdateCursor` object against the point feature class that stores the potential store locations, and update the applicable fields with the enrichment information:

```
cntr = 1
for rslt in decoded['results']:
    lstFeatures = rslt['value']['FeatureSet'][0]['features']
    for ftr in lstFeatures:
        sixStar = (ftr['attributes']['MP29083a_B'])
        yrOutBreakfast = (ftr['attributes']['X1147_X'])
        siteID = (ftr['attributes']['myID'])

        whereClause = 'NAME = ' + "\'" + siteID + "\'"
        with arcpy.da.UpdateCursor(in_features, ("NAME",
"AvgStar6", "TotBfSpend"), whereClause) as cursor:
            for row in cursor:
                row[1] = sixStar
                row[2] = yrOutBreakfast
                cursor.updateRow(row)
                arcpy.AddMessage("Record number: " + str(cntr) + "
written to feature class")
                cntr = cntr + 1
```

29. You can check your work by examining the `C:\ArcGIS_Blueprint_Python\solutions\ch7\EnrichPotentialStores.py` solution file. Refer to the `getParameterInfo()` and `execute()` methods.

30. Save the file and exit your Python development environment.

31. In the **Catalog** view of **ArcMap**, double-click on the **Enrich Potential Stores** tool to test your code. The tool dialog box shown in the following screenshot should appear. Select the **PotentialCoffeeStores** feature class created in the previous step as **Layer to Enrich** and leave the default **Drive Time** of 5 minutes as shown in the following screenshot:

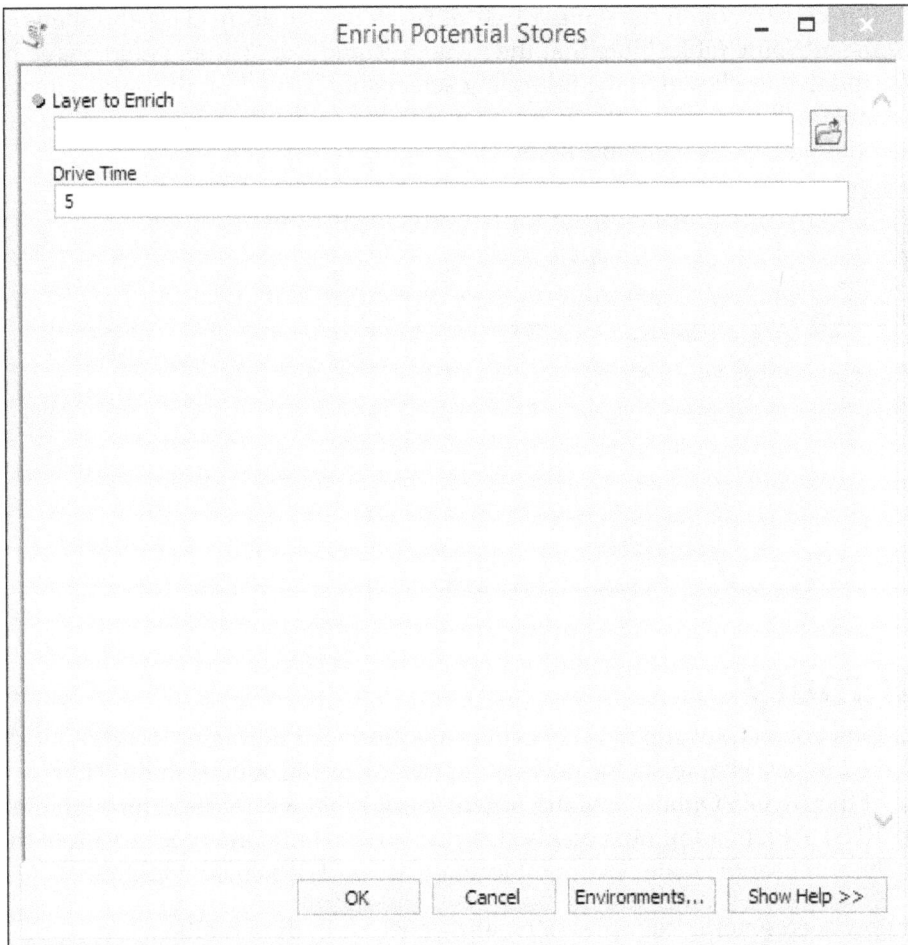

32. Click on **OK** to execute the tool. If there aren't any errors in your tool, you should see progress information indicating that the features have been updated.

33. If necessary, open the `PotentialCoffeeStores` feature class and then open the attribute table. Note that the `AvgStar6` and `TotBfSpend` fields have been updated as shown in the following screenshot. Note that the values in your table will probably differ from mine unless you have created the exact same potential coffee store locations.

OBJECTID *	SHAPE *	NAME	DESCRIPTION	AvgStar6	TotBfSpend
1	Point	Site 1	Near a major shopping area	4509	3912861
2	Point	Site 2	<Null>	5239	4980351
3	Point	Site 3	<Null>	8713	7875676

(0 out of 3 Selected)

Potential Coffee Stores Sites

Summary

This chapter covered several new concepts in addition to reinforcing several skills that we covered in past chapters. One new concept that was introduced in this chapter is the use of the ArcGIS Online GeoEnrichment service that we accessed through the ArcGIS REST API. In addition, you also learned how to build an interactive tool that allows the end user to create new point locations as part of a custom tool.

We've now seen several examples of creating custom Python toolboxes in ArcGIS Desktop, so you should be familiar with these by now and have a good understanding of how they work. You should also be familiar with the Python requests module that is used to make requests to external web services as we've seen this in action on several occasions as well. Finally, we used the **ArcPy** mapping and data access modules in this chapter once again in order to support several operations, including applying symbology to a layer and using cursor objects to query and update the data in a feature class.

In the next chapter, you will build a set of tools to support search and rescue operations. This will include tools that are used to define the last known location of a subject and the search area definition boundaries as well as a tool that will enable the data to be visualized in Google Earth.

8

Supporting Search and Rescue Operations with ArcPy, Python Add-Ins, and simplekml

Search and Rescue (SAR) operations are inherently geographic in nature and can benefit from GIS tools designed to support search efforts. The identification of the **Last Known Position (LKP)** is an essential first step in the process. Based on this LKP, additional tools can support **SAR** operations through the identification of potential search areas and the definition of search sectors for the assignment of search teams. The visualization of these datasets through a tool such as Google Earth is extremely helpful in order to understand how terrain can affect the likely movement patterns of lost individuals. For example, people are much more likely to follow well-defined trails and move toward areas of lower elevation as opposed to areas of higher elevation.

The application built in this chapter will support **SAR** operations through the inclusion of several custom tools in an **ArcGIS Python Toolbox** as well as Python add-ins. The Python add-in for ArcMap will include a tool to define the last known position (LKP) and a tool to define polygon search sectors. In addition, a custom **ArcGIS Python Toolbox** will contain a tool to define buffer distances around the LKP, a tool to extract datasets to the Google Earth format, and a tool to assign attribute data to the LKP.

In this chapter, we will cover the following topics:

- Creating tools for Python add-ins
- Combining add-ins with custom tools
- Converting data to the Google Earth **Keyhole Markup Language** (**KML**) format with `simplekml`
- Using the `arcpy.mapping` module

Design

The design of this application will include the creation of three custom tools inside an **ArcGIS Desktop Python Toolbox** as well as a Python add-in for ArcGIS Desktop that includes two tools.

The two tools, that are part of the Python add-in, include a tool that can be used to define a point location that indicates the last known position of the lost individual, and a tool that can be used to sketch search sectors.

For the **ArcGIS Desktop Python Toolbox**, there is a tool that will be used to assign attributes to the LKP, a tool to define buffer polygons representing distances around the LKP, and a tool to export data to the Google Earth KML format. The following diagram shows how it works:

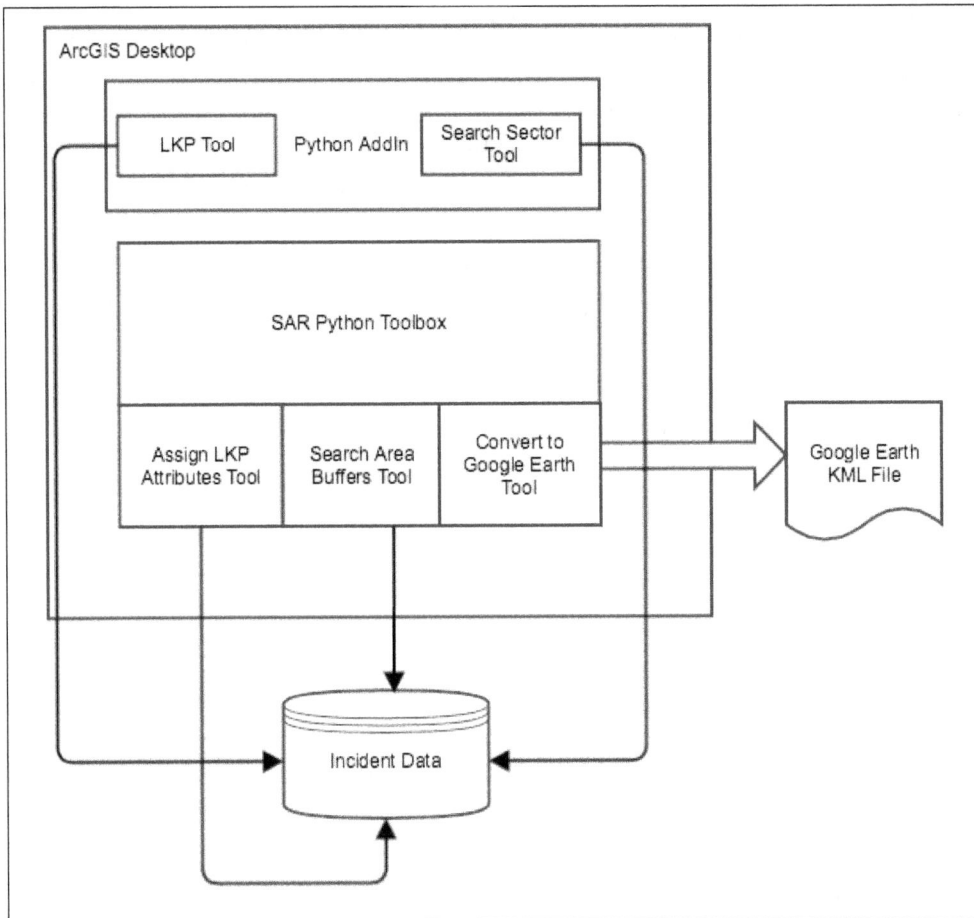

Let's get started and build the application.

Creating the Last Known Position tool

In this step, you'll create an **ArcGIS Python Add-In** for the purpose of allowing an analyst to define the LKP of the individual in the need of rescue. This LKP tool will be created inside a toolbar container that will eventually house a second tool that will be used to create polygon search sectors. A screenshot of the final toolbar is shown as follows:

The LKP tool will enable the selection of a point location on the map. After selecting the LKP, the tool will then display a dialog that collects attribute information about the location, including the subject name, incident date, gender, weight, hair color, and other attributes. This information will be collected via a custom tool inside an **ArcGIS Python Toolbox**, which will be created in this section as well.

The user interface for the attribute tool will appear as seen in the following screenshot:

Follow these to create the **ArcGIS Python Add-In** and the custom ArcGIS tool to capture the location and attributes of the LKP of the subject. You should already be familiar with the process of using the **ArcGIS Python Add-In Wizard** to create the structure of an ArcGIS Desktop add-in, so I won't go into much detail on this subject. If you need a refresher, refer to previous chapters:

1. Open **ArcMap** with the `C:\ArcGIS_Blueprint_Python\ch8\SAR.mxd` file and examine the contents to get a better knowledge of the database, that will be used in this application.

2. Open the **ArcGIS Python Add-In Wizard** and go to `C:\ArcGIS_Blueprint_Python\ch8` folder.

3. Click on the **Make New Folder** button with `ch8` folder selected and name the folder `add-in`.

4. Click on **OK**.

5. In the **ArcGIS Python Add-In Wizard** on the **Project Settings** tab, give the project a name of `Search and Rescue Addin`. You can leave the version as **0.1**.

6. Click on the **Add-In Contents** tab.

7. Create a new toolbar by right-clicking on **TOOLBARS** and selecting **New Toolbar**.

8. Type `Search and Rescue` toolbar in the **Caption** textbox.

9. Right-click on the **New Toolbar** and select **New Tool**.

10. Enter `LKP` for **Caption** and `LKPClass` for **Class Name**.

11. Right-click on **Search and Rescue** toolbar and select **New Tool**.

12. Give the tool a **Caption of Search Sector**, a **Class Name** of `SearchSectorClass`, and an **ID** of `addin_addin.tool_1`.

13. Click on the **Save** button.

14. Click on the **Open Folder** button.

15. Open the `Install` folder and open the `addin_addin.py` script in your Python development environment.

16. This tool will use only the `onMouseDownMap()` method, so remove all methods in the `LKP` class with the exception of `__init__` and `onMouseDownMap()`.

17. Import the `pythonaddins` and `os` modules:

```
import arcpy
import pythonaddins
import os
```

18. Inside the `__init__` method, make sure that `self.shape` is set to `NONE`, as shown in the following code, and `self.cursor` is set to a value of `4`:

```
def __init__(self):
    self.enabled = True
    self.cursor = 4
    self.shape = "NONE"
```

19. Find the `onMouseDown()` method and inside it, use the `SaveDialog()` method to capture the path and the filename, the feature class name of the layer that will store the geometry and attributes of the `LKP`:

```
def onMouseDownMap(self, x, y, button, shift):
    #path to the output feature class
    fullPath = pythonaddins.SaveDialog("Save LKP")
    path = os.path.split(fullPath)[0]
    layerName = os.path.split(fullPath)[1]
```

20. Use the `OpenDialog()` method to capture the feature class that will contain the attribute schema of the input feature class:

```
def onMouseDownMap(self, x, y, button, shift):
    #path to the output feature class
    fullPath = pythonaddins.SaveDialog("Save LKP")
    path = os.path.split(fullPath)[0]
    layerName = os.path.split(fullPath)[1]


    #schema feature class
    schemaFC = pythonaddins.OpenDialog("Feature Class Template")
```

21. Define the output spatial reference:

```
def onMouseDownMap(self, x, y, button, shift):
    #path to the output feature class
    fullPath = pythonaddins.SaveDialog("Save LKP")
    path = os.path.split(fullPath)[0]
    layerName = os.path.split(fullPath)[1]


    #schema feature class
    schemaFC = pythonaddins.OpenDialog("Feature Class Template")

    sr = arcpy.SpatialReference(26911)
```

22. Create a new feature class using the information collected from the user:

```
def onMouseDownMap(self, x, y, button, shift):
    #path to the output feature class
    fullPath = pythonaddins.SaveDialog("Save LKP")
    path = os.path.split(fullPath)[0]
    layerName = os.path.split(fullPath)[1]


    #schema feature class
    schemaFC = pythonaddins.OpenDialog("Feature Class Template")

    sr = arcpy.SpatialReference(26911)


    #create a new feature class
    arcpy.CreateFeatureclass_management(path, layerName, "POINT",
schemaFC, spatial_reference=sr)
```

23. Use the `InsertCursor` method to insert the point location of the `LKP`:

```
def onMouseDownMap(self, x, y, button, shift):
    #path to the output feature class
    fullPath = pythonaddins.SaveDialog("Save LKP")
    path = os.path.split(fullPath)[0]
    layerName = os.path.split(fullPath)[1]


    #schema feature class
    schemaFC = pythonaddins.OpenDialog("Feature Class Template")

    sr = arcpy.SpatialReference(26911)

    #create a new feture class
    arcpy.CreateFeatureclass_management(path, layerName, "POINT",
schemaFC)

    #insert the record
    row_value = [(x,y)]
    with arcpy.da.InsertCursor(fullPath, ["SHAPE@XY"]) as cursor:
        cursor.insertRow(row_value)
```

24. Use the `GPToolDialog()` method to execute the `AssignLKPAttributes` custom tool. We are yet to create the `AssignLKPAttributes` tool, but we will do this in the next few steps. This tool will be used to capture attribute information about the `LKP`. For now, just put a placeholder for the path to the tool. Later, we'll return to this line of code and insert the actual path:

```
def onMouseDownMap(self, x, y, button, shift):
    #path to the output feature class
    fullPath = pythonaddins.SaveDialog("Save LKP")
    path = os.path.split(fullPath)[0]
    layerName = os.path.split(fullPath)[1]


    #schema feature class
def onMouseDownMap(self, x, y, button, shift):
    #path to the output feature class
    fullPath = pythonaddins.SaveDialog("Save LKP")
    path = os.path.split(fullPath)[0]
    layerName = os.path.split(fullPath)[1]


    #schema feature class
```

```
schemaFC = pythonaddins.OpenDialog("Feature Class Template")

sr = arcpy.SpatialReference(26911)

#create a new feture class
arcpy.CreateFeatureclass_management(path, layerName, "POINT",
schemaFC)

#insert the record
row_value = [(x,y)]
with arcpy.da.InsertCursor(fullPath, ["SHAPE@XY"]) as cursor:
    cursor.insertRow(row_value)

pythonaddins.GPToolDialog("C:\Users\Eric Pimpler\AppData\
Roaming\ESRI\Desktop10.3\ArcToolbox\My Toolboxes\SAR.pyt",
"AssignLKPAttributes")
```

25. Save the Python script and close the development environment.

26. Open **ArcMap** and display the **ArcCatalog** pane. In the **Toolboxes** folder under **My Toolboxes**, create a new **Python Toolbox** and call it SAR.pyt.

27. Open the code for the **SAR** toolbox in your Python development environment by right-clicking on the toolbox and selecting **Edit**.

28. Rename the Tool class to AssignLKPAttributes. Also, update the self. label and self.description properties, as shown in the following code:

```
class AssignLKPAttributes(object):
    def __init__(self):
        """Define the tool (tool name is the name of the
class)."""
        self.label = "Assign Last Known Point Attributes"
        self.description = "Assign Last Known Point Attributes"
```

29. Add the tool to the self.tools list in the Toolbox class:

```
self.tools = [AssignLKPAttributes]
```

30. Find the getParameterInfo() method and add the following parameters to capture the input feature class, incident date, description, name, and other attributes of the LKP and add them to the parameter list:

```
def getParameterInfo(self):
    """Define parameter definitions"""
    outFC = arcpy.Parameter(displayName = "Output Feature
Class", \
                        name="outFC", \
                        datatype="DEFeatureClass",\
```

```
                              parameterType="Required",\
                              direction="Input")

     ## begin date for import
     incidentDate = arcpy.Parameter(
         displayName="Incident Date",
         name="incidentDate",
         datatype="GPDate",
         parameterType="Optional",
         direction="Input")
     incidentDate.value = "01/01/2015"

     description = arcpy.Parameter(
         displayName="Description",
         name="description",
         datatype="GPString",
         multiValue="False",
         parameterType="Optional",
         direction="Input")

     incidentName = arcpy.Parameter(
         displayName="Incident Name",
         name="incidentName",
         datatype="GPString",
         multiValue="False",
         parameterType="Optional",
         direction="Input")

     subjectName = arcpy.Parameter(
         displayName="Name",
         name="subjectName",
         datatype="GPString",
         parameterType="Optional",
         direction="Input",
         multiValue = False)

     gender = arcpy.Parameter(
         displayName="Gender",
         name="gender",
         datatype="GPString",
         parameterType="Optional",
         direction="Input",
         multiValue = False)
```

```
gender.filter.list = ["MALE", "FEMALE"]

weight = arcpy.Parameter(
    displayName="Weight",
    name="weight",
    datatype="GPString",
    parameterType="Optional",
    direction="Input",
    multiValue = False)

hairColor = arcpy.Parameter(
    displayName="Hair Color",
    name="hairColor",
    datatype="GPString",
    parameterType="Optional",
    direction="Input",
    multiValue = False)

other = arcpy.Parameter(
    displayName="Other",
    name="other",
    datatype="GPString",
    parameterType="Optional",
    direction="Input",
    multiValue = False)

height = arcpy.Parameter(
    displayName="Height",
    name="height",
    datatype="GPString",
    parameterType="Optional",
    direction="Input",
    multiValue = False)

clothing = arcpy.Parameter(
    displayName="Clothing",
    name="clothing",
    datatype="GPString",
    parameterType="Optional",
    direction="Input",
    multiValue = False)

age = arcpy.Parameter(
    displayName="Age",
```

```
                name="age",
                datatype="GPString",
                parameterType="Optional",
                direction="Input",
                multiValue = False)
        params = [outFC, incidentDate, description, incidentName,
    subjectName, gender, weight, hairColor, other, height, clothing,
    age]
```

31. Find the `execute()` method and add the following code to capture the input parameters supplied by the end user:

```
def execute(self, parameters, messages):
        """The source code of the tool."""
        outFC = parameters[0].valueAsText
        incidentDate = parameters[1].valueAsText
        description = parameters[2].valueAsText
        incidentName = parameters[3].valueAsText
        subjectName = parameters[4].valueAsText
        gender = parameters[5].valueAsText
        weight = parameters[6].valueAsText
        hairColor = parameters[7].valueAsText
        other = parameters[8].valueAsText
        height = parameters[9].valueAsText
        clothing = parameters[10].valueAsText
        age = parameters[11].valueAsText
```

32. Use `UpdateCursor` to update the attributes:

```
def execute(self, parameters, messages):
        """The source code of the tool."""
        outFC = parameters[0].valueAsText
        incidentDate = parameters[1].valueAsText
        description = parameters[2].valueAsText
        incidentName = parameters[3].valueAsText
        subjectName = parameters[4].valueAsText
        gender = parameters[5].valueAsText
        weight = parameters[6].valueAsText
        hairColor = parameters[7].valueAsText
        other = parameters[8].valueAsText
        height = parameters[9].valueAsText
        clothing = parameters[10].valueAsText
        age = parameters[11].valueAsText

        with arcpy.da.UpdateCursor(outFC, ("Date", "Description",
    "Incident_Name", "Name", "Gender", "Weight", "Hair_Color",
    "Other", "Height", "Clothing", "Age")) as cursor:
            for row in cursor:
```

```
row[0] = incidentDate
row[1] = description
row[2] = incidentName
row[3] = subjectName
row[4] = gender
row[5] = weight
row[6] = hairColor
row[7] = other
row[8] = height
row[9] = clothing
row[10] = age
cursor.updateRow(row)
```

33. Save your code and exit the development environment.

34. In **ArcMap**, go to the **Catalog** window and right-click on the SAR.pyt toolbox to display the **SAR Properties** dialog box, as shown in the following screenshot:

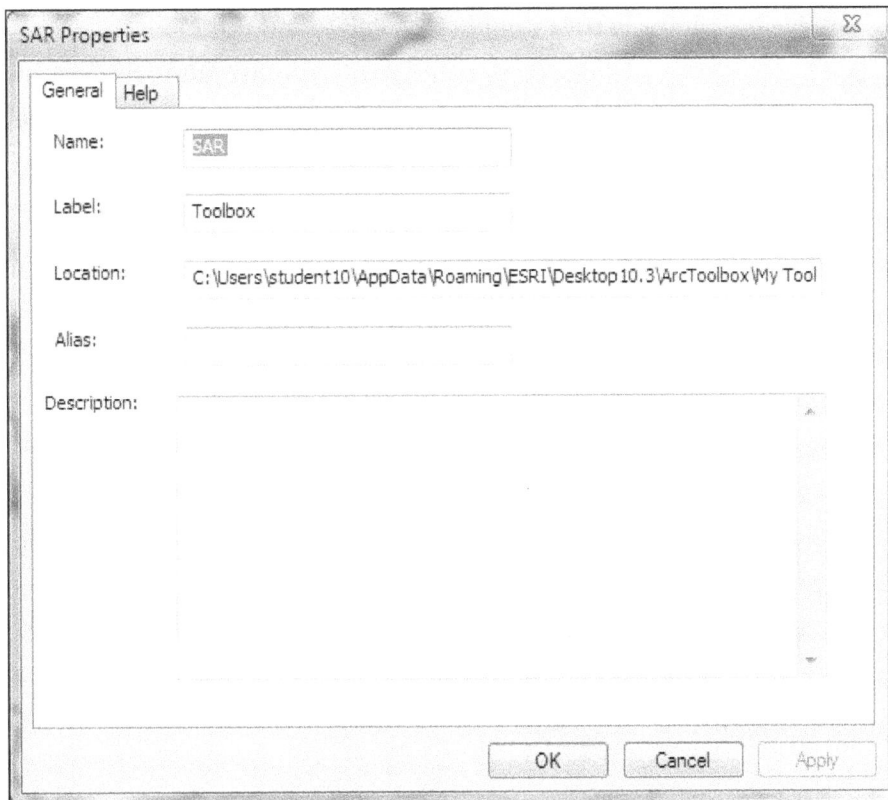

35. Highlight and copy the path in the **Location** textbox.

36. Return to the `addin_addin.py` file in the `Install` folder of the Python add-in and open it in your Python development environment.

37. Find the line of code that sets `GPToolDialog` and paste the path as the first parameter. An example is shown here. Your path will differ from mine:

```
pythonaddins.GPToolDialog("C:\Users\Eric Pimpler\AppData\
Roaming\ESRI\Desktop10.3\ArcToolbox\My Toolboxes\SAR.pyt",
"AssignLKPAttributes")
```

38. You can check your work by examining the solution files in the `C:\ArcGIS_Blueprint_Python\solutions\ch8` folder. Open `addin_addin.py` and review the `LKP` class. Open the `ConvertToGoogleEarth.py` file and examine the `AssignLKPAttributes` class.

39. Save the file and exit the development environment.

40. In the working directory of this add-in, double-click on the `makeaddin.py` script to create a new **Esri Addin** file called `addin.esriaddin`.

41. Double-click on the **Esri Addin** file to install the add-in.

42. Test the work by opening **ArcMap** and `SAR.mxd` and adding **SAR** toolbar.

43. If required, open the **Search and Rescue** toolbar by going to **Customize | Toolbars | Search and Rescue** toolbar.

44. Click on the `LKP` tool and then click on a location on the map to define a new point location that represents a missing person. This will display the **Save LKP** dialog box, as shown in the following screenshot. The layers that appear in your dialog box may differ from what you see in this screenshot:

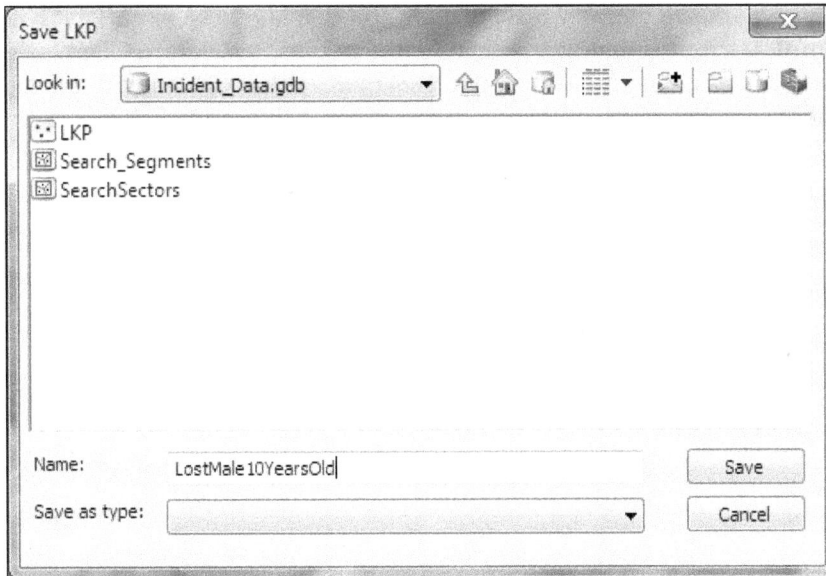

45. Navigate to the `C:\ArcGIS_Blueprint_Python\data\SAR\Incident_Data.gdb` geodatabase and name your feature class `LostMale10YearsOld`. You can apply whatever name you'd like. Click on the **Save** button.

46. The **Feature Class Template** dialog box will then be displayed. Select `C:\ArcGIS_Blueprint_Python\data\SAR\Incident_Data.gdb\LKP`, as shown in the following screenshot. This defines the `feature` class schema to be used for the creation of the new feature class. Click on the **Add** button:

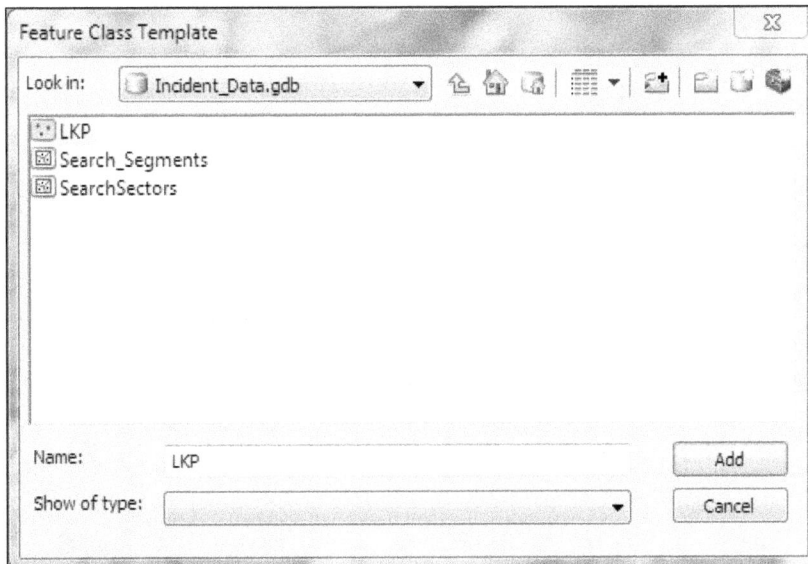

47. This will create the new feature class and display the **Assign Last Known Point Attributes** dialog box seen here. For **Output Feature Class**, select the feature class that you just created (`LostMale10YearsOld`) with the `LKP` tool, fill in as many of the other attributes as you'd like, and click on `OK`, as shown in the following screenshot:

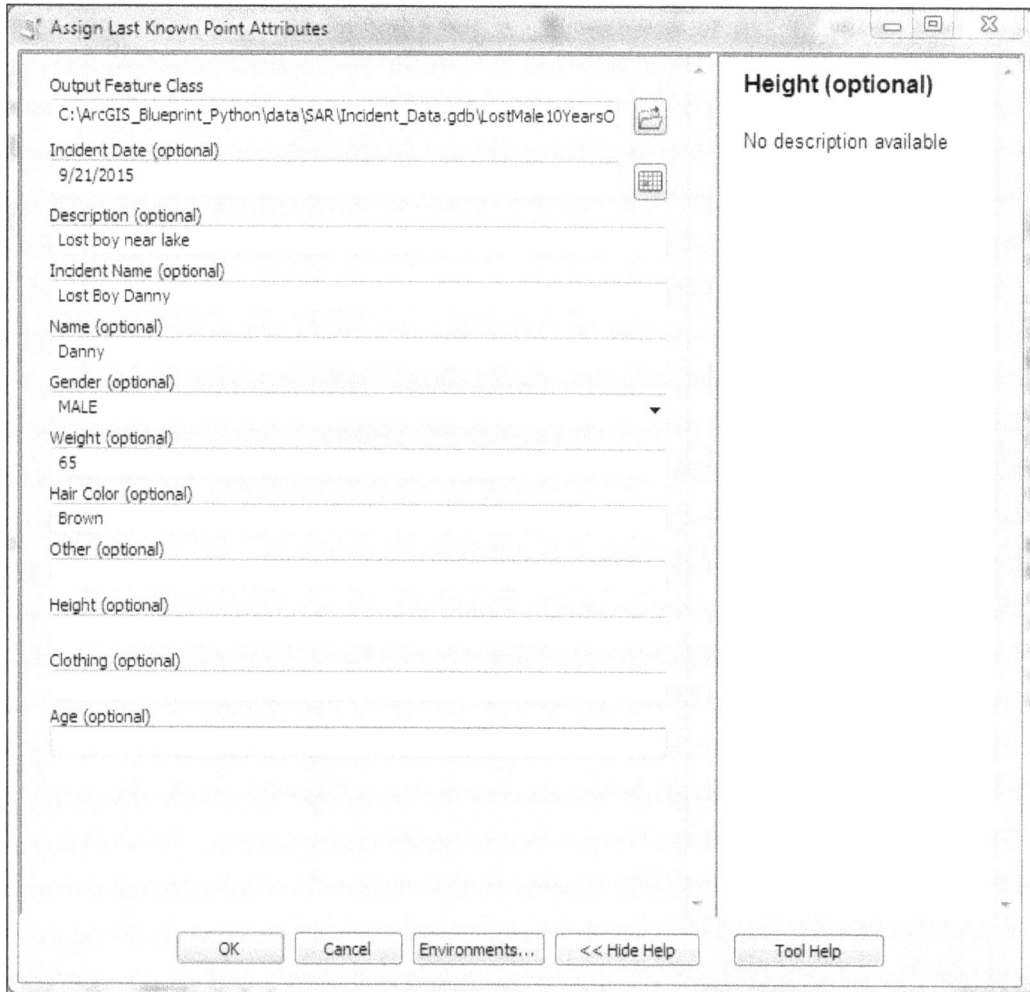

48. Close the progress dialog box when the processing is done.

Creating the Search Area Buffers tool

The search and rescue operations team can benefit from the knowledge of the potential search radius in which they'll need to operate. To assist with this task, the next tool that will be added to the **SAR** toolbox will be a **Search Area Buffers** tool that calculates multiple buffers based on the LKP template created in the previous step. This tool will be a variant of the existing **Multi-Ring Buffer** tool provided by ArcGIS Desktop. The tool will be simplified to include only distances in miles.

The following steps will help you to create the **Search Area Buffers** tool:

1. In **ArcMap**, right-click on the SAR.pyt custom toolbox in the **Catalog** view and select **Edit** to open the code in the Python development environment.

2. Copy and paste the existing AssignLKPAttributes class inside the SAR.pyt file.

3. Rename the newly pasted AssignLKPAttributes class to SearchAreaBuffers class.

4. Update the self.label and self.description properties, as shown in the following code:

```
class SearchAreaBuffers(object):
    def __init__(self):
        """Define the tool (tool name is the name of the
class)."""
        self.label = "Search Area Buffers"
        self.description = "Search Area Buffers"
```

5. Add the tool to the self.tools list in the Toolbox class:

```
self.tools = [AssignLKPAttributes, SearchAreaBuffers]
```

6. Find the getParameterInfo() method and remove the existing parameters that were copied. Next, add the following parameters so that your code appears as follows:

```
def getParameterInfo(self):
        """Define parameter definitions"""
        inFC = arcpy.Parameter(displayName = "Input Feature
Class", \
                    name="inFC", \
                    datatype="GPFeatureLayer",\
                    parameterType="Required",\
                    direction="Input")
```

```
outFC = arcpy.Parameter(displayName = "Output Feature
Class", \
                        name="outFC", \
                        datatype="DEFeatureClass",\
                        parameterType="Required",\
                        direction="Output")

distanceVals = arcpy.Parameter(displayName = "Distances in
Miles", \
                        name="distanceVals", \
                        datatype="GPDouble",\
                        parameterType="Required",\
                        multiValue="True",\
                        direction="Input")

params = [inFC, outFC, distanceVals]
return params
```

7. Find the `execute()` method and remove the existing code. Add the following code to capture the input parameters and call the `MultiRingBuffer` tool:

```
def execute(self, parameters, messages):
        """The source code of the tool."""
        inFC = parameters[0].valueAsText
        outFC = parameters[1].valueAsText
        distanceVals = parameters[2].valueAsText

        arcpy.MultipleRingBuffer_analysis(inFC, outFC,
distanceVals)
```

8. You can check your work by examining the solution file in the `C:\ArcGIS_Blueprint_Python\solutions\ch8` folder. Open the `ConvertToGoogleEarth.py` file and examine the `SearchAreaBuffers` class.

9. Save the Python script and exit the development environment.

10. Double-click on the new `Search Area Buffers` tool from the **SAR** toolbox. This will display the dialog box as shown in the following screenshot. Navigate to **LostMale10YearsOld** as the input feature class along with an output feature class called `SearchAreaBuffers`, which should be saved in the `Incident_Data` geodatabase as well as various output distances:

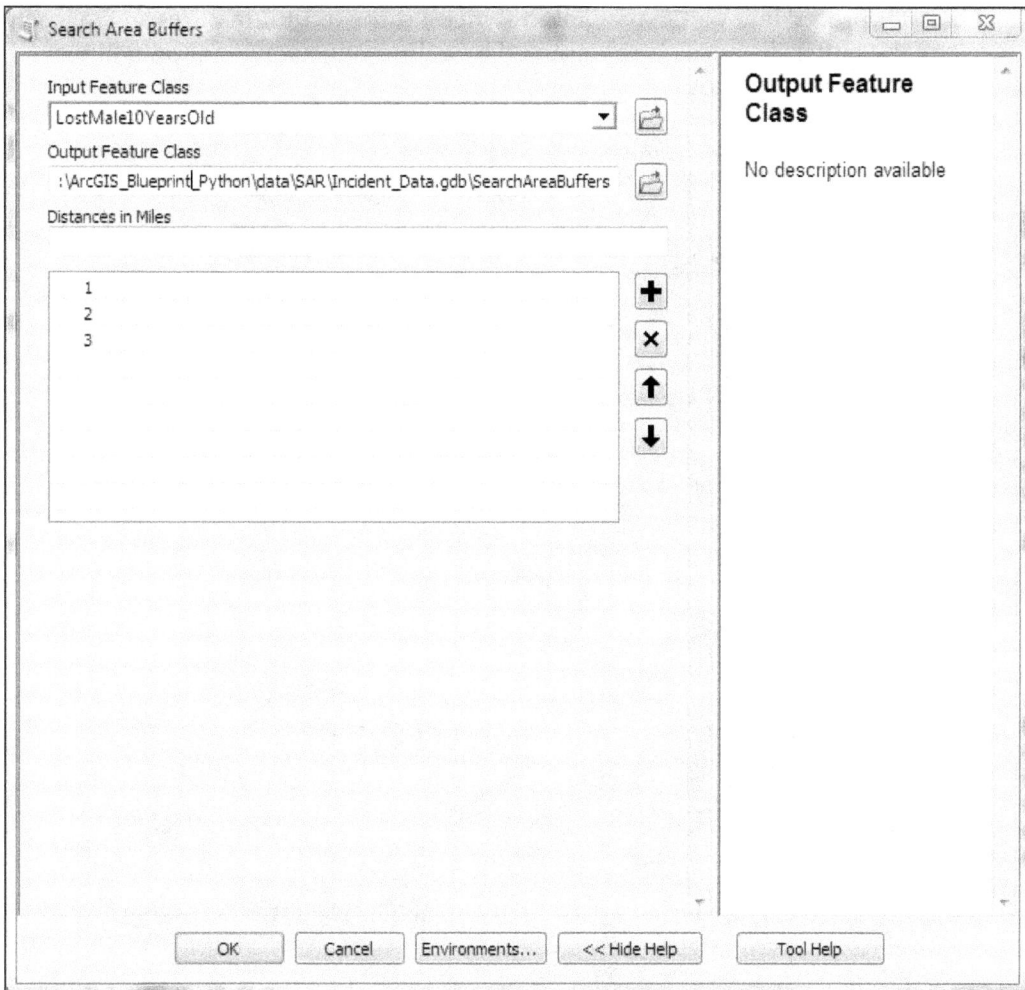

11. Click on **OK**.

12. After processing, you may want to symbolize the output features so that each of the buffer distances is represented by a different color. As shown in the following screenshot:

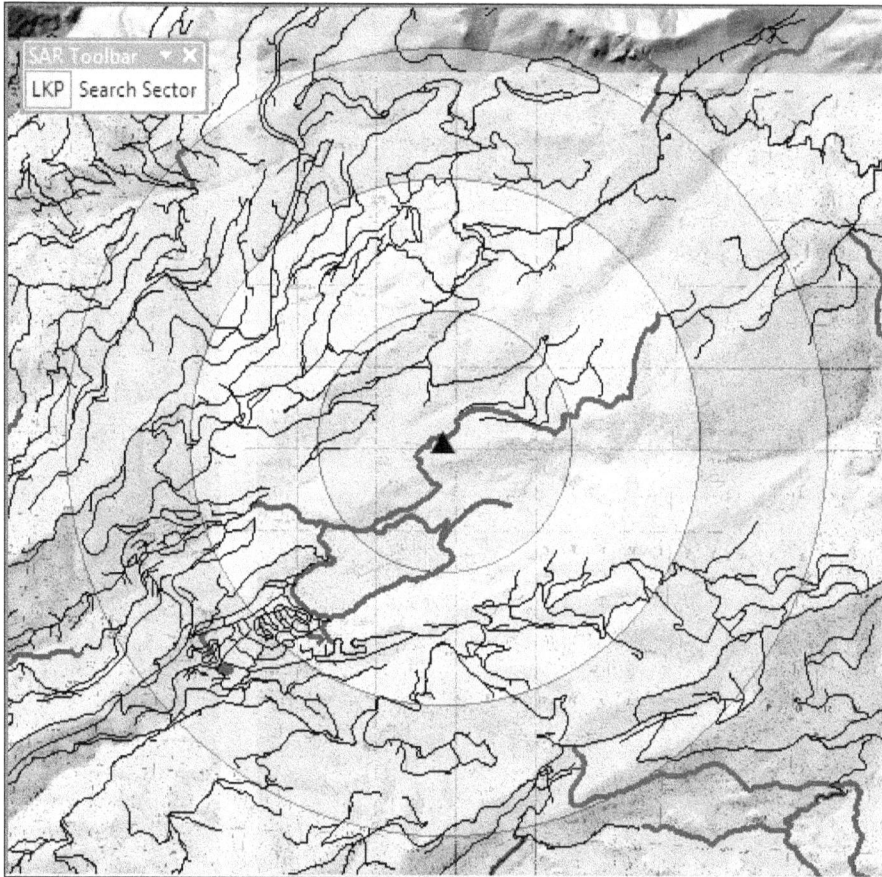

In the next section of the chapter, we'll create a Search Sector tool that will enable Search and Rescue personnel to draw search sector polygons.

Creating the Search Sector tool

Efficient search and rescue operations will require that groups of personnel be assigned to specific areas. In this section, we'll create a tool that will allow **Search and Rescue** analysts to sketch polygon features that represent specific areas that will then be assigned to search groups. The following steps will help you to create **Search Sector** tool:

1. Close **ArcMap** if required.

2. In Windows Explorer, return to the working directory that stores the Python add-in for this project and open the addin_addin.py file from the Install folder in your Python development environment.

3. Find the SearchSector class and remove all the methods with the exception of the __init__ and onLine() methods.

4. In the __init__ method, set the self.shape property to LINE, as shown here, along with a cursor type:

```
self.shape = "LINE"
self.cursor = 3
```

5. In the onLine() method, set the workspace environment variable:

```
def onLine(self, line_geometry):
    arcpy.env.workspace = "C:\ArcGIS_Blueprint_Python\data\SAR\
Incident_Data.gdb"
```

6. In the next code block that you'll add to the onLine() method, create an arcpy Array object, retrieve the line geometry that is sketched on the map, add the vertices of the line to Array, add the first vertices as the last point in the array, and create a new Polygon object from Array. The creation of the last vertices using the first point is necessary in order to close the polygon:

```
def onLine(self, line_geometry):
    arcpy.env.workspace = "C:\ArcGIS_Blueprint_Python\data\SAR\
Incident_Data.gdb"

    array = arcpy.Array()
    part = line_geometry.getPart(0)
    for pt in part:
        array.add(pt)
    array.add(line_geometry.firstPoint)
    polygon = arcpy.Polygon(array)
```

7. Create an `InsertCursor` object that references the `SearchSectors` feature class that is provided for you in the `Incident_Data` geodatabase. Insert the new polygon and refresh the view:

```
def onLine(self, line_geometry):
    arcpy.env.workspace = "C:\ArcGIS_Blueprint_Python\data\SAR\
Incident_Data.gdb"

    array = arcpy.Array()
    part = line_geometry.getPart(0)
    for pt in part:
        array.add(pt)
    array.add(line_geometry.firstPoint)
    polygon = arcpy.Polygon(array)

    with arcpy.da.InsertCursor("SearchSectors", ("SHAPE@")) as
cursor:
        cursor.insertRow((polygon,))
        arcpy.RefreshActiveView()
```

8. You can check your work by examining the solution files in the `C:\ArcGIS_Blueprint_Python\solutions\ch8` folder. Open `addin_addin.py` and review the `SearchSector` class.

9. Save the script and exit the Python development environment.

10. In Windows Explorer, go to the `C:\ArcGIS_Blueprint_Python\ch8\addin` folder and delete the existing `addin.esriaddin` file.

11. Double-click on `makeaddin.py` to recreate this file with the changes you just made.

12. Double-click on the `addin.esriaddin` file and reinstall the `add-in`.

13. Open the `SAR.mxd` file found in `C:\ArcGIS_Blueprint_Python\ch8`.

14. If required, add the `SearchSectors` feature class from the `Incident_Data` geodatabase.

15. If required, display the **SAR** toolbar by going to **Customize | Toolbars | SAR** toolbar.

16. Click on the **Search Sector** tool on the toolbar and start drawing a polygon using your mouse. When you're done drawing, double-click with the mouse to create and store the polygon. The following screenshot displays a completed polygon:

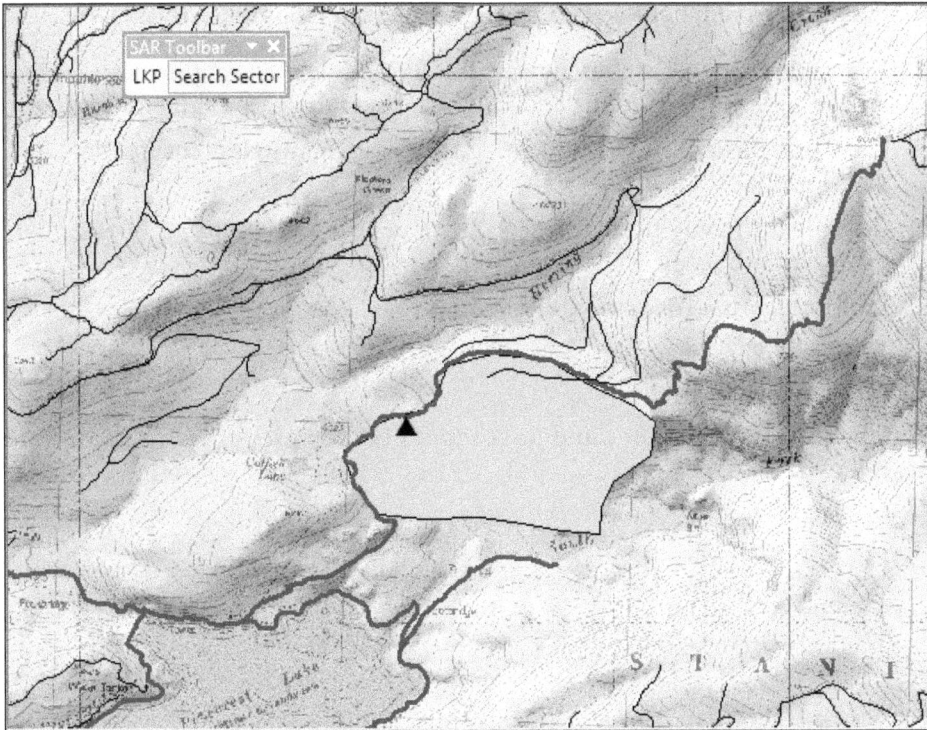

In the next section, a tool will be created to generate Google Earth format files from these newly created datasets.

Creating the Convert to Google Earth tool

Because visualization is very important to **Search and Rescue** personnel, the ability to see these datasets in a 3D environment is helpful. In this section, we'll create a custom tool that can be used to export these datasets to Google Earth's Keyhole Markup Language (KML) format.

To complete this section, you will need to install the `simplekml` Python module. You can use `pip` to install this module by using the following command from the command prompt:

```
pip install simplekml
```

The following steps will help you to create **Convert to Google Earth** tool:

1. In **ArcMap**, open **Catalog** view, right-click on the SAR.pyt custom toolbox, and select **Edit** to open the code in the Python development environment.

2. Copy and paste the existing AssignLKPAttributes class inside the SAR.pyt file.

3. Rename the newly pasted AssignLKPAttributes class to ConvertToGoogleEarth class.

4. Import the simplekml module:

```
import arcpy
import simplekml
```

5. Update the self.label and self.description properties, as shown in the following code:

```
class ConvertToGoogleEarth(object):
    def __init__(self):
        """Define the tool (tool name is the name of the
class)."""
        self.label = "Convert to Google Earth"
        self.description = "Convert to Google Earth"
```

6. Add the tool to the self.tools list in the Toolbox class:

```
self.tools = [AssignLKPAttributes, SearchAreaBuffers,
ConvertToGoogleEarth]
```

7. Find the getParameterInfo() method and remove the existing parameters that were copied. Next, add the following parameters. These parameters will be used to capture the input feature class to be converted to the KML format along with an output filename. Your code should appear as follows:

```
def getParameterInfo(self):

    """Define parameter definitions"""
    in_fc = arcpy.Parameter(
                name='in_features',
                displayName='Input Features',
                datatype='GPFeatureLayer',
                direction='Input',
                parameterType='Required')

    out_file = arcpy.Parameter(
                name='out_file',
                displayName='Output KML File',
                datatype='DEFile',
```

```
                    direction='Output',
                    parameterType='Required')
    out_file.filter.list = ['kml']

    params = [in_fc, out_file]
    return params
```

8. Find the `execute()` method and remove the existing parameters that were copied.

9. In the `execute()` method, add code to capture the input parameters:

```
def execute(self, parameters, messages):
    """The source code of the tool."""
    in_fc = parameters[0].valueAsText
    out_file = parameters[1].valueAsText
```

10. Create a new KML file:

```
def execute(self, parameters, messages):
    """The source code of the tool."""
    in_fc = parameters[0].valueAsText
    out_file = parameters[1].valueAsText

    kml = simplekml.Kml()
    kml.document.name = out_file
```

11. Create a new `SpatialReference` object that will be used to define the output parameters in a WGS84 coordinate system:

```
def execute(self, parameters, messages):
    """The source code of the tool."""
    in_fc = parameters[0].valueAsText
    out_file = parameters[1].valueAsText

    kml = simplekml.Kml()
    kml.document.name = out_file

    sr = arcpy.SpatialReference(4326)
```

12. Using the `Describe()` function, add the code that will determine the shape type of the input feature class. For this particular tool, we will only convert points and polygons, but you can easily add another code block to process polylines:

```
def execute(self, parameters, messages):
    """The source code of the tool."""
    in_fc = parameters[0].valueAsText
```

```
    out_file = parameters[1].valueAsText

    kml = simplekml.Kml()
    kml.document.name = out_file

    sr = arcpy.SpatialReference(4326)

    desc = arcpy.Describe(in_fc)
    shapeType = desc.shapeType
    if shapeType == 'Point':
```

13. In the code block for point feature classes, create a `SearchCursor` class that will return the geometry of the point along with the contents of the `incident_name` and name fields for the input layer:

```
def execute(self, parameters, messages):
    """The source code of the tool."""
    in_fc = parameters[0].valueAsText
    out_file = parameters[1].valueAsText

    kml = simplekml.Kml()
    kml.document.name = out_file

    sr = arcpy.SpatialReference(4326)

    desc = arcpy.Describe(in_fc)
    shapeType = desc.shapeType
    if shapeType == 'Point':
        with arcpy.da.SearchCursor(in_fc,('SHAPE@','Incident_
Name','Name'),spatial_reference=sr) as cursor:
```

14. Loop through the `SearchCursor` object and retrieve the *x* and *y* coordinates along with the `Incidentname` and `Name` fields:

```
def execute(self, parameters, messages):
    """The source code of the tool."""
    in_fc = parameters[0].valueAsText
    out_file = parameters[1].valueAsText

    kml = simplekml.Kml()
    kml.document.name = out_file

    sr = arcpy.SpatialReference(4326)

    desc = arcpy.Describe(in_fc)
```

```
        shapeType = desc.shapeType
        if shapeType == 'Point':
            with arcpy.da.SearchCursor(in_fc,('SHAPE@','Incident_
Name','Name'),spatial_reference=sr) as cursor:
                for row in cursor:
                    x = row[0].firstPoint.X
                    y = row[0].firstPoint.Y
                    incidentName = row[1]
                    name = row[2]
```

15. Create a new `kml` point object using the `simplekml` library and assign the
    `name`, `description`, and `coordinates` properties:

```
def execute(self, parameters, messages):
        """The source code of the tool."""
        in_fc = parameters[0].valueAsText
        out_file = parameters[1].valueAsText

        kml = simplekml.Kml()
        kml.document.name = out_file

        sr = arcpy.SpatialReference(4326)

        desc = arcpy.Describe(in_fc)
        shapeType = desc.shapeType
        if shapeType == 'Point':
            with arcpy.da.SearchCursor(in_fc,('SHAPE@','Incident_
Name','Name'),spatial_reference=sr) as cursor:
                for row in cursor:
                    x = row[0].firstPoint.X
                    y = row[0].firstPoint.Y
                    incidentName = row[1]
                    name = row[2]
                    pnt = kml.newpoint()
                    pnt.name = name
                    pnt.description = incidentName
                    pnt.coords = [(x,y)]
```

16. Next, we'll turn our attention to coding the KML for polygon layers. Inside
    the `elif` statement, create `SearchCursor` method:

```
elif shapeType == 'Polygon':
        with arcpy.da.SearchCursor(in_fc, ('OID@','SHAPE@'),spatial_
reference=sr) as cursor:
```

17. Loop through the rows and retrieve the vertices for each polygon:

```
elif shapeType == 'Polygon':
    with arcpy.da.SearchCursor(in_fc, ('OID@','SHAPE@'),spatial_
reference=sr) as cursor:
        for row in cursor:
            listVertices = []
            # Step through each part of the feature
            for part in row[1]:
                # Step through each vertex in the feature
                for pnt in part:
                    if pnt:
                        # get x,y coordinates of current point
                        newVertex = pnt.X, pnt.Y
                        listVertices.append(newVertex)
```

18. Create a new `kml` polygon object and assign the `outerboundaryis` property:

```
elif shapeType == 'Polygon':
    with arcpy.da.SearchCursor(in_fc, ('OID@','SHAPE@'),spatial_
reference=sr) as cursor:
        for row in cursor:
            listVertices = []
            # Step through each part of the feature
            for part in row[1]:
                # Step through each vertex in the feature
                for pnt in part:
                    if pnt:
                        # get x,y coordinates of current point
                        newVertex = pnt.X, pnt.Y
                        listVertices.append(newVertex)
                        poly = kml.newpolygon()
                        poly.outerboundaryis = listVertices
```

19. Save the output KML file:

```
def execute(self, parameters, messages):
        """The source code of the tool."""
        in_fc = parameters[0].valueAsText
        out_file = parameters[1].valueAsText

        kml = simplekml.Kml()
```

```
kml.document.name = out_file

sr = arcpy.SpatialReference(4326)

desc = arcpy.Describe(in_fc)
shapeType = desc.shapeType
if shapeType == 'Point':
    with arcpy.da.SearchCursor(in_fc,('SHAPE@','Incident_
Name','Name'),spatial_reference=sr) as cursor:
        for row in cursor:
            x = row[0].firstPoint.X
            y = row[0].firstPoint.Y
            incidentName = row[1]
            name = row[2]
            pnt = kml.newpoint()
            pnt.name = name
            pnt.description = incidentName
            pnt.coords = [(x,y)]
    elif shapeType == 'Polygon':
        with arcpy.da.SearchCursor(in_fc, ('OID@','SHAPE@'),
spatial_reference=sr) as cursor:
            for row in cursor:
                listVertices = []

                # Step through each part of the feature
                for part in row[1]:

                    # Step through each vertex in the feature
                    for pnt in part:
                        if pnt:
                            # get x,y coordinates of current
point

                            newVertex = pnt.X, pnt.Y
                            listVertices.append(newVertex)
                            poly = kml.newpolygon()
                            poly.outerboundaryis =
listVertices
        kml.save(out_file)
```

20. You can check your work by examining the solution files in the
    `C:\ArcGIS_Blueprint_Python\solutions\ch8` folder. Open
    the `ConvertToGoogleEarth.py` file and examine the
    `ConvertToGoogleEarth` class.

21. Save your script and exit the Python development environment.

22. Double-click on the **Convert to Google Earth** tool in the **SAR** toolbox to display the tool, as shown in the following screenshot:

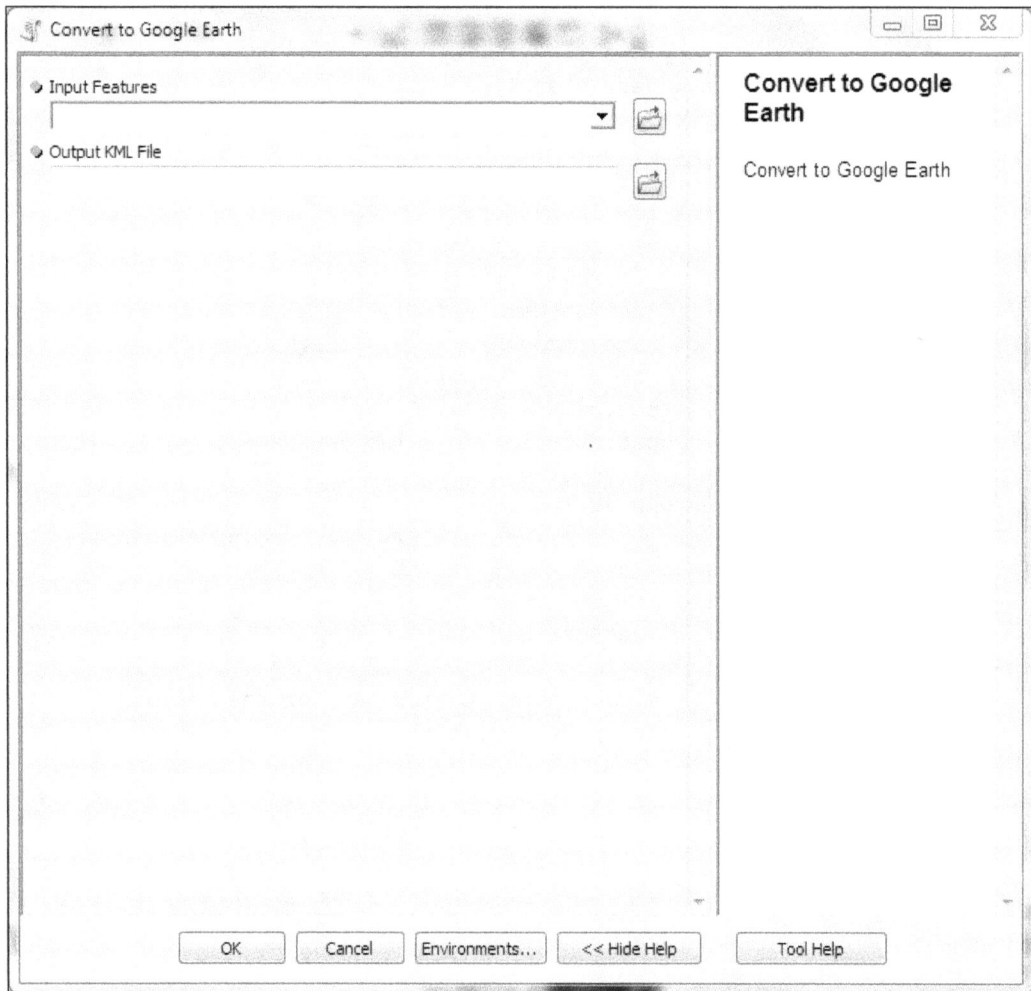

23. To test the tool, select the `LostMale10YearsOld` feature class and define the output as `C:\ArcGIS_Blueprint_Python\ch8\LostMale10YearsOld.kml`. Click on the **OK** button.

24. For this next step, you'll need to install Google Earth Pro application on your computer. Google Earth Pro is a free product that you can install by downloading the software from `http://www.google.com/earth/download/gep/agree.html`. The license key is `GEPFREE`.

25. Open Google Earth Pro and navigate to **File | Open**. Then, go to `C:\ArcGIS_Blueprint_Python\ch8` and open `LostMale10YearsOld.kml`.

26. Your location will most likely vary, but you should see a point location similar to what is shown in the following screenshot:

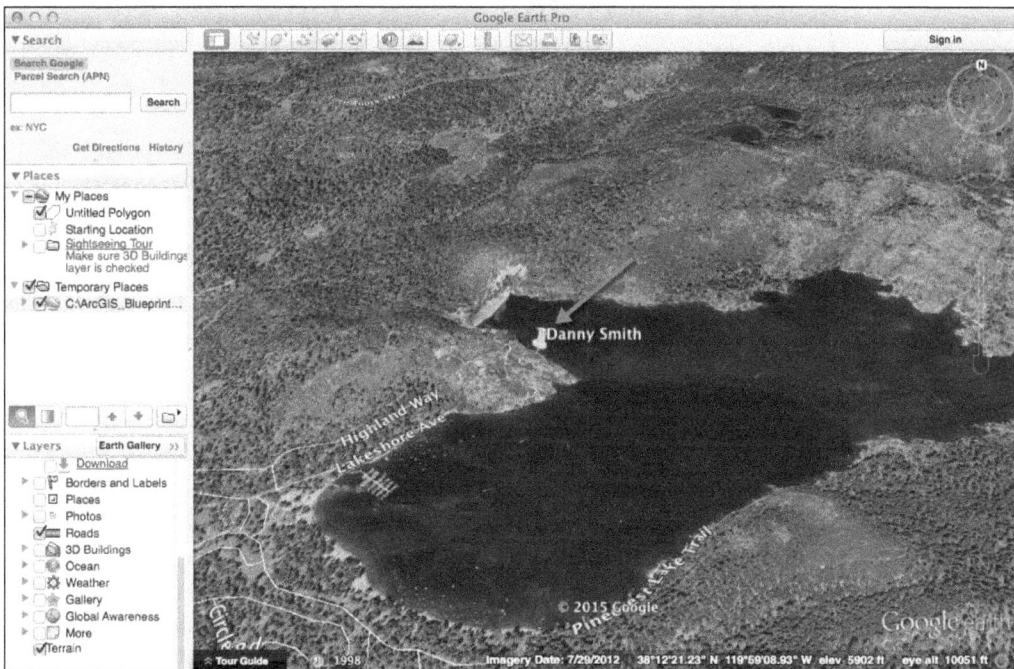

27. Repeat the process with the `SearchSectors` layer to see something similar to what is shown in the following screenshot:

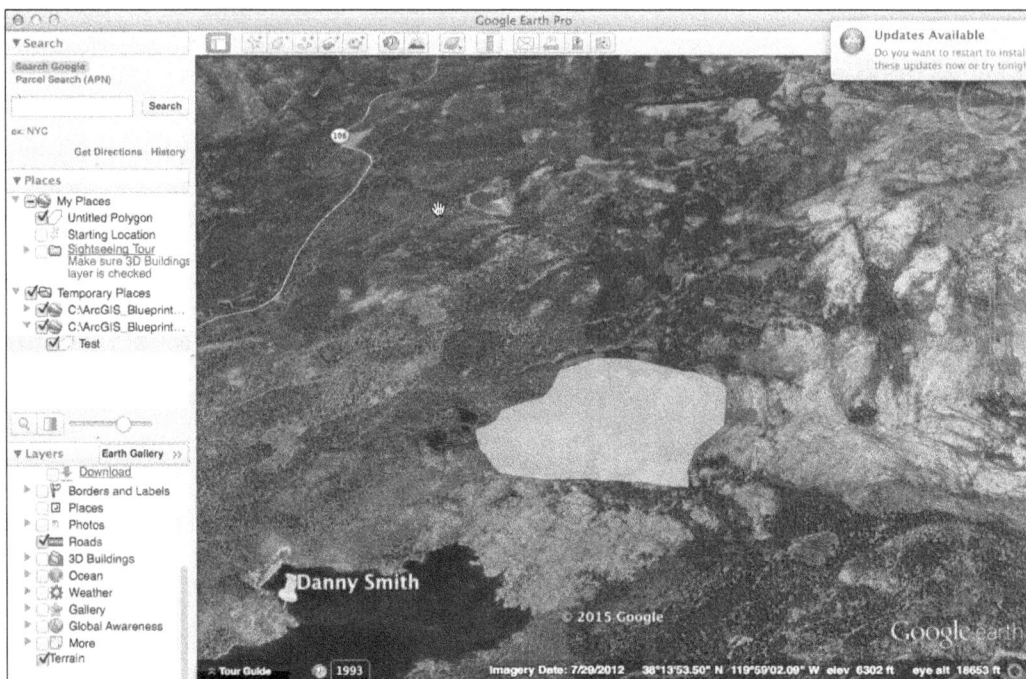

Summary

This chapter covered several topics that we've seen in other chapters, including the creation of Python add-ins as well as custom Python toolboxes and tools. However, we introduced several new topics, including the inclusion of the `shapekml` module to convert ArcGIS data to the Google Earth format, the use of **ArcGIS Python Add-In** tool to draw polygons, and the integration of Python add-ins with custom Python toolboxes through the `pythonaddins` module.

In the next chapter, you will learn how to use Python to create a real-time social media application using the `tweepy` Python module.

Real-Time Twitter Mapping with Tweepy, ArcPy, and the Twitter API

Events and news stories generate a massive amount of social media attention. Whether it's something as important as a natural disaster or there is social strife or something purely for entertainment, there is a large volume of information that needs to be mined. As GIS practitioners, we are primarily interested in geographic analysis, and some social media channels provide geographic context related to their information streams. For example, Twitter users can make their tweets location-enabled. Location-enabled Twitter accounts will ensure that each tweet will be tagged with the current geographic coordinates of the user. Many Twitter users have not made their accounts location-enabled due to security reasons or because they are unaware of this feature or are simply not interested. In fact, only about 2% of tweets contain location information.

The application built in this chapter will mine a live stream of tweets containing specific terms and hash tags. Tweets that contain geographic coordinates will be written to a local geodatabase for further analysis. In addition, several tools will be created to enable the analysis of this social media data. Finally, the results will be shared with the public through ArcGIS Online.

In this chapter, we will cover the following topics:

- Creating a Python script that uses the `tweepy` library to mine a live stream of Twitter data
- The analysis of social media data through geostatistical analysis tools
- Scheduling Python scripts with Windows Task Scheduler

Design

The design of this application is quite simple and involves only the creation of a single Python script that uses the tweepy module and arcpy to mine a live stream of tweets and write the information to a local feature class. The script will also be scheduled using Windows Task Scheduler:

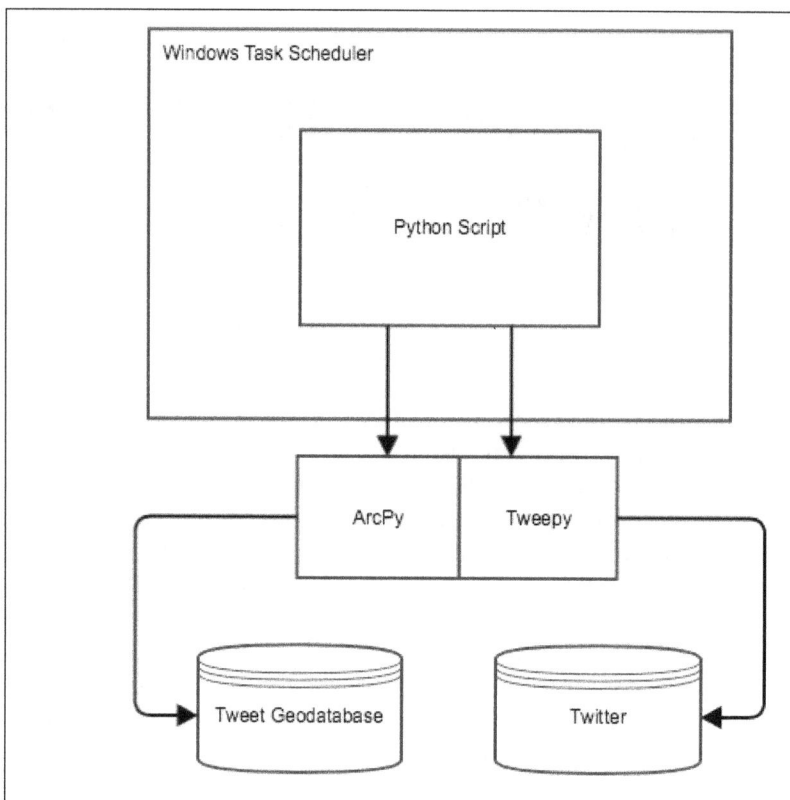

Let's get started and build the application.

Extracting Tweet geographic coordinates with tweepy

In this step, you'll create a Python script that uses the Python tweepy module to extract tweets containing geographic coordinates and write this information to a local feature class. The tweepy module is designed to provide access to all the Twitter REST API methods. This will be a standalone script that can be run from the command line or the Python development environment. It will create a listener style environment that will continue to be executed, looking for live tweets until you kill the process.

Follow these steps to create the script:

1. Before completing this section of the chapter, you will need to download and install the tweepy Python module using pip. Issue the following command from Command Prompt. The instruction to install pip can be found in previous chapter:

   ```
   pip install tweepy
   ```

2. Tweepy supports the OAuth authentication required to connect with Twitter. The first step is to register your client application with Twitter. In this case, the client application is simply a desktop Python script, but this one-time step is necessary.

3. Log in to your Twitter account.

4. Go to https://apps.twitter.com.

5. Click on the **Create New App** button shown in the following screenshot:

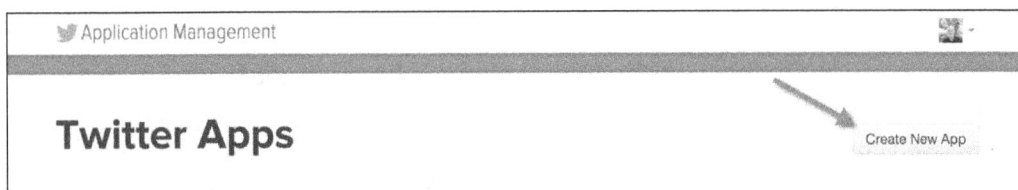

6. Add the application details. In this chapter, we'll create a geographic analysis of tweets about **Southeastern Conference (SEC)** football during a one-weekend time period. However, if you'd like to explore a different topic, you are certainly welcome to do that. You'll need to provide a URL to a website that is publicly accessible:

Create an application

Application Details

Name *

> SEC College Football

Your application name. This is used to attribute the source of a tweet and in user-facing authorization screens. 32 characters max.

Description *

> Pulls tweets about SEC college football games

Your application description, which will be shown in user-facing authorization screens. Between 10 and 200 characters max.

Website *

> http://<your domain>|

Your application's publicly accessible home page, where users can go to download, make use of, or find out more information about source attribution for tweets created by your application and will be shown in user-facing authorization screens.

(If you don't have a URL yet, just put a placeholder here but remember to change it later.)

7. Click on the **Create your Twitter application** button at the bottom of this form.

8. This will return a new page with the application details. Click on the **Keys** and **Access Tokens** tab. Make a note of the **Consumer Key (API Key)** and **Consumer Secret (Secret Key)** seen here. Mine have been obscured, but you should see a long sequence of characters in both the cases. Keep these handy because you'll need these details in a later step:

9. Open your Python development environment. A shell window will work fine in this particular case.

10. Import the `tweepy` objects from the `tweepy` module:

```
import tweepy
```

11. Create an instance of `OAuthHandler` by passing in the consumer key which is `consumer_key` and consumer secret which is `consumer_secret` as strings:

```
auth = tweepy.OAuthHandler("<consumer_key>", "<consumer_secret>")
```

12. Get the redirected URL by issuing the following command:

`redirect_url = auth.get_authorization_url()`

13. Print the redirected URL:

```
print(redirect_url)
```

You should see something similar to what is shown in here, but keep in mind that your token will not be the same as mine, so if you go to the following URL, you will see request token invalid message on the webpage: https://api.twitter.com/oauth/authorize?oauth_token=LAfKfgAAAAAh8FgAAABUELLojA.

14. Open a web browser and copy and paste your URL. Hit the *Enter* key, and you should see something similar to what is shown in the following screenshot. You may be required to log in to your Twitter account:

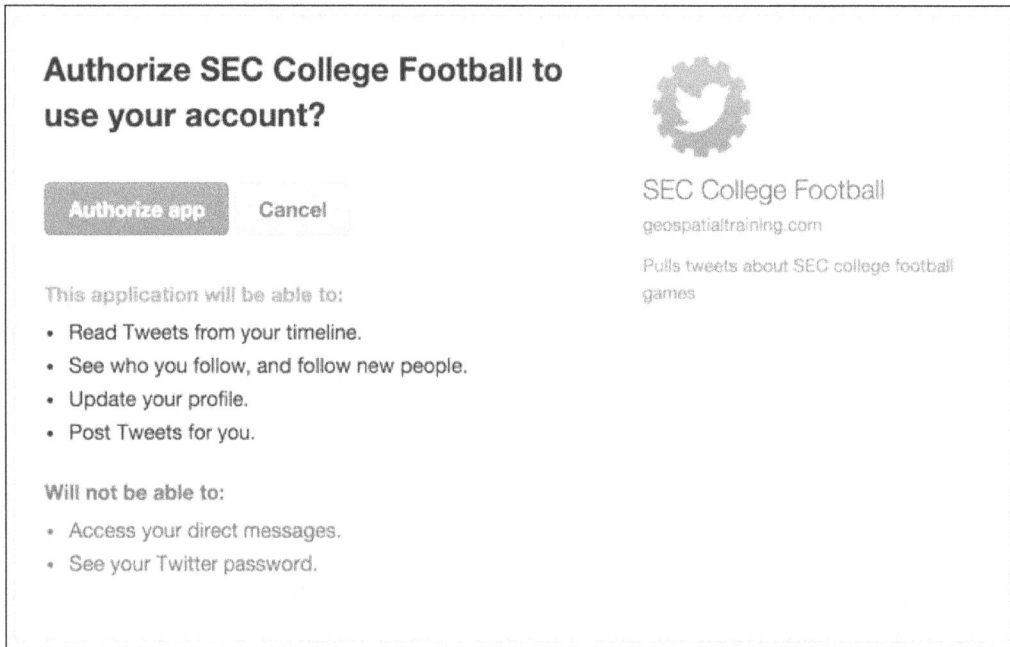

Authorize SEC College Football to use your account?

Authorize app   Cancel

SEC College Football
geospatialtraining.com

Pulls tweets about SEC college football games

This application will be able to:

- Read Tweets from your timeline.
- See who you follow, and follow new people.
- Update your profile.
- Post Tweets for you.

Will not be able to:

- Access your direct messages.
- See your Twitter password.

15. Click on the **Authorize app** button, and you should see something similar to what is shown in the following screenshot. Keep in mind that the number you see will not be the same:

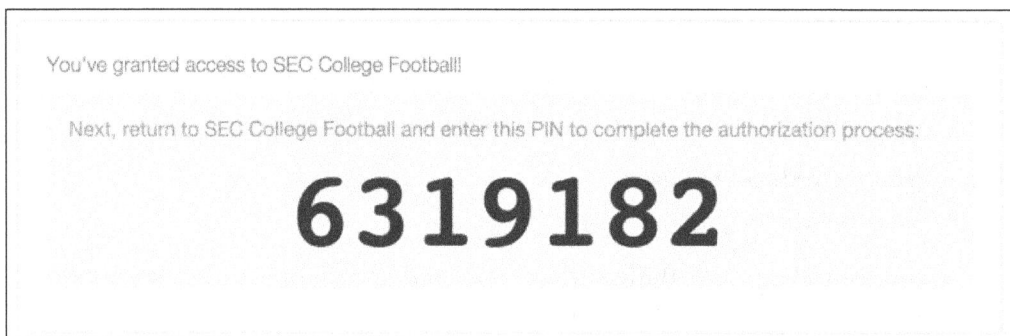

You've granted access to SEC College Football!

Next, return to SEC College Football and enter this PIN to complete the authorization process:

6319182

16. Return to the Python shell environment, type the following line of code, and hit *Enter*:

```
verifier = raw_input('Verifier:')
```

17. Type in the verification code you just generated, as shown here, and click on *Enter*, but make sure you enter the number you generated and not the example:

```
Verifier: 6319182
```

18. In the Python shell, type the following:

```
print(auth.get_access_token(verifier))
```

19. You will see something similar to the following, but keep in mind that your values will, again, not be the same as mine. The first value is the token key and the second is the token secret. Write this information down or save it somewhere because you will need it when you write the script:

```
(u'20646590-fUYjN8BuSqLjfOpODxye799rG2WF6wBMAsUddyIwC', u'GVAJTeOS
klHmnG0dDsy5kpPBk8yBXuQWRSqrFIWJAFlQy')
```

20. In your Python development environment, create a new script.

21. Save the script as `C:\ArcGIS_Blueprint_Python\ch9\tweepy_stream.py`.

22. Import the `StreamListener`, `OAuthHandler`, and `Stream` objects from the `tweepy` library:

```
from tweepy.streaming import StreamListener
from tweepy import OAuthHandler
from tweepy import Stream
```

23. Also, import the `arcpy`, `sys`, and `time` modules:

```
import arcpy
import sys
import time
```

24. Create `global` variables to hold the consumer key and secret (the Twitter application information), as well as the Twitter token key and secret. These should be the values you gathered up until this point:

```
#global variables
consumer_key = '<consumer_key>'
consumer_secret = '<consumer_secret>'
token_key = '<token_key>'
token_secret = '<token_secret>'
```

25. Create `global` variables to hold the start time of the script as well as the current workspace:

```
#global variables
consumer_key = '<consumer_key>'
consumer_secret = '<consumer_secret>'
token_key = '<token_key>'
token_secret = '<token_secret>'

start_time = time.time()
arcpy.env.workspace = r'c:\ArcGIS_Blueprint_Python\data\Twitter\
TweetInformation.gdb'
```

26. Create the `main()` function:

```
def main():


if __name__ == '__main__':
    main()
```

27. Inside the `main()` function, create a `try`/`except` block and create variables to hold the input feature class and time for the monitoring of tweets. The `sys.argv[]` list is used to capture the command-line input of the script when executed from Command Prompt. The `featureClass` variable will hold the input feature class name, and the `monitorTime` variable will hold the amount of time for which the script will monitor tweets. This value is in seconds; the input value will be input in hours and then converted to seconds:

```
def main():
    try:  #new
        featureClass = sys.argv[1]
        monitorTime = sys.argv[2]
        monitorTime = monitorTime * 3600


    except Exception as e:
        print(e.message)
```

28. Create the spatial reference for the output feature class, enable the script to overwrite an existing feature class, and call the `CreateFeatureClass` tool:

```
try:  #new
    featureClass = sys.argv[1]
```

```
monitorTime = sys.argv[2]
monitorTime = monitorTime * 3600

sr = arcpy.SpatialReference(4326)
arcpy.env.overwriteOutput = True
arcpy.CreateFeatureclass_management(arcpy.env.workspace,
featureClass, "POINT", spatial_reference=sr)
```

29. Inside the `main()` function, create a new instance of `OAuthHandler` by passing in the consumer key and consumer token, and set the Twitter access token by passing in the token key and token secret:

```
try:  #new
    featureClass = sys.argv[1]
    monitorTime = sys.argv[2]
    monitorTime = monitorTime * 3600

    sr = arcpy.SpatialReference(4326)
    arcpy.env.overwriteOutput = True
    arcpy.CreateFeatureclass_management(arcpy.env.workspace,
featureClass, "POINT", spatial_reference=sr)

    auth = OAuthHandler(consumer_key, consumer_secret)
    auth.set_access_token(token_key, token_secret)
```

30. Create a new `Stream` object by passing in instances of `OAuthHandler` and `StdOutListener`:

```
try:  #new
    featureClass = sys.argv[1]
    monitorTime = sys.argv[2]
    monitorTime = monitorTime * 3600

    sr = arcpy.SpatialReference(4326)
    arcpy.env.overwriteOutput = True
    arcpy.CreateFeatureclass_management(arcpy.env.workspace,
featureClass, "POINT", spatial_reference=sr)

    auth = OAuthHandler(consumer_key, consumer_secret)
    auth.set_access_token(token_key, token_secret)

    stream = Stream(auth, StdOutListener(start_time, featureClass,
time_limit=monitorTime))
```

31. Set the filter for the `Stream` object. The filter includes terms that will be used in the search for tweets. I've included a number of terms related to searching tweets of SEC conference football teams, including Alabama, Texas A&M, Auburn, LSU, and Georgia. However, you can feel free to include search terms that are relevant to any topic that you'd like to monitor and map. I've commented out a second filter that can be used to monitor the wildfire information to give you another option:

```
try:   #new
    featureClass = sys.argv[1]
    monitorTime = sys.argv[2]
    monitorTime = monitorTime * 3600

    sr = arcpy.SpatialReference(4326)
    arcpy.env.overwriteOutput = True
    arcpy.CreateFeatureclass_management(arcpy.env.workspace,
featureClass, "POINT", spatial_reference=sr)

    auth = OAuthHandler(consumer_key, consumer_secret)
    auth.set_access_token(token_key, token_secret)

    stream = Stream(auth, StdOutListener(start_time, featureClass,
time_limit=monitorTime))   #172800
    stream.filter(track=['#SEC', '#SECFootball', '#RollTide',
'#GigEm', '#Bama', '#UGABulldogs','#Dawgs', '#GeorgiaBulldogs',
'##A&MFootball', '#KyleField', '#Aggies', '#gigem','#LSUFootball',
'#LSUFB', '#WarEagle','#AuburnFootball' ])
```

32. Above the `main()` function and below the `global` variable, create a new class called `StdOutListener`. Objects created from this class will listen for tweets on the `Stream` object and are filtered based on the terms that are provided:

```
#global variables
consumer_key = 'x9jRE3KQm1LlEFcHsL6bP4TRa'
consumer_secret =
'8VVzPzY0DJbbgbBk5bgWCrBADzLEqdnATNbw1z0LUWF5MWuu4g'
token_key = '2997753385-nCFmNPAo2LOt7LLF311Kw0JdsAhcNSq8yQThxtO'
token_secret = '0Dck37JE7HV56Rs5t5GUkbW3C61qepG4fi070RiP4SNdm'

class StdOutListener(StreamListener):
```

33. Create an `__init__` method, as shown in the following code. The `__init__` method is a constructor for the class and will be used to set various properties:

```
class StdOutListener(StreamListener):
    def __init__(self, start_time, featureClass, time_limit):
        super(StdOutListener, self).__init__()
```

```
                    self.time = start_time
                    self.limit = time_limit
                    self.featureClass = featureClass
```

34. Create methods called `on_status`, `on_error`, and `on_timeout` inside the `StdOutListener` class:

```
class StdOutListener(StreamListener):

    def on_status(self, status):

    def on_error(self, status):

    def on_timeout(self):
```

35. Inside the `on_status` method, add a while loop that will execute as long as the time has not exceeded the amount of time to monitor the script execution. You'll recall that the amount to be monitored is input from the command line when the script is initiated:

```
def on_status(self, status):
    while (time.time() - self.time) < self.limit:
```

36. Inside the `on_status` method, add an `if-else` statement that tests to see whether the `status.geo` property has been set. Keep in mind that most people do not enable the location status on their Twitter profile, so the majority of tweets will not contain coordinate information. Therefore, we need an `if-else` statement that will branch the code depending upon whether or not the coordinate information is available:

```
def on_status(self, status):
    while (time.time() - self.time) < self.limit:
        if status.geo is not None:

        else:
            print "No coordinates found"
            return True
```

37. Inside the `if` statement, get the `status.geo` property, which is returned as a Python dictionary object, and pull out the value associated with the coordinates key. This will be a Python list containing the latitude and longitude coordinates:

```
if status.geo is not None:
    dictCoords = status.geo
    listCoords = dictCoords['coordinates']
    latitude = listCoords[0]
```

```
        longitude = listCoords[1]
else:
    print "No coordinates found"
    return True
```

38. Create an `InsertCursor` object and insert a new row:

```
if status.geo is not None:
    dictCoords = status.geo
    listCoords = dictCoords['coordinates']
    latitude = listCoords[0]
    longitude = listCoords[1]

    cursor = arcpy.da.InsertCursor(self.featureClass,("SHAPE@XY"))
    cursor.insertRow([(longitude,latitude)])

    print(str(listCoords[0]) + "," + str(listCoords[1]))
    return True
```

39. In the `on_error` method, add the code block shown here to print out any error messages that may occur while the script is executing:

```
def on_error(self, status):
    print('Error...')
    print status
    return True
```

40. In the `on_timeout` method, access the following code block to print out any messages related to a timeout condition:

```
def on_timeout(self):
    print('Timeout...')
    return True
```

41. Save your script.

42. You can check your work by examining the solution script found at `C:\ArcGIS_Blueprint_Python\solutions\ch9\tweepy_stream.py`.

Scheduling the script

In this section, you'll learn how to schedule the script using the Windows Task Scheduler. Scheduling the script will require the creation of a batch file. Batch files can contain scripts and operating system commands. The batch file will then be added to the Windows Task Scheduler to run at a specific time interval. To do this follow these steps:

1. Open Notepad.

2. Add the following lines of text to the file. These lines of code will switch the directory where the Python script is stored and execute the script, passing in a name for the output feature class and the number of hours to be monitored for tweets. You can change this value if you'd like:

```
cd c:\ArcGIS_Blueprint_Python\ch9
python tweepy_stream.py Tweets 48
```

3. Save the file to your desktop as `MonitorTweets.bat`. Make sure you change the **Save As Type** drop-down list to **All Files**; otherwise, you'll end up with a file called `MonitorTweets.bat.txt`.

4. Open the Windows Task Scheduler by navigating to **Start** | **All Programs** | **Accessories** | System **Tools** | **Control Panel** | **Administrative Tools**. Select **Task Scheduler**. The scheduler should appear as shown in the following screenshot:

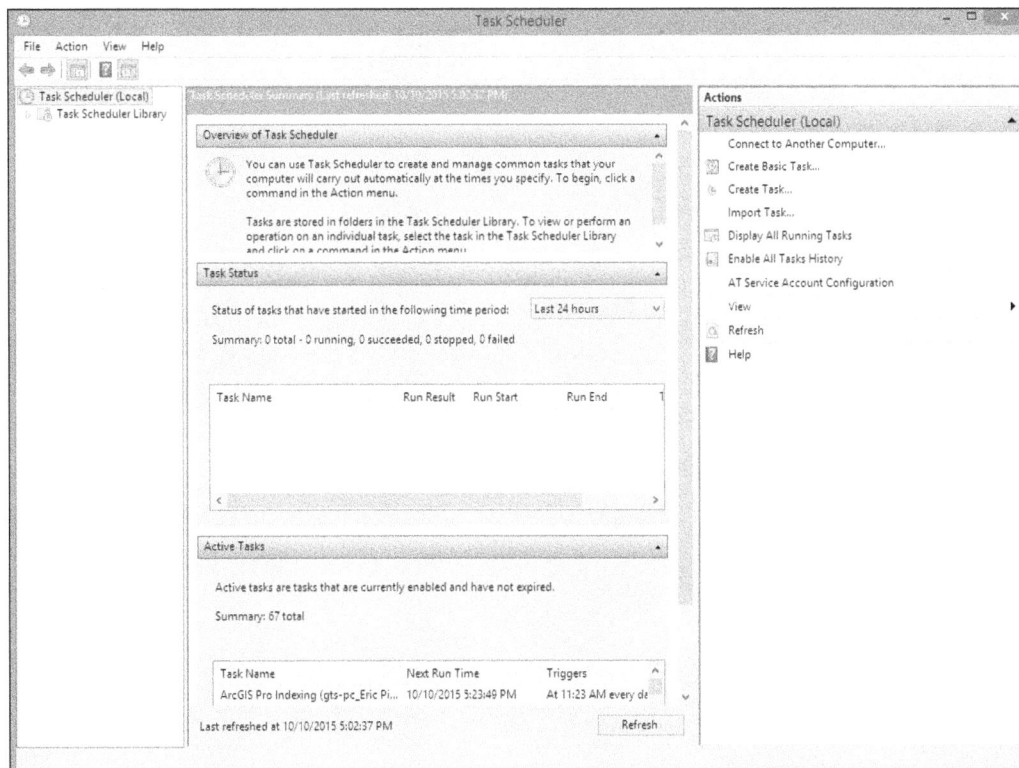

5. Select the Action menu item and then select Create a Basic Task to display the Create Basic Task Wizard dialog box, as shown in the following screenshot:

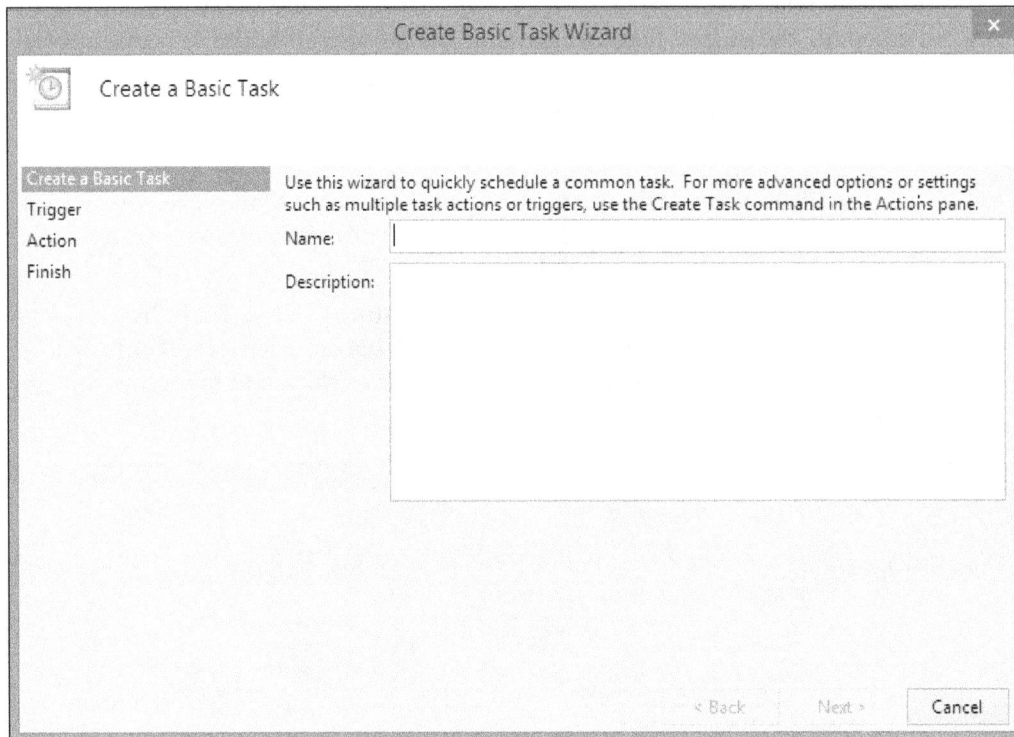

6. Give you task a name. In this case, we will call it Monitor Tweets. Then, click on **Next**:

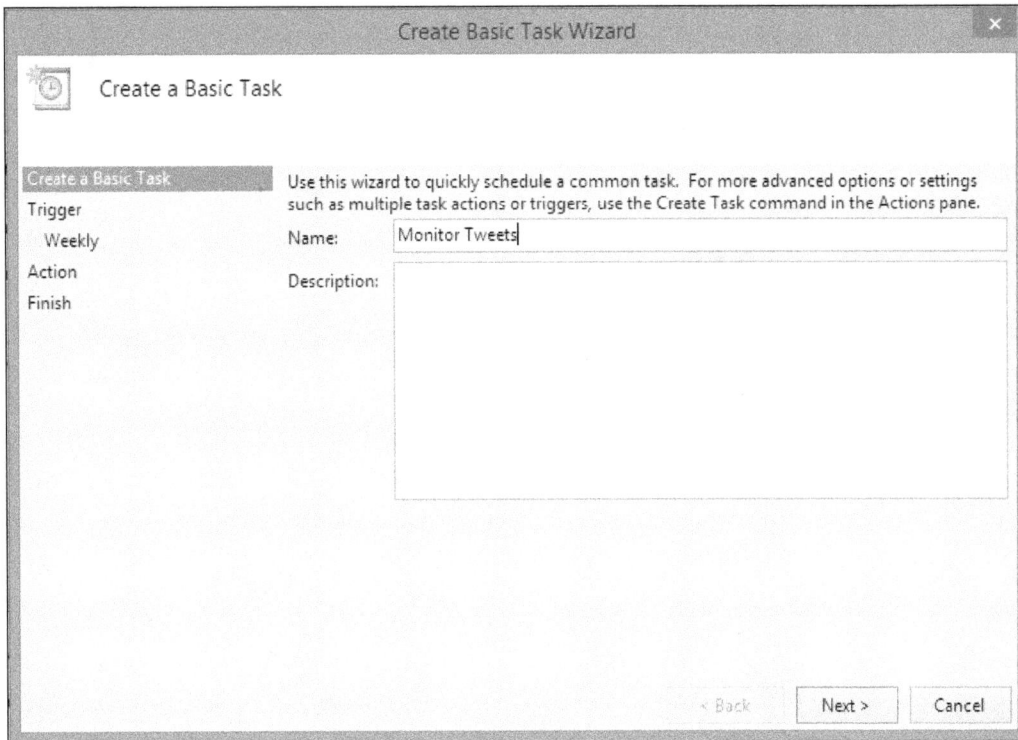

7. Select a trigger for when the task should be executed. Select **Weekly** as the trigger and click on **Next**:

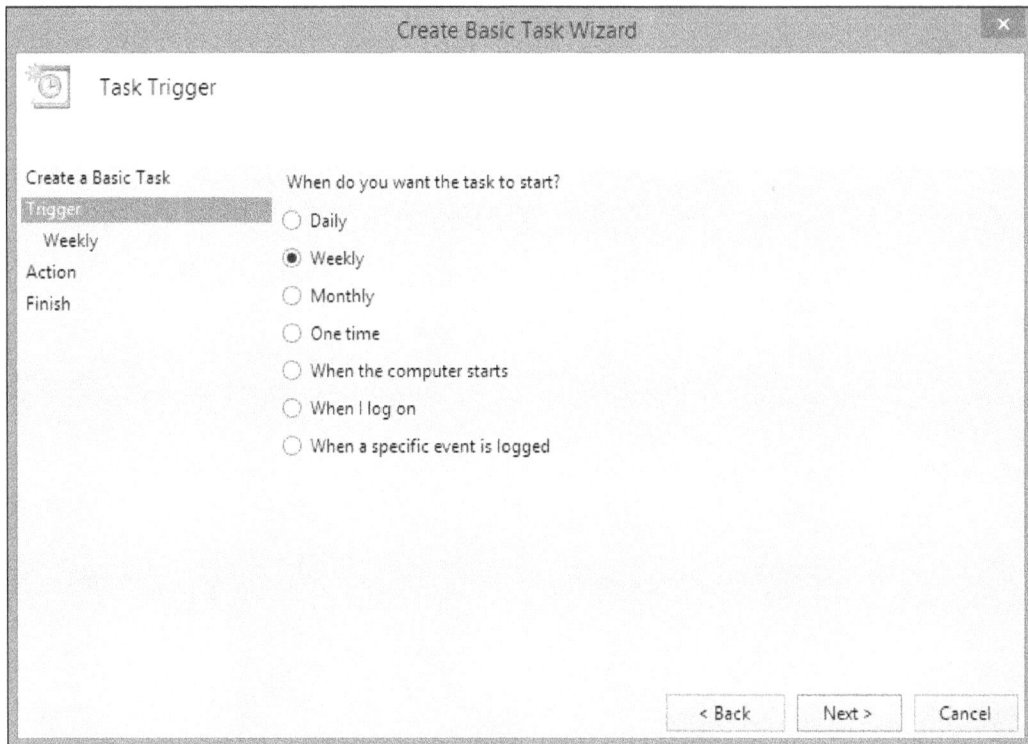

8. Select a **Start** date and time along with a recurrence interval. In this exercise, we're monitoring tweets from SEC football games, so the day on this date would be Saturday morning. Click on **Next**:

9. Select **Start a Program** as the action on the next dialog.

10. Go to the script and click on **Next**:

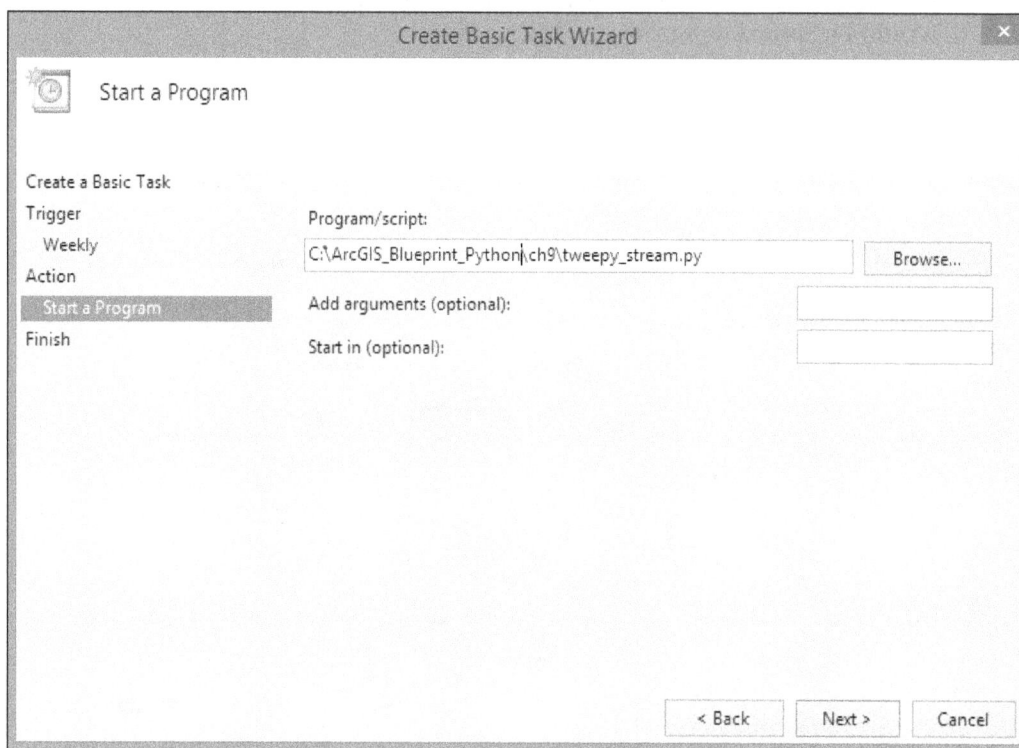

11. Click on **Finish**.

12. The task should be displayed in the list of active tasks.

Creating the heatmap

In this section, you'll create a heatmap of the Twitter feed using the **Optimized Hot Spot Analysis** tool found in the **Spatial Statistics Tools** toolbox:

1. If required, open **ArcMap** and create a new map document file.

2. Add a basemap like **Dark Gray Canvas** or **Light Gray Canvas** work well to display a heatmap. The **Light Gray Canvas** basemap is displayed in the following screenshot:

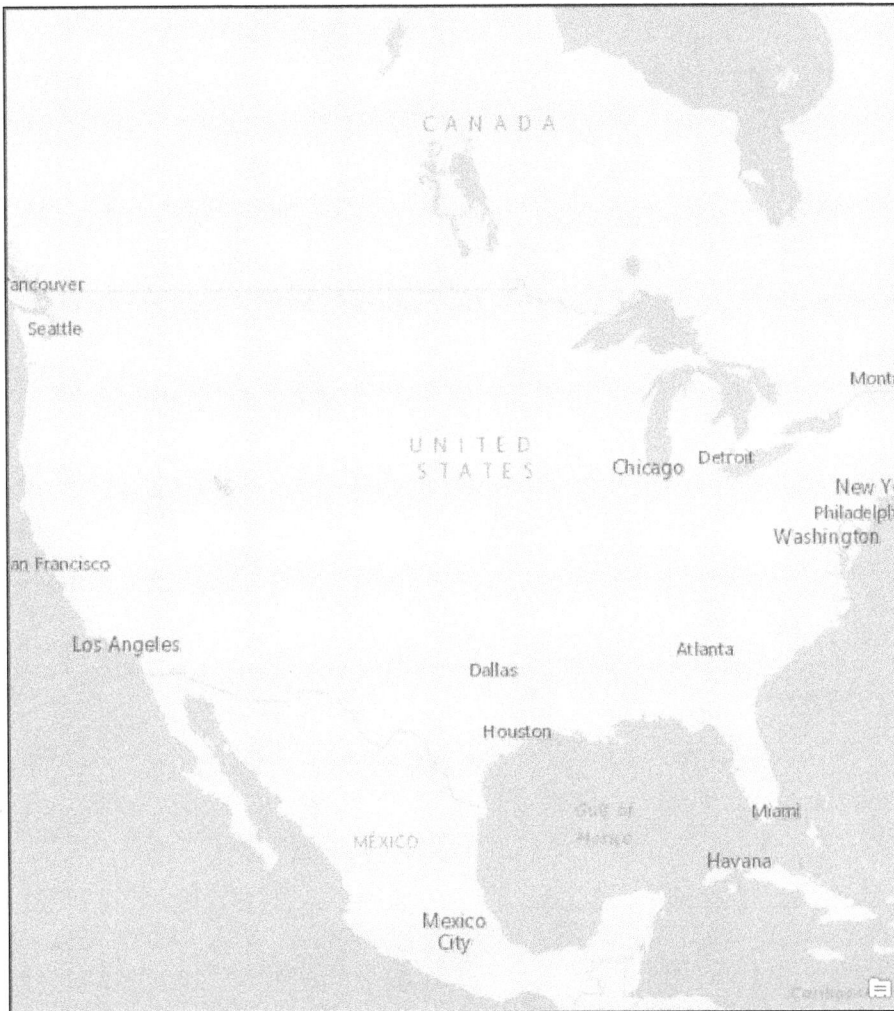

3. Add the Tweets `feature` class as a layer to the display. The distribution of your tweet points will not be the same as what is displayed in the following screenshot:

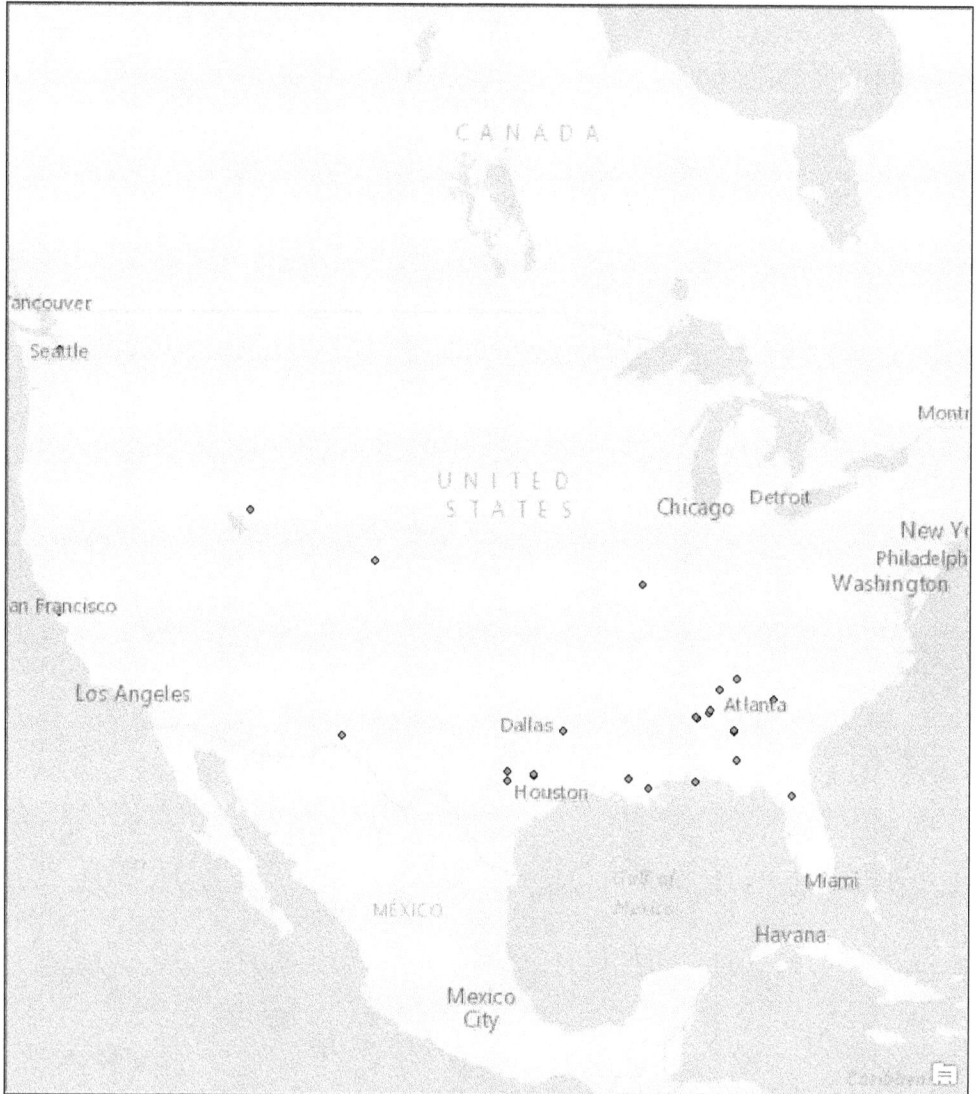

4. If you zoom in to a specific area, you should see some clustering of the points, as shown in the following screenshot:

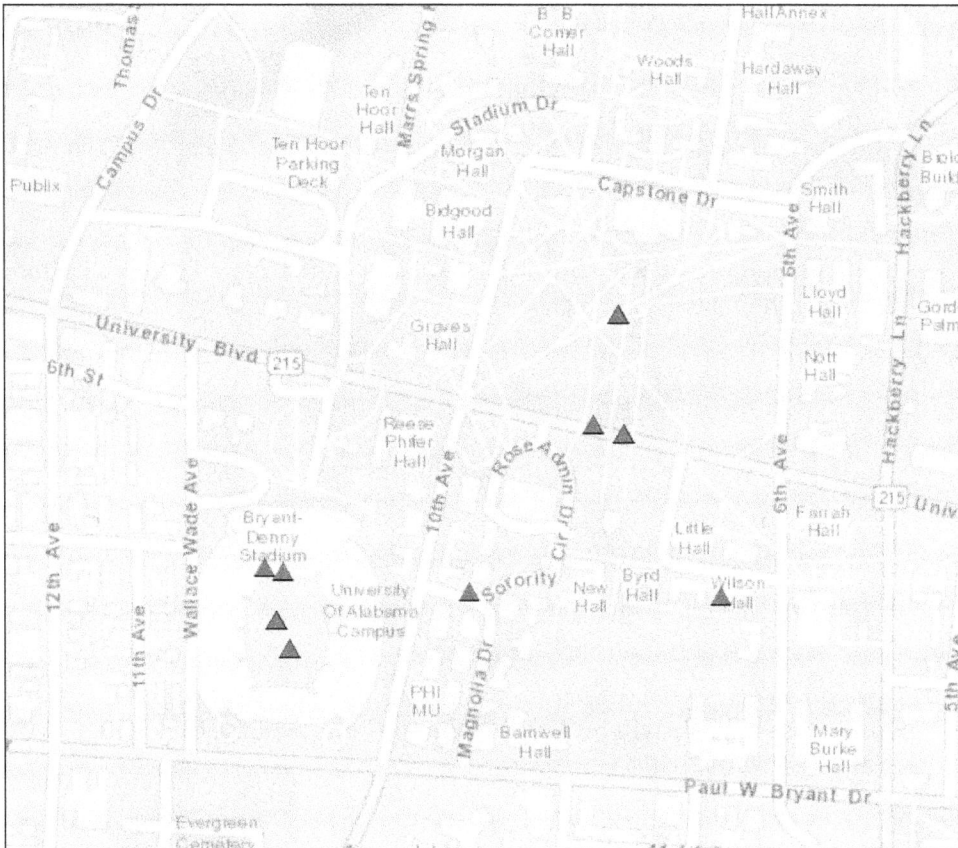

Over the next few steps, you'll create a bounding box polygon to confine the distribution of the points. For this particular exercise, the tweet activity for SEC football games examined. The geographic distribution of the tweet activity is relatively confined to the south and south-eastern parts of the United States. However, there will certainly be outliers. The bounding box will limit the geographic area that's used to define the hotspot analysis.

5. Add the `BoundingPolygon` feature class as a layer to the display. It should be empty.

6. Display the **Editor** toolbar in **ArcMap**.

7. Navigate to **Editor | Start Editing** and click on the **Create Features** button shown in the following screenshot:

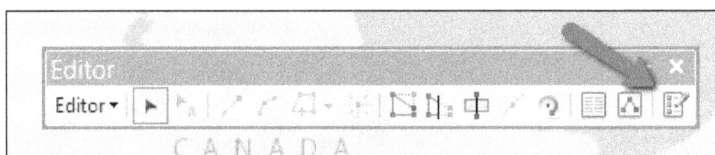

8. In the **Create Features** dialog, select **Organize Templates**.

9. Select **New Template** and then select the `BoundingPolygon` layer.

10. Click on **Finish** and then click on **Close**.

11. Click on `BoundingPolygon` from the **Create Features** dialog, and the **Construction Tools** will be displayed at the bottom of the dialog as shown in the following screenshot:

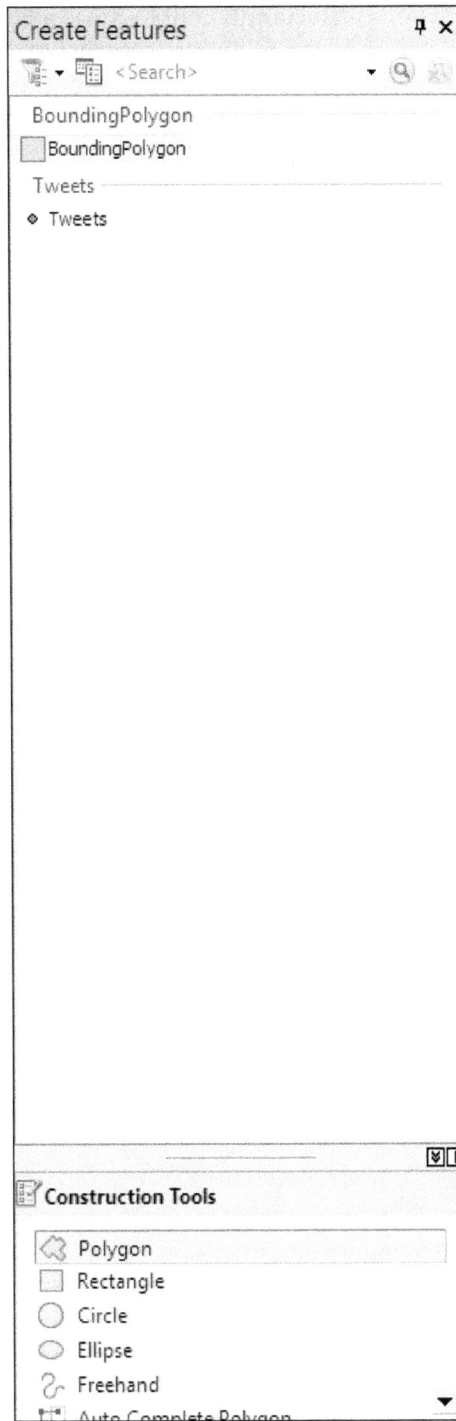

Create Features   ⊓ ✕

▤ ▾ ▦ <Search>   ▾ ⊕ ⚏

BoundingPolygon

☐ BoundingPolygon

Tweets

◇ Tweets

⊻ ▯

▤ **Construction Tools**

☌ Polygon

☐ Rectangle

◯ Circle

◯ Ellipse

ᔐ Freehand

Ⱶ Auto Complete Polygon

12. Use either the **Polygon** or **Rectangle** tool to draw a bounding polygonal area similar to what is shown in the following screenshot. Your bounding polygon may be totally different than mine depending upon the distribution of your data. However, you'd want to limit the geographic area that will be analyzed in the hotspot analysis.

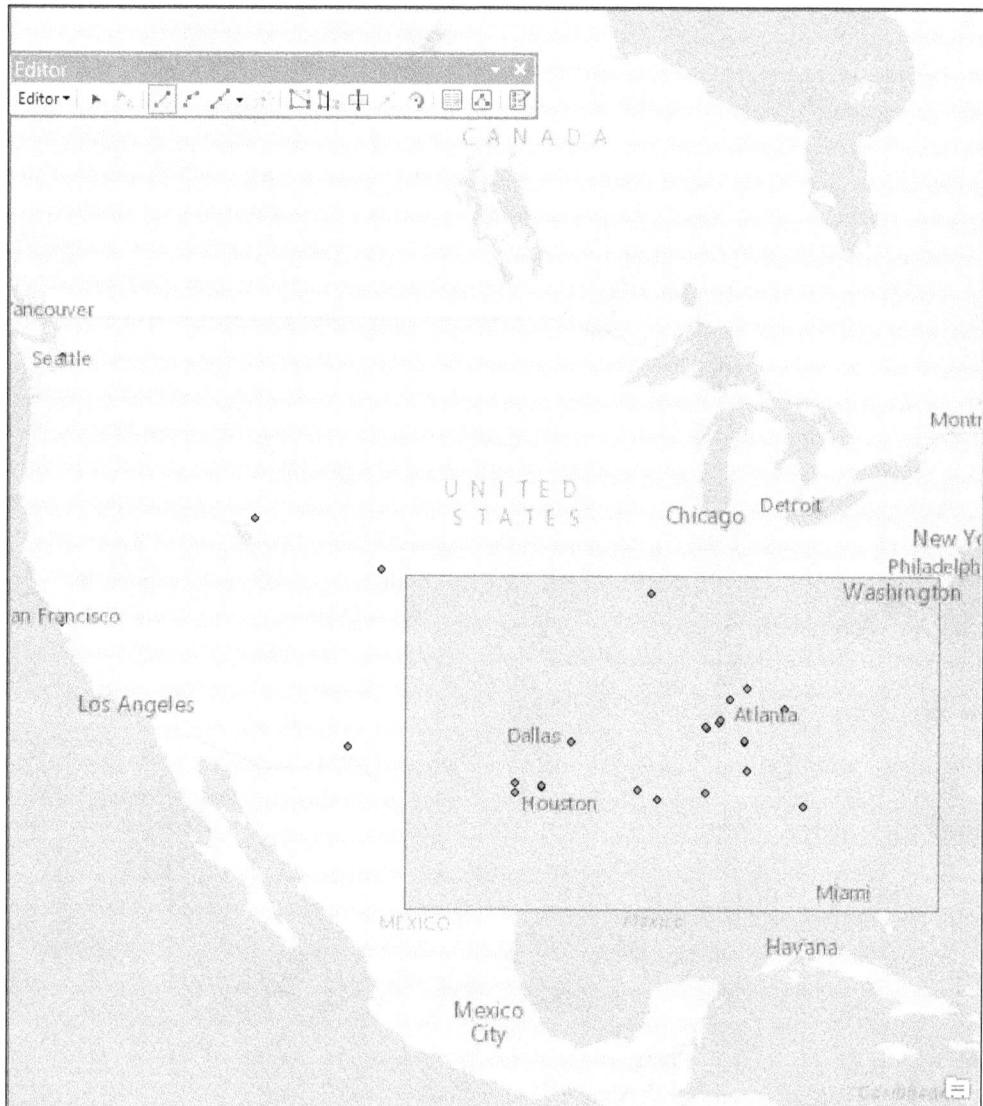

13. Navigate to **Editor | Save Edits** and then go to **Editor | Stop Editing** from the **Editor** toolbar.

14. Close the **Editor** toolbar.

15. Open **ArcToolbox** and find the **Optimized Hot Spot Analysis** tool in the **Spatial Statistics Tools** toolbox. It should be inside the **Mapping Clusters** toolset.

16. Double-click on **Optimized Hot Spot Analysis** to display the dialog shown here:

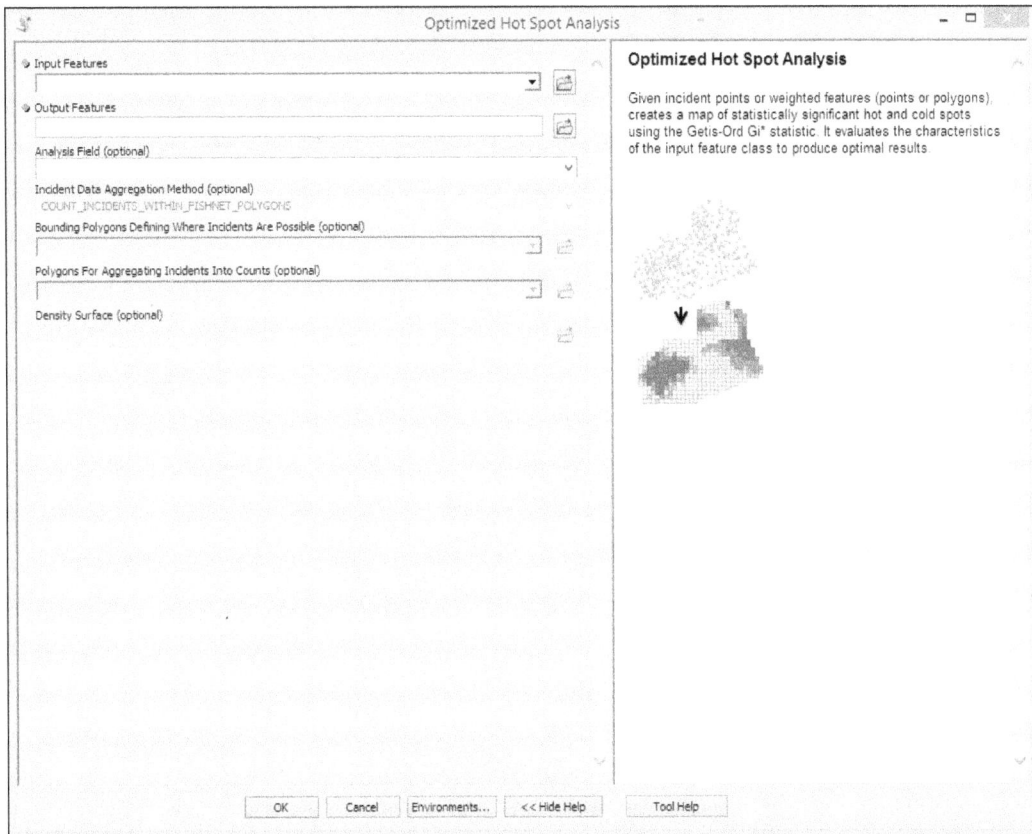

17. Select `Tweets` as **Input Features**, `HeatMap` as **Output Features**, and `BoundingPolygon` as **Bounding Polygons Defining Where Incidents are Possible**. The remainder of the parameters can be left as the defaults, or they can remain undefined. Take a look at the following screenshot to verify your parameters:

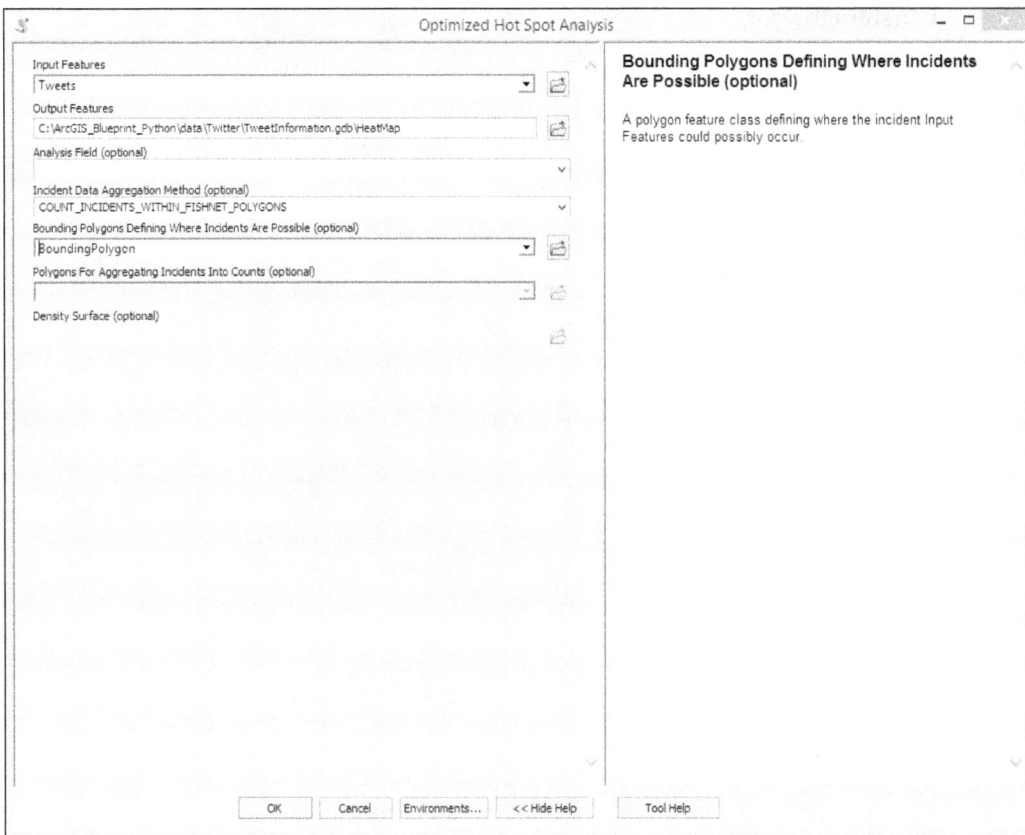

18. Click on OK to execute the tool. The progress dialog should display information as it proceeds, as shown in the following screenshot, and will ultimately create an output feature class:

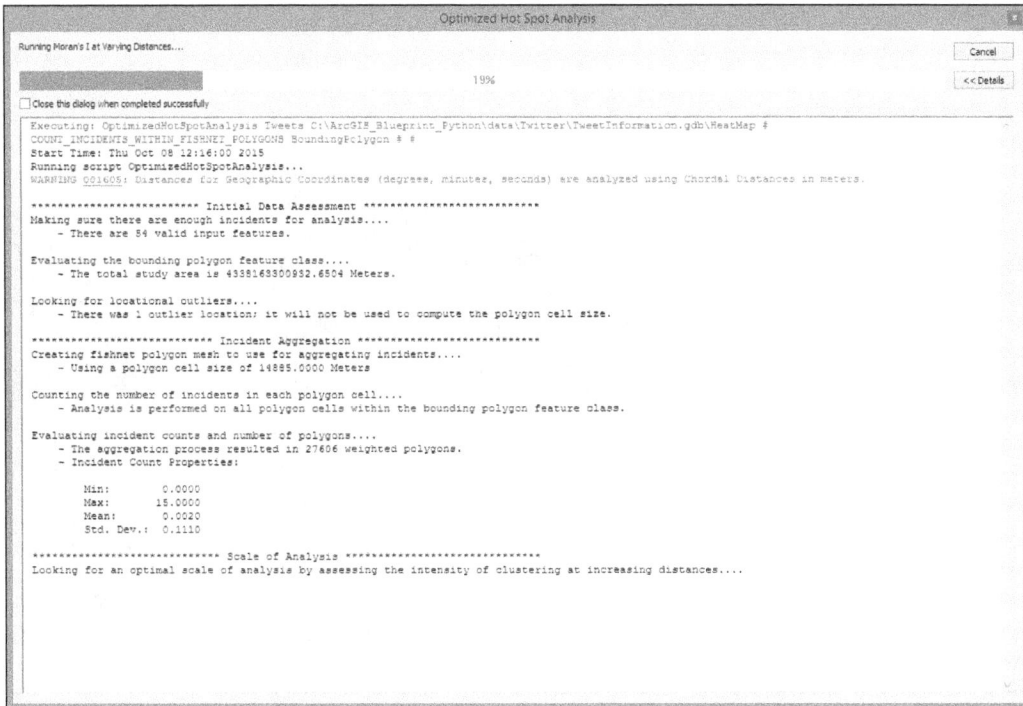

19. The output feature class will be symbolized to show cold spots, hotspots, and areas that are not significant. An example can be seen in the following screenshot. Note that the output is clipped to the bounding polygon:

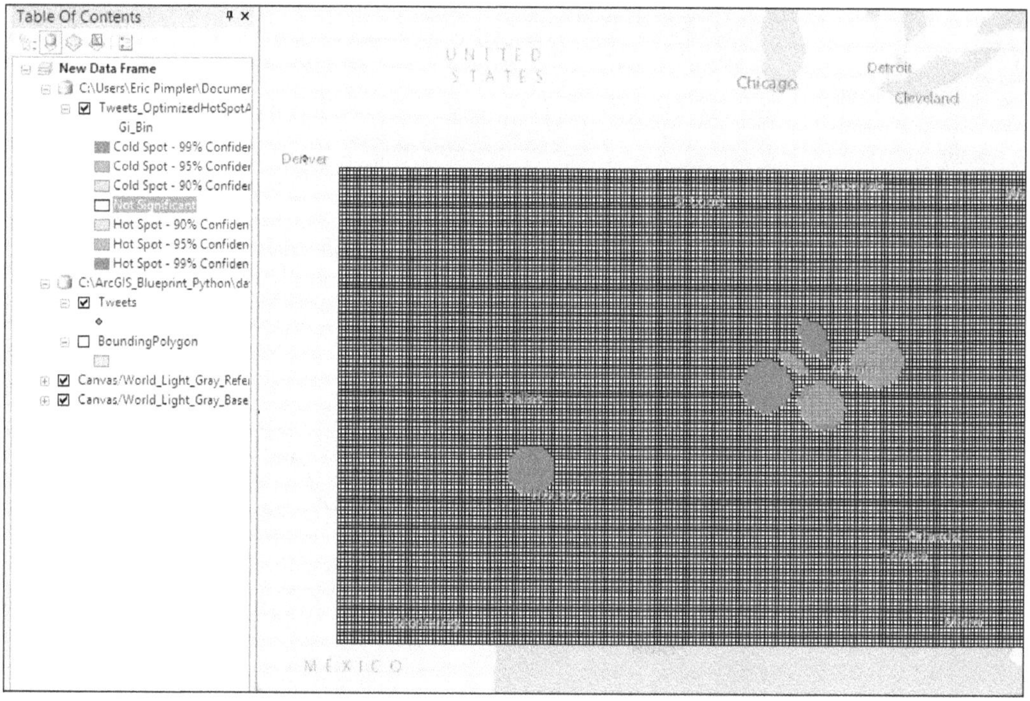

20. In the data frame containing your output `HeatMap` feature class, click on the symbol used to represent **Not Significant**.

21. In the **Symbol Selector** dialog, change the current symbol to **Hollow** and set **Outline Width** to 0. Click on **OK**.

22. The result should be something similar to what is shown in the following
    screenshot. Keep in mind that your data will differ:

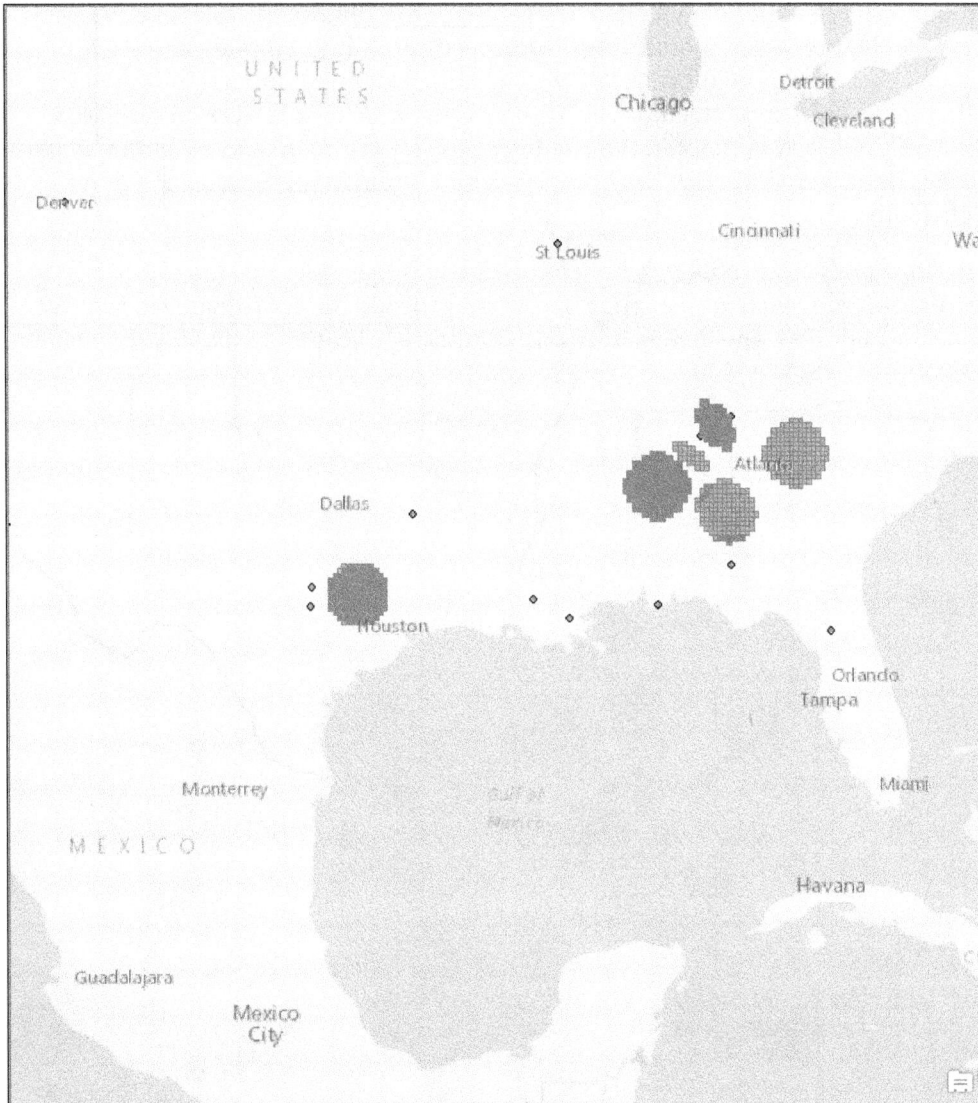

Summary

This chapter introduced several new topics, including the `tweepy` module used to monitor live Twitter data feeds and the use of the Windows Task Scheduler to automate the process of monitoring Twitter activity. Although only about 2% of tweets include location information, we can still get a good understanding of the spatial patterns of social media when monitoring large events over an extended period of time. In this chapter, the live tweets were written to a local feature class and then mapped to the **Hot Spot Analysis** tool found in the **Spatial Statistics Tools** toolbox.

In the next chapter, you'll learn how to use Python to extract the geographic coordinates from smartphone photos, reverse geocode the coordinates to retrieve the nearest address, and create an ArcGIS Online application to display the results.

10
Integrating Smartphone Photos with ArcGIS Desktop and ArcGIS Online

Today, almost everyone uses a smartphone. These phones have many capabilities, including the ability to take photos and videos. Because they also include a GPS, photos can be location enabled so that the geographic coordinates of each photo are captured and stored with the metadata that accompanies the photos. Photo metadata is stored in an **Exchangeable Image File Format** (**EXIF**). The Python's **Python Imaging Library** (**PIL**) module can be used to extract this information, including latitude and longitude coordinates. Using this extracted coordinate information, a reverse geocoding process can then be applied to each coordinate to determine the nearest address of the photo. This can be extremely useful for organizations, such as property managers, real estate agents, local government organizations, and more. Employees can be sent into the field with a smartphone to capture photos of properties or other assets without having to be concerned about capturing address and GPS information.

In this chapter, we will create a real estate application that reads photo metadata, extracts the coordinate information, retrieves the nearest address to the photo, and writes this information to a local feature class. In addition, the photos will be copied to a Dropbox account using the Python `dropbox` module so that the photos can be accessed through a web application. Finally, the property feature class will be uploaded to ArcGIS Online, integrated with the Dropbox photos, and shared as a web-based map.

In this chapter we will cover the following topics:

- Extracting geographic coordinates from smartphone photos with the Python `PIL` module
- Writing extracted coordinate information to a feature class with `ArcPy`
- Reverse geocoding smartphone photos to obtain nearest address
- Copying smartphone photos to Dropbox with the Python `dropbox` module

Design

The design of this application involves quite a few moving parts. Photo metadata information will be extracted using the Python `PIL` module. The extracted information will include geographic coordinates. The coordinate information can then be passed to the Esri World Geocoding service as a reverse geocoding operation to obtain the nearest address to the photo. The coordinate and address information can then be written to a local point feature class using the `ArcPy Data Access` module. The photos will also be copied to Dropbox so that they can be accessed through a web-based application. The final step in this chapter will be to upload the local file geodatabase to ArcGIS Online where it will be configured alongside the Dropbox photos to display property locations and photos that can be shared in a web application:

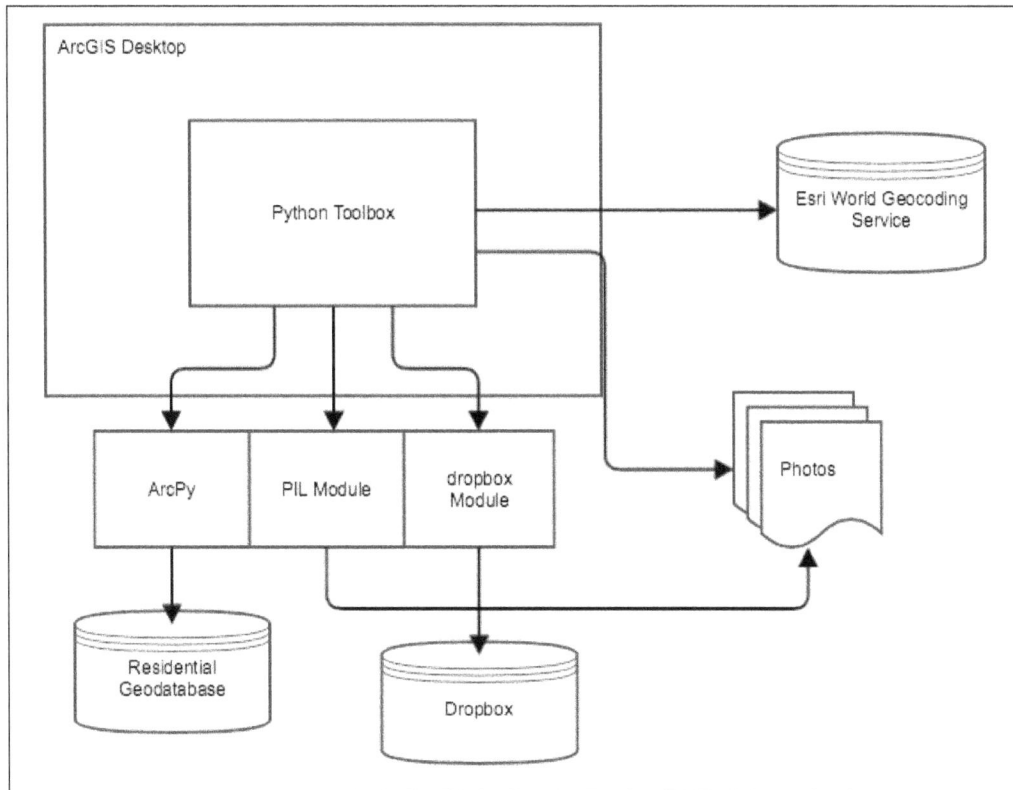

Let's get started building the application.

Taking photos

For this exercise, a number of photos have been provided for you to use. They are located in the `C:\ArcGIS_Blueprint_Python\ch10\photos` folder. However, you can use your own photos if you'd prefer. The code for this application does require that you use an iPhone or iPad device to take the photos. If you have an Android or other device, the metadata created with the photos will be different and require that your code be altered to account for the differences.

Photos taken with the camera application on an iPhone can store geographic coordinates in the metadata associated with each photo. However, you will need to turn on **Location Services**. The steps to do so are provided as follows:

1. Open the **Settings** app on your iPhone.
2. Select **Privacy**.

3. You should see **Location Services** at the top of the **Privacy** dialog as seen in the screenshot here:

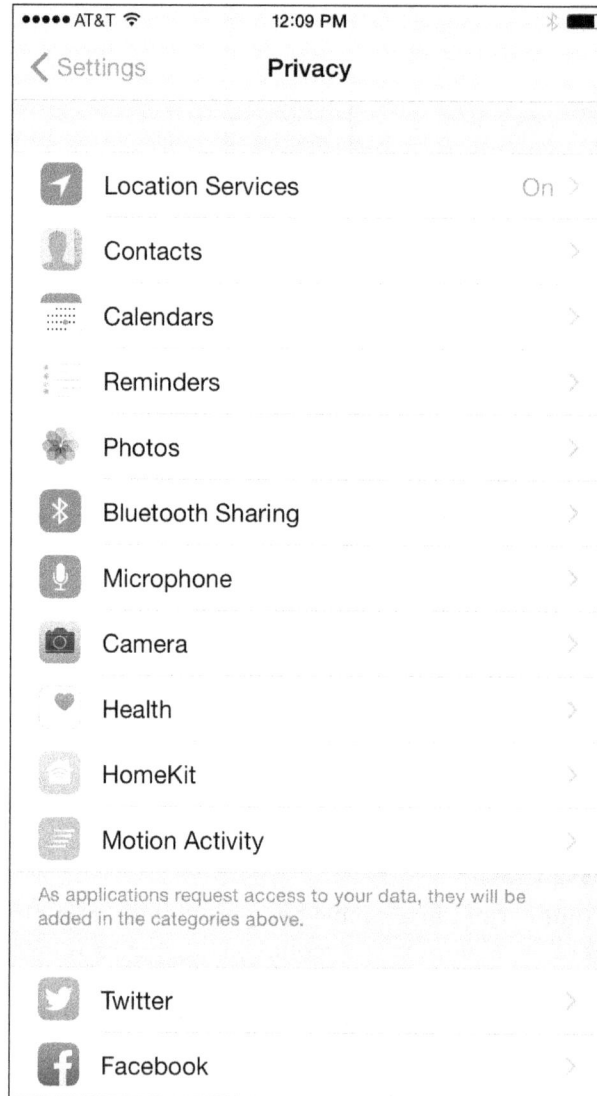

4. Click **Location Services**, find the **Camera** app, and select **While Using**, as shown in the following image:

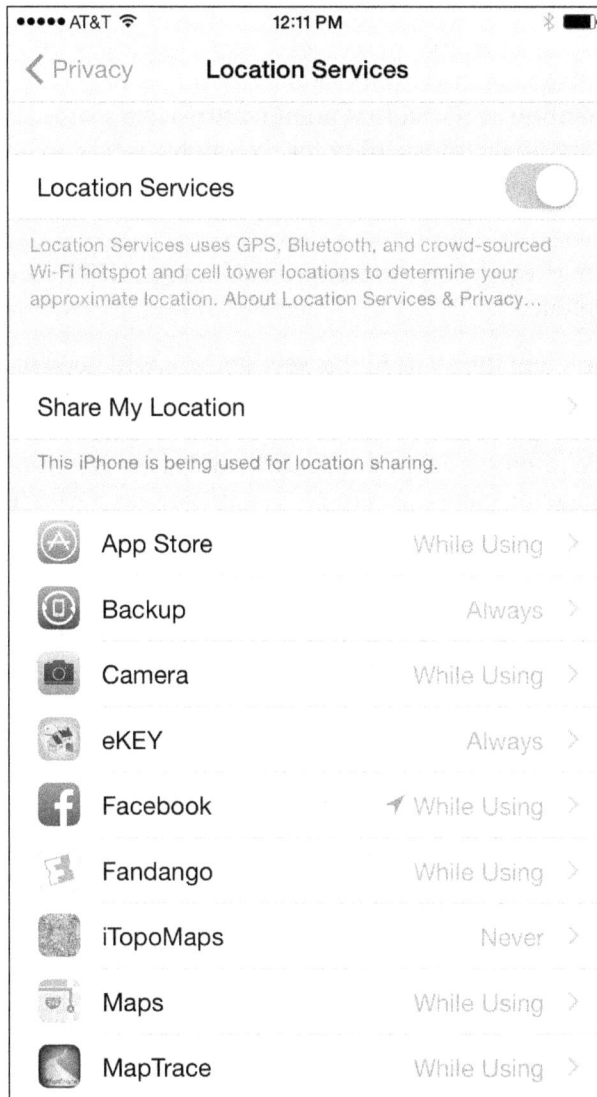

This will ensure that any photos taken with the **Camera** app will include geographic coordinates.

Converting iPhone photos to a feature class

In this step, you'll write a tool that processes a series of photos taken with an Apple iPhone. The tool will extract the latitude and longitude coordinates of each photo and write the information as individual point features in a feature class stored in a file geodatabase. Coordinate information for the photos can be extracted using the Python PIL module. In a later step, we'll update the script to also copy the photos to Dropbox.

Perform the following steps given, to create a custom **ArcGIS Python Toolbox** and tool to process the photos:

1. Before completing the steps in this section, you will need to download and install the Python PIL module. Open a Command Prompt, type the following command, and press *Enter*. This assumes that you have already installed pip from the previous chapter. Windows installers can also be found at the project downloads page at http://www.pythonware.com/products/pil/. We have the following code:

   ```
   pip install PIL
   ```

2. Open **Catalog** view in **ArcMap** and create a new **Python Toolbox** inside **Toolboxes | My Toolboxes**. Rename the default **Toolbox** to ProcessingPhotos.pyt.

3. Open your Python development environment by right-clicking the new toolbox you created and selecting **Edit**.

4. Import the shutil, os, sys, and arcpy modules along with several objects from the PIL module as follows:

   ```
   import shutil, os, sys, arcpy
   from PIL import Image   #PIL: http://www.pythonware.com/products/
   pil/
   from PIL.ExifTags import TAGS, GPSTAGS
   ```

5. Change the Tool class to ConvertPhotosToGeodatabase and set the label and description properties as follows:

   ```
   class ConvertPhotosToGeodatabase(object):
       def __init__(self):
           """Define the tool (tool name is the name of the
   class)."""
           self.label = "Convert Photos to Geodatabase"
           self.description = "Convert Photos to Geodatabase"
           self.canRunInBackground = False
   ```

6. Add the `ConvertPhotosToGeodatabase` tool to the tools list inside the toolbox:

```
self.tools = [ConvertPhotosToGeodatabase]
```

7. Define a `label` and `description` for the tool as follows:

```
class ConvertPhotosToGeodatabase(object):
    def __init__(self):
        """Define the tool (tool name is the name of the
class)."""
        self.label = "Convert Photos to Geodatabase"
        self.description = "Convert Photos to Geodatabase"
        self.canRunInBackground = False
```

8. This tool will require four input parameters. The first will provide a path to the folder containing the photos. The second and third parameters will provide the path and name of the output file geodatabase to be created, and the final parameter will be the output feature class name. Add the following code block to the `getParameterInfo()` method:

```
def getParameterInfo(self):
    """Define parameter definitions"""
    param0 = arcpy.Parameter(displayName = "Path to Pictures", \
                  name="folderToImport", \
                  datatype="DEFolder", \
                  parameterType="Required",\
                  direction="Input")

    param1 = arcpy.Parameter(displayName = "Path to File
Geodatabase", \
                  name="pathToFGDB", \
                  datatype="DEFolder", \
                  parameterType="Required",\
                  direction="Input")

    param2 = arcpy.Parameter(displayName = "File Geodatabase
Name", \
                  name="fgdbName", \
                  datatype="GPString", \
                  parameterType="Required",\
                  direction="Input")

    param3 = arcpy.Parameter(displayName = "Output Feature Class
Name", \
                  name="output_fc", \
```

```
                                  datatype="GPString",\
                                  parameterType="Required",\
                                  direction="Input")

        params = [param0, param1, param2, param3]
        return params
```

9. In the `execute()` method, capture the input parameters as follows:

```
def execute(self, parameters, messages):
    """The source code of the tool."""
    inPicFolder = parameters[0].valueAsText
    path = parameters[1].valueAsText
    fgdb = parameters[2].valueAsText
    fc = parameters[3].valueAsText
```

10. Next, call the `makeGISpointsFromPics()` function as seen here. We haven't created this method yet, but we will do so in the next step:

```
def execute(self, parameters, messages):
    """The source code of the tool."""
    inPicFolder = parameters[0].valueAsText
    path = parameters[1].valueAsText
    fgdb = parameters[2].valueAsText
    fc = parameters[3].valueAsText

    makeGISpointsFromPics(inPicFolder, path, fgdb, fc)
```

11. Create a new function named `makeGISpointsFromPics()`, as seen here. Make sure that you unindent the definition of this method so that it is completely left justified as follows:

```
def execute(self, parameters, messages):
    """The source code of the tool."""
    inPicFolder = parameters[0].valueAsText
    path = parameters[1].valueAsText
    fgdb = parameters[2].valueAsText
    fc = parameters[3].valueAsText

    makeGISpointsFromPics(inPicFolder, path, fgdb, fc)


def makeGISpointsFromPics(inPicFolder, path, fgdb, fc):
```

12. The `makeGISpointsFromPics()` function will control the extraction of the geographic coordinates from each photo and will also copy each of the photos into the file geodatabase. Add a start message and call a function named `_createFGDB()`. This function hasn't been created yet. We have the following code:

```
def makeGISpointsFromPics(inPicFolder, path, fgdb, fc):

    '''Top level function... '''
    arcpy.AddMessage( "----Beginning to build a GIS Point data for
geotagged photos----")

    #build the geodatabase
    _createFGDB(path,fgdb, fc)
```

13. Create a new function named `_createFGDB()`, as seen here. This function will create the new file geodatabase and feature class. Also, the function should be completely left justified:

```
def _createFGDB(path, fgdbName, fcName):
```

14. Inside the `_createFGDB()` function, create a new file geodatabase:

```
def _createFGDB(path, fgdbName, fcName):

    '''create a fgdb and a fc with a field to contain the path to
a picture'''
    if arcpy.Exists(path + "\\" + fgdbName):
        arcpy.AddMessage( "...the file gdb already exists")
        pass
    else:
        arcpy.CreateFileGDB_management(path, fgdbName)
        arcpy.AddMessage( "...created the file gdb")
```

15. Next, add a code block that creates a new feature class as follows:

```
def _createFGDB(path, fgdbName, fcName):

    '''create a fgdb and a fc with a field to contain the path to
a picture'''
    if arcpy.Exists(path + "\\" + fgdbName):
        arcpy.AddMessage( "...the file gdb already exists")
        pass
    else:
        arcpy.CreateFileGDB_management(path, fgdbName)
        arcpy.AddMessage( "...created the file gdb")

    if arcpy.Exists(path + "\\" + fgdbName + "\\" + fcName):
```

```
            arcpy.AddMessage( "...the fc already exists")
            pass
        else:
            spRef = r"Coordinate Systems\Geographic Coordinate
Systems\World\WGS 1984.prj"

            arcpy.AddMessage(path + "\\" + fgdbName)
            arcpy.AddMessage(fcName)

            arcpy.CreateFeatureclass_management(path + "\\" + fgdbName
+ ".gdb", fcName, "POINT", "#", "#", "#", spRef)
            arcpy.AddMessage( "...made fc")

            arcpy.AddField_management(path + "\\" + fgdbName + ".gdb"
+ "\\" + fcName, "name", "TEXT", "#", "#", "255","#", "#", "#",
"#")
            arcpy.AddField_management(path + "\\" + fgdbName + ".gdb"
+ "\\" + fcName, "pic_url", "TEXT", "#", "#", "255","#", "#", "#",
"#")
            arcpy.AddField_management(path + "\\" + fgdbName + ".gdb"
+ "\\" + fcName, "PicName", "TEXT", "#", "#", "100","#", "#", "#",
"#")
            arcpy.AddMessage( "...added fields")
```

16. Return to the `makeGISpointsFromPIcs()` function and execute the
    `Describe()` function on the new feature class as follows:

    ```
    def makeGISpointsFromPics(inPicFolder, path, fgdb, fc):

        '''Top level function... '''
        arcpy.AddMessage( "----Beginning to build a GIS Point data for
    geotagged photos----")

        #build the geodatabase
        _createFGDB(path,fgdb, fc)

        dsc = arcpy.Describe(path +  "\\" + fgdb + ".gdb" + "\\" + fc)
        shpFld = dsc.ShapeFieldName
    ```

17. Get a list of `.jpg` photos in the specified folder:

    ```
    def makeGISpointsFromPics(inPicFolder, path, fgdb, fc):

        '''Top level function... '''
    ```

```
arcpy.AddMessage( "----Beginning to build a GIS Point data for
geotagged photos----")

    #build the geodatabase
    _createFGDB(path,fgdb, fc)

    dsc = arcpy.Describe(path +  "\\" + fgdb + ".gdb" + "\\" + fc)
    shpFld = dsc.ShapeFieldName

    pics = os.listdir(inPicFolder)
    pics = [p for p in pics if p.endswith(".JPG") or p.endswith(".
jpg")]
```

18. Set up a looping structure for each of the photos:

```
def makeGISpointsFromPics(inPicFolder, path, fgdb, fc):

    '''Top level function... '''
    arcpy.AddMessage( "----Beginning to build a GIS Point data for
geotagged photos----")

    #build the geodatabase
    _createFGDB(path,fgdb, fc)

    dsc = arcpy.Describe(path +  "\\" + fgdb + ".gdb" + "\\" + fc)
    shpFld = dsc.ShapeFieldName

    pics = os.listdir(inPicFolder)
    pics = [p for p in pics if p.endswith(".JPG") or p.endswith(".
jpg")]

    i = len(pics)
    for pic in pics:
        try:

        except Exception as e:
            arcpy.AddMessage( e.message )
```

19. Inside the try block, add a call to the get_exif_data() function. We'll create
    this function in the next step:

```
try:

    exif_data = get_exif_data(inPicFolder + "\\" + pic)

except Exception as e:
    arcpy.AddMessage( e.message )
```

20. Create a new function named `get_exif_data()`. This function will extract the metadata from a photo. Exchangeable Image File Format (EXIF), is a standard that specifies the formats for images, sounds, and other tags used by digital cameras:

```
def get_exif_data(fn):
    """Returns a dictionary from the exif data of an PIL Image
item. Also converts the GPS Tags"""
    image = Image.open(fn)

    exif_data = {}
    info = image._getexif()
    if info:
        for tag, value in info.items():
            decoded = TAGS.get(tag, tag)
            if decoded == "GPSInfo":
                gps_data = {}
                for t in value:
                    sub_decoded = GPSTAGS.get(t, t)
                    gps_data[sub_decoded] = value[t]

                exif_data[decoded] = gps_data
            else:
                exif_data[decoded] = value

    return exif_data
```

21. Return to the `try` block in the `makeGISpointsFromPics()` function and add some messaging information about the photo to the ArcGIS progress dialog:

```
try:

    exif_data = get_exif_data(inPicFolder + "\\" + pic)

    arcpy.AddMessage("\n")
    arcpy.AddMessage(pic)
    arcpy.AddMessage(exif_data.get("GPSInfo"))
```

22. Call the `get_lat_lon()` function and pass in the `exif_data` variable. We'll create the `get_lat_lon()` function in the next step:

```
try:

    exif_data = get_exif_data(inPicFolder + "\\" + pic)

    arcpy.AddMessage("\n")
```

```
arcpy.AddMessage(pic)
arcpy.AddMessage(exif_data.get("GPSInfo"))

coordinates = get_lat_lon(exif_data)
```

23. Create the `get_lat_lon()` function. Inside the function, create an if
    statement that tests the `exif_data` dictionary variable for the presence of the
    `GPSInfo` key as follows:

```
def get_lat_lon(exif_data):
    """Returns the latitude and longitude, if available, from the
provided exif_data (obtained through get_exif_data above)"""
    lat = None
    lon = None

    if "GPSInfo" in exif_data:
```

24. Inside the `if` statement, pull out the information related to the latitude,
    longitude, and reference:

```
if "GPSInfo" in exif_data:
    gps_info = exif_data["GPSInfo"]

    gps_latitude = _get_if_exist(gps_info, "GPSLatitude")
    gps_latitude_ref = _get_if_exist(gps_info, 'GPSLatitudeRef')
    gps_longitude = _get_if_exist(gps_info, 'GPSLongitude')
    gps_longitude_ref = _get_if_exist(gps_info, 'GPSLongitudeRef')
```

25. Note that we called a function named `_get_if_exist()`. This function
    doesn't exist yet, but we'll create it in the coming steps along with a function
    named `_convert_to_degrees`. Convert the metadata to latitude and
    longitude coordinates:

```
def get_lat_lon(exif_data):
    """Returns the latitude and longitude, if available, from the
provided exif_data (obtained through get_exif_data above)"""
    lat = None
    lon = None

    if "GPSInfo" in exif_data:
        gps_info = exif_data["GPSInfo"]

        gps_latitude = _get_if_exist(gps_info, "GPSLatitude")
        gps_latitude_ref = _get_if_exist(gps_info,
'GPSLatitudeRef')
        gps_longitude = _get_if_exist(gps_info, 'GPSLongitude')
```

```
        gps_longitude_ref = _get_if_exist(gps_info,
'GPSLongitudeRef')


        if gps_latitude and gps_latitude_ref and gps_longitude and
gps_longitude_ref:
            lat = _convert_to_degrees(gps_latitude)
            if gps_latitude_ref != "N":
                lat = 0 - lat

            lon = _convert_to_degrees(gps_longitude)
            if gps_longitude_ref != "E":
                lon = 0 - lon


    return lat, lon
```

26. Create a function named _get_if_exists() as follows and this simply retrieves a value associated with a key, if the key exists in the dictionary:

```
def _get_if_exist(data, key):
    if key in data:
        return data[key]


    return None
```

27. Create the _convert_to_degrees() function seen as follows:

```
def _convert_to_degrees(value):
    """Helper function to convert the GPS coordinates stored in
the EXIF to degrees in float format"""
    d0 = value[0][0]
    d1 = value[0][1]
    d = float(d0) / float(d1)

    m0 = value[1][0]
    m1 = value[1][1]
    m = float(m0) / float(m1)

    s0 = value[2][0]
    s1 = value[2][1]
    s = float(s0) / float(s1)

    return d + (m / 60.0) + (s / 3600.0)
```

28. Return to the `makeGISpointsFromPics()` function. Inside the `try` block, get the latitude and longitude coordinates as follows:

```
try:

    exif_data = get_exif_data(inPicFolder + "\\" + pic)

    arcpy.AddMessage("\n")
    arcpy.AddMessage(pic)
    arcpy.AddMessage(exif_data.get("GPSInfo"))

    coordinates = get_lat_lon(exif_data)
    latitude = coordinates[0]
    longitude = coordinates[1]
```

29. We want to automatically derive the address where each picture was taken. This can be accomplished through the use of reverse geocoding. Reverse geocoding is used to find the nearest address to a given point. Because we have the latitude and longitude coordinates for each photo, we should be able to obtain the nearest address by calling a reverse geocoding service. First, add references to the `json` and `requests` modules:

```
import shutil, os, sys, arcpy
from PIL import Image   #PIL: http://www.pythonware.com/products/pil/
from PIL.ExifTags import TAGS, GPSTAGS
import json, requests
```

30. Create a new function named `getAddress()`, as seen here and create a variable that references the Esri World Geocoding service. The Esri World Geocoding service can perform reverse geocoding operations. For more information on this service please refer to `https://developers.arcgis.com/rest/geocode/api-reference/overview-world-geocoding-service.htm`:

```
def getAddress(latitude, longitude):
    agisurl = "http://geocode.arcgis.com/arcgis/rest/services/World/GeocodeServer/reverseGeocode?"
```

31. Append the latitude and longitude coordinates passed to the `getAddress()` function along with the desired output format and a distance value (in meters). The nearest address within this distance will be returned. If no address is found within this distance, an address will not be assigned:

```
def getAddress(latitude, longitude):
    agisurl = "http://geocode.arcgis.com/arcgis/rest/services/World/GeocodeServer/reverseGeocode?"

    agisurl = agisurl + "location=" + str(longitude) + "," + str(latitude) + "&f=pjson&distance=5000"
```

32. Pass the `agisurl` to the `requests.get()` method and return a response. Convert the returned `json` format to a Python dictionary:

```
def getAddress(latitude, longitude):
    agisurl = "http://geocode.arcgis.com/arcgis/rest/services/
World/GeocodeServer/reverseGeocode?"

    agisurl = agisurl + "location=" + str(longitude) + "," +
str(latitude) + "&f=pjson&distance=5000"

    r = requests.get(agisurl)
    decoded = json.loads(r.text)
```

33. The `json` response that is returned will be formatted similar to the code you see later. Note that we want to retrieve the value associated with the `Address` key, so we'll need to first retrieve the `address` key and then the `Address` key. When this `json` format data is converted to a Python dictionary, the keys will include address and location. Both of these keys contain values that are also Python dictionaries. So to retrieve the `Address` key, we'll need to drill down to the value of the Address key, which is itself a value of the `address` key. I know that it's a little confusing:

```
{
 "address": {
  "Address": "6 Avenue Gustave Eiffel",
  "Neighborhood": "7e Arrondissement",
  "City": "Paris",
  "Subregion": "Paris",
  "Region": "Île-de-France",
  "Postal": "75007",
  "PostalExt": null,
  "CountryCode": "FRA",
  "Loc_name": "FRA.PointAddress"
 },
 "location": {
  "x": 2.2946500041892821,
  "y": 48.857489996304814,
  "spatialReference": {
   "wkid": 4326,
   "latestWkid": 4326
  }
 }
}
```

34. Retrieve the `address` information and return this value to the calling function:

```
def getAddress(latitude, longitude):
    agisurl = "http://geocode.arcgis.com/arcgis/rest/services/
World/GeocodeServer/reverseGeocode?"

    agisurl = agisurl + "location=" + str(longitude) + "," +
str(latitude) + "&f=pjson&distance=5000"

    r = requests.get(agisurl)
    decoded = json.loads(r.text)
    address = decoded["address"]["Address"]
    return address
```

35. Return to the `makeGISpointsFromPics()` method and call the `getAddress()` function as follows:

```
try:

    exif_data = get_exif_data(inPicFolder + "\\" + pic)

    arcpy.AddMessage("\n")
    arcpy.AddMessage(pic)
    arcpy.AddMessage(exif_data.get("GPSInfo"))

    coordinates = get_lat_lon(exif_data)
    latitude = coordinates[0]
    longitude = coordinates[1]

    address = getAddress(latitude, longitude)
```

36. Create a new `arcpy.Point` object from the latitude and longitude as follows:

```
try:

    exif_data = get_exif_data(inPicFolder + "\\" + pic)

    arcpy.AddMessage("\n")
    arcpy.AddMessage(pic)
    arcpy.AddMessage(exif_data.get("GPSInfo"))

    coordinates = get_lat_lon(exif_data)
    latitude = coordinates[0]
    longitude = coordinates[1]
    address = getAddress(latitude, longitude)

    pnt = arcpy.Point(longitude,latitude) #pnt is now an "object"
that arcmap recognizes
```

37. Create `InsertCursor` and `Row` objects for the new record as follows:

```
pnt = arcpy.Point(longitude,latitude) #pnt is now an "object" that
arcmap recognizes
rows = arcpy.InsertCursor(path + "\\" + fgdb + ".gdb" + "\\" + fc)
row = rows.newRow()
```

38. Set the values for the new row, including the `geometry` and `attributes`. Insert the row into the `feature` class:

```
pnt = arcpy.Point(longitude,latitude) #pnt is now an "object" that
arcmap recognizes
rows = arcpy.InsertCursor(path + "\\" + fgdb + ".gdb" + "\\" + fc)
row = rows.newRow()
row.setValue(shpFld, pnt)

row.name = address
row.PicName = pic

rows.insertRow(row)
```

39. Update the `progress` dialog as follows:

```
try:

    exif_data = get_exif_data(inPicFolder + "\\" + pic)

    arcpy.AddMessage("\n")
    arcpy.AddMessage(pic)
    arcpy.AddMessage(exif_data.get("GPSInfo"))

    coordinates = get_lat_lon(exif_data)
    latitude = coordinates[0]
    longitude = coordinates[1]

    address = getAddress(latitude, longitude)

    pnt = arcpy.Point(longitude,latitude) #pnt is now an "object"
that arcmap recognizes
    rows = arcpy.InsertCursor(path + "\\" + fgdb + ".gdb" + "\\" +
fc)
    row = rows.newRow()

    row.setValue(shpFld, pnt)

    row.name = address
```

```
row.PicName = pic
rows.insertRow(row)

arcpy.AddMessage( "...added a point, " + str(i - 1) + " to
go.")
    i -= 1
```

40. You can review your code by examining the solution code in the `C:\ArcGIS_ Blueprint_Python\solutions\ch10\ProcessingPhotos.py` file.

41. Save your work and close the Python development environment.

42. In **ArcMap**, open `Ch10.mxd` found in `C:\ArcGIS_Blueprint_Python\Ch10`.

43. In the **Catalog** window, go to **Toolboxes | My Toolboxes |** `ProcessingPhotos.pyt` and double-click on the **Convert Photos to Geodatabase** tool to display the dialog shown in the following screenshot. Add the parameter information as seen:

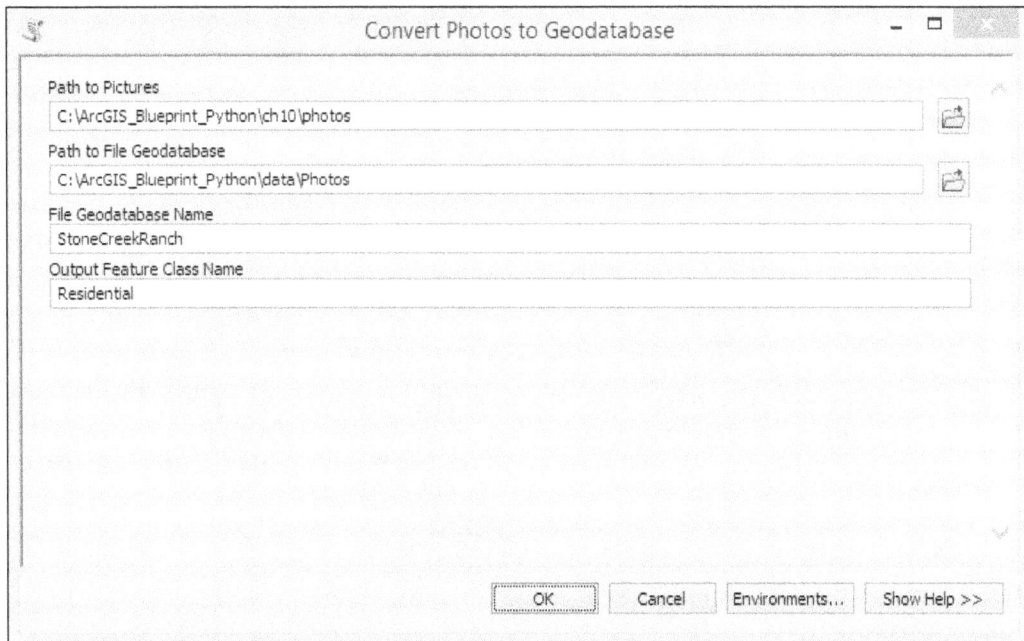

44. Click on **OK** to execute the tool.

45. Add the **Residential** feature class to the display. You should see a handful of point locations, as seen in the following screenshot:

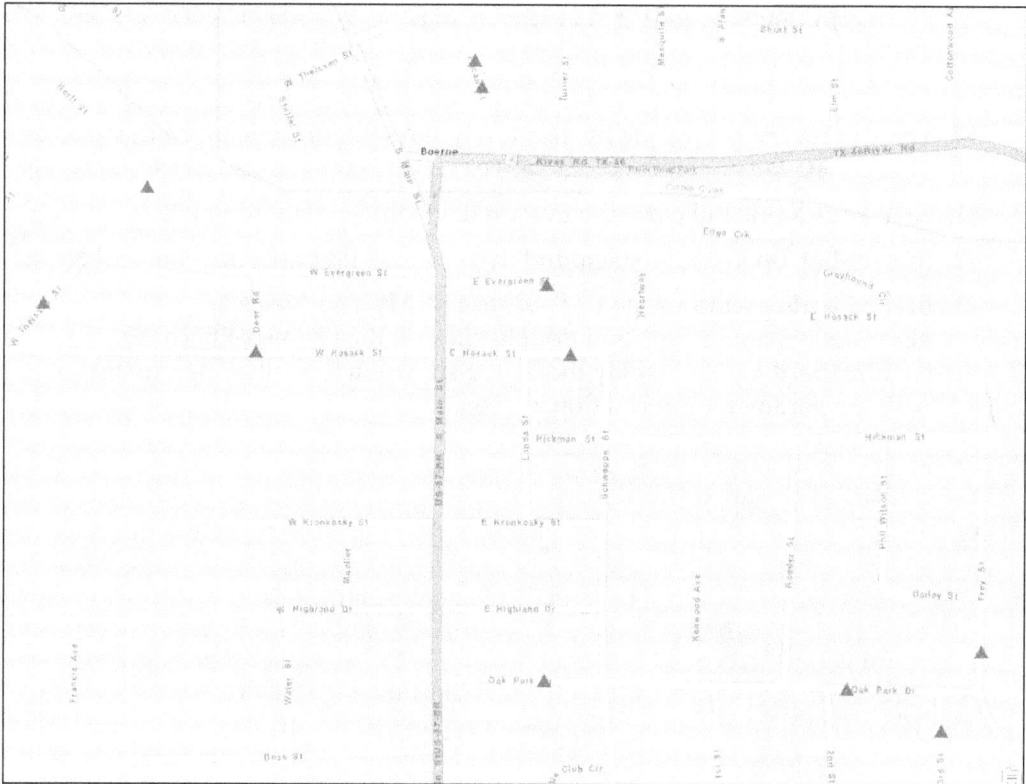

46. Identify one of the point locations to see the attribute information stored with each feature, as seen in the screenshot. Note that your attributes may differ from mine, but you should have some attribute information for the name and `PicName` fields:

```
Identify                                                    □ ×

Identify from:    │ <Top-most layer>                    │ ▼ │

⊟‥Residential
    └‥‥ 341 Ammann Rd


                                                          ⌃□

Location:    │ -10,979,593.088  3,475,060.344 Meters    │ ▼

Field        Value

OBJECTID     14
Shape        Point
name         341 Ammann Rd
PicName      IMG_0707.JPG




<                                                      >

Identified 1 feature
```

47. Copying the photos to Dropbox.

48. In this step, we will add code to copy the photos from the local computer to a
    Dropbox account so that the photos can be accessed from a web application.
    You will need a Dropbox account to complete this step. For more information
    and to create an account, you can visit `http://dropbox.com`.

Follow the steps here to add code that will copy the photos to Dropbox:

1. Use `pip` to install the Python `dropbox` module. You can get more information on the **Python SDK for Dropbox** at `https://www.dropbox.com/developers-v1/core/sdks/python`:

   ```
   pip install dropbox
   ```

2. Before writing the code to store the photos in Dropbox, you will need to create an **App** inside Dropbox. This will allow you to read and write files in Dropbox. Open a web browser and navigate to `http://www.dropbox.com/developers-v1`.

3. Log in to your Dropbox account.

4. The Dropbox API contains a core API that allows you to read and write files in Dropbox. This Core API is based on HTTP and OAuth to provide low-level calls and access a user's Dropbox account. You'll first need to register a new app using the **App Console**. In your browser navigate to `http://www.dropbox.com/developers-v1/apps`.

5. You should see something similar to the screenshot here. Click on the **Create App** button to get started as follows:

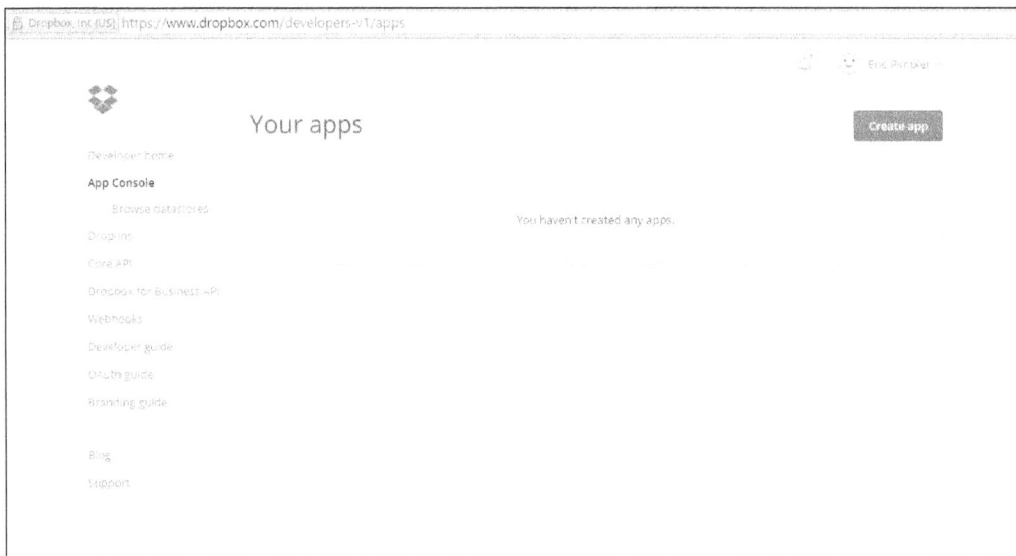

6. Select **Dropbox API app** and enter the parameters, as seen in the screenshot here. Please note that the Dropbox interface changes periodically, so your view may differ somewhat from the screenshot:

Create a new Dropbox Platform app

What type of app do you want to create?

Drop-ins app
Chooser or Saver

Dropbox API app

To create a Dropbox for Business app, visit the Dropbox for Business app creation page.

Can your app be limited to its own folder?

Yes — My app only needs access to files it creates.

No — My app needs access to files already on Dropbox.

Provide an app name, and you're on your way.

RealEstatePhotos

Terms of Service

I agree to Dropbox API Terms and Conditions

Create app

7. Click on the **Create app** button, and you should see a detailed screen similar to that shown in the screenshot here. You will want to note the **App key** and **App secret**. You'll use these values in just a few moments:

RealEstatePhotos

Settings	Branding	Analytics

Status	Development	Apply for production
Development users	Only you	Enable additional users
Permission type	App folder	
App folder name	RealEstatePhotos	Change
App key	████████████	
App secret	████████████	
OAuth 2	**Redirect URIs**	
	https:// (http allowed for localhost)	Add
	Allow implicit grant	
	Allow	

8. Open a Python shell window. The IDLE shell window will work fine for this.

9. Import the dropbox module and set variable for the app key and app secret. These are the values generated in the previous screenshot:

```
import dropbox
app_key = "<your app key here>"
app_secret = <your app secret here>
```

10. Create an instance of the `DropboxOAuth2FlowNoRedirect` object by passing in the app key and secret as follows:

```
flow = dropbox.client.DropboxOAuth2FlowNoRedirect(app_key, app_secret)
```

11. Start the `flow` object.

```
authorize_url = flow.start()
```

12. Have the user sign in and authorize the token by using the URL you get like `https://www.dropbox.com/1/oauth2/authorize?response_type=code&client_id=a62sar870yn7dsa`. this might show you error, because this is specific to each user:

```
print '1. Go to: ' + authorize_url
```

13. Open a web browser and copy and paste the URL printed in step 12 into the address bar and press Enter. You should see something similar to the following screenshot:

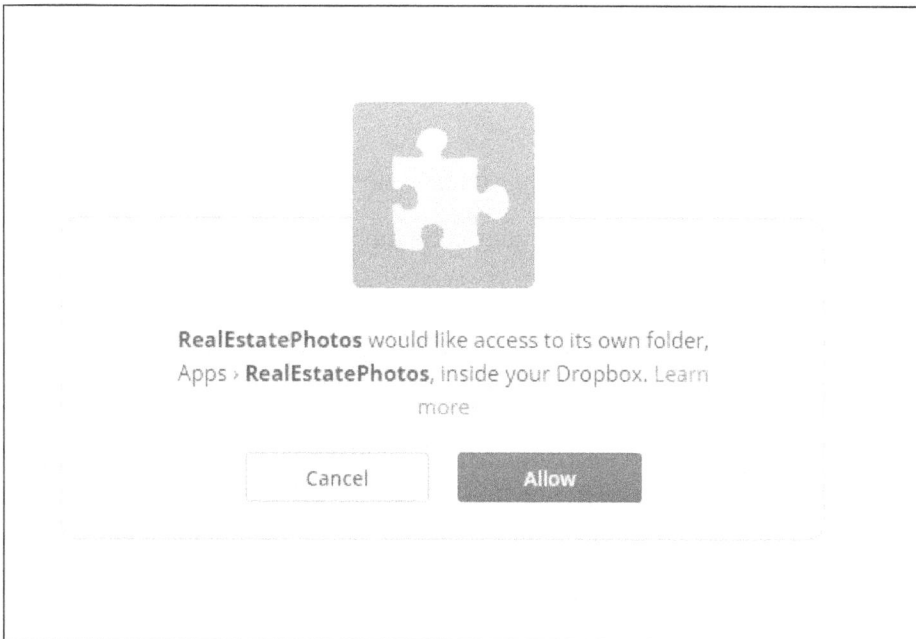

14. Click on **Allow**, sometimes you might have to log in first.

15. After clicking, you should be presented with a screen similar to what is displayed in the following screenshot, and your code will not be the same:

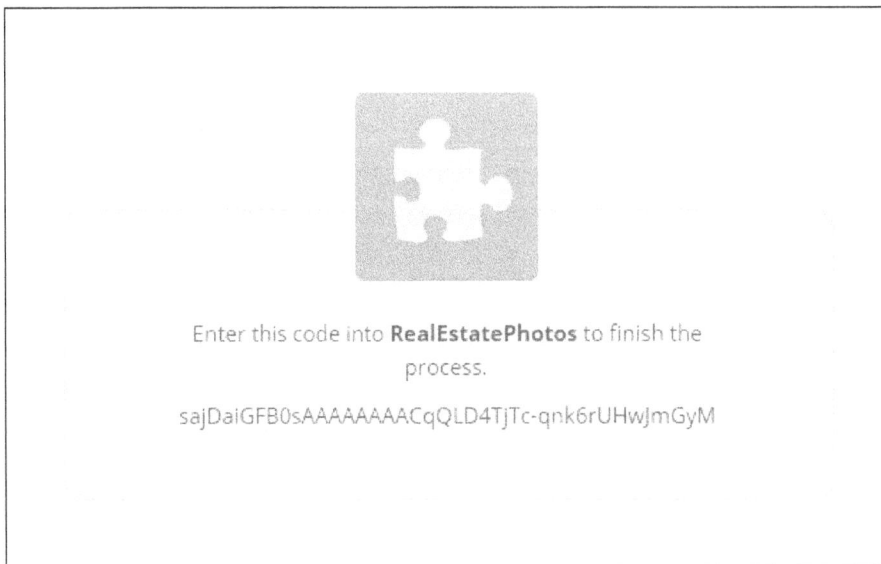

Enter this code into **RealEstatePhotos** to finish the process.

sajDaiGFB0sAAAAAAACqQLD4TjTc-qnk6rUHwJmGyM

16. Copy the authorization code.'

```
code = raw_input("Enter the authorization code here: ").strip()
```

17. Enter the authorization code, your code might be different:

```
sajDaiGFB0sAAAAAAACqQLD4TjTc-qnk6rUHwJmGyM
```

18. Generate the token by entering the following line of code:

```
# This will fail if the user enters an invalid authorization code
access_token, user_id = flow.finish(code)
```

19. Print out the access token and user_id. You'll want to save the access token and user ID. This series of steps only has to be done once if you save the token:

```
print access_token
<The access token will be printed out. Save the token so you can
use it for your requests>
print user_id
<The user id will be printed out>
```

20. You can test the access token by entering the following lines of code. You should get some sort of response similar to that shown as follows:

```
client = dropbox.client.DropboxClient(access_token)
print 'linked account: ', client.account_info()
```

```
linked account: {u'referral_link': u'https://db.tt/
PIgQ4g9F', u'display_name': u'Eric Pimpler', u'uid': 37354582,
u'locale': u'en', u'email_verified': True, u'email': u'eric@
geospatialtraining.com', u'is_paired': False, u'team': None,
u'name_details': {u'familiar_name': u'Eric', u'surname':
u'Pimpler', u'given_name': u'Eric'}, u'country': u'US', u'quota_
info': {u'datastores': 0, u'shared': 567215708, u'quota':
1101927546880L, u'normal': 2872036212L}}
```

21. The entire section of code should look something like the screenshot here, and you can ignore the warning message:

```
Python 2.7.8 Shell

File  Edit  Shell  Debug  Options  Windows

Python 2.7.8 (default, Jun 30 2014, 16:03:49) [MSC v.1500 32 bit (Intel)] on win
32
Type "copyright", "credits" or "license()" for more information.
>>> import dropbox
>>> app_key = '            '
>>> app_secret = '            '
>>> flow = dropbox.client.DropboxOAuth2FlowNoRedirect(app_key, app_secret)
>>> authorize_url = flow.start()
>>> print '1. Go to: ' + authorize_url
1. Go to: https://www.dropbox.com/1/oauth2/authorize?response_type=code&client_i
d=a62sar870yn7dsa
>>> code = raw_input("Enter the authorization code here: ").strip()
Enter the authorization code here:
>>> access_token, user_id = flow.finish(code)

Warning (from warnings module):
  File "C:\Python27\ArcGIS10.3\lib\site-packages\urllib3\util\ssl_.py", line 100
    InsecurePlatformWarning
InsecurePlatformWarning: A true SSLContext object is not available. This prevent
s urllib3 from configuring SSL appropriately and may cause certain SSL connectio
ns to fail. For more information, see https://urllib3.readthedocs.org/en/latest/
security.html#insecureplatformwarning.
>>> print access_token

>>> print user_id

>>> client = dropbox.client.DropboxClient(access_token)
>>> print 'linked account: ', client.account_info()
linked account: {u'referral_link': u'https://db.tt/PIgQ4g9F', u'display_name':
u'Eric Pimpler', u'uid': 37354582, u'locale': u'en', u'email_verified': True, u'
email': u'eric@geospatialtraining.com', u'is_paired': False, u'team': None, u'na
me_details': {u'familiar_name': u'Eric', u'surname': u'Pimpler', u'given_name':
u'Eric'}, u'country': u'US', u'quota_info': {u'datastores': 0, u'shared': 567215
708, u'quota': 1101927546880L, u'normal': 2872036212L}}
```

22. Now return to the Python development environment for the `ProcessingPhotos.pyt` Python toolbox.

23. Import the `dropbox` module.

```
import shutil, os, sys, arcpy
from PIL import Image   #PIL: http://www.pythonware.com/products/
pil/
from PIL.ExifTags import TAGS, GPSTAGS
import json, requests
import dropbox
```

24. Create a new function named `sendPhotoDropbox()` and create a new instance of `DropboxClient` by passing in the access token you generated in the previous step. The `sendPhotoDropbox()` function should accept two parameters including the full path to the image file that will be uploaded to Dropbox as well as the filename that will be created. Use the following code:

```
def sendPhotoDropbox(fullPath,fn):
    client = dropbox.client.DropboxClient('<your access token.')
```

25. Copy the file to Dropbox as follows:

```
def sendPhotoDropbox(fullPath,fn):
    client = dropbox.client.DropboxClient('sajDaiGFB0sAAAAAAAACqp
vO-oHvp3R33aNNPFIeOhaYH3o8qECbYUSyZs0MUU-S')
    f = open(fullPath, 'rb')
    response = client.put_file('/' + fn, f)
```

26. Create a shared link for the photo that we'll write to the output feature class and return this value to the calling function. By default, the end of the URL contains some extraneous characters that we chop off using `[:-5]`:

```
def sendPhotoDropbox(fullPath,fn):
    client = dropbox.client.DropboxClient('sajDaiGFB0sAAAAAAAACqp
vO-oHvp3R33aNNPFIeOhaYH3o8qECbYUSyZs0MUU-S')
    f = open(fullPath, 'rb')
    response = client.put_file('/' + fn, f)
    shared_img = client.share('/' + fn, short_url=False)
    return shared_img["url"][:-5]
```

27. Return to the `makeGISpointsFromPics()` method, and inside the `try` statement, add the following line of code:

```
try:

    exif_data = get_exif_data(inPicFolder + "\\" + pic)

    arcpy.AddMessage("\n")
    arcpy.AddMessage(pic)
    arcpy.AddMessage(exif_data.get("GPSInfo"))

    coordinates = get_lat_lon(exif_data)
    latitude = coordinates[0]
    longitude = coordinates[1]

    address = getAddress(latitude, longitude)
    url = sendPhotoDropbox(inPicFolder + "\\" +  pic, pic)
```

28. Add a new line that inserts the URL into the feature class.

```
row.name = address
row.pic_url = url
row.PicName = pic
```

29. Save your work.

30. You can review your code by examining the solution code in the `C:\ArcGIS_Blueprint_Python\solutions\ch10\ProcessingPhotos.py` file.

31. Open **ArcMap** with the `Ch10.mxd` file. Run the tool again, creating a new file geodatabase and feature class.

32. Add the output **Residential** feature class to the map and identify a point feature to see the output. It should look similar to the screenshot here. Note the content of the `pic_url` field:

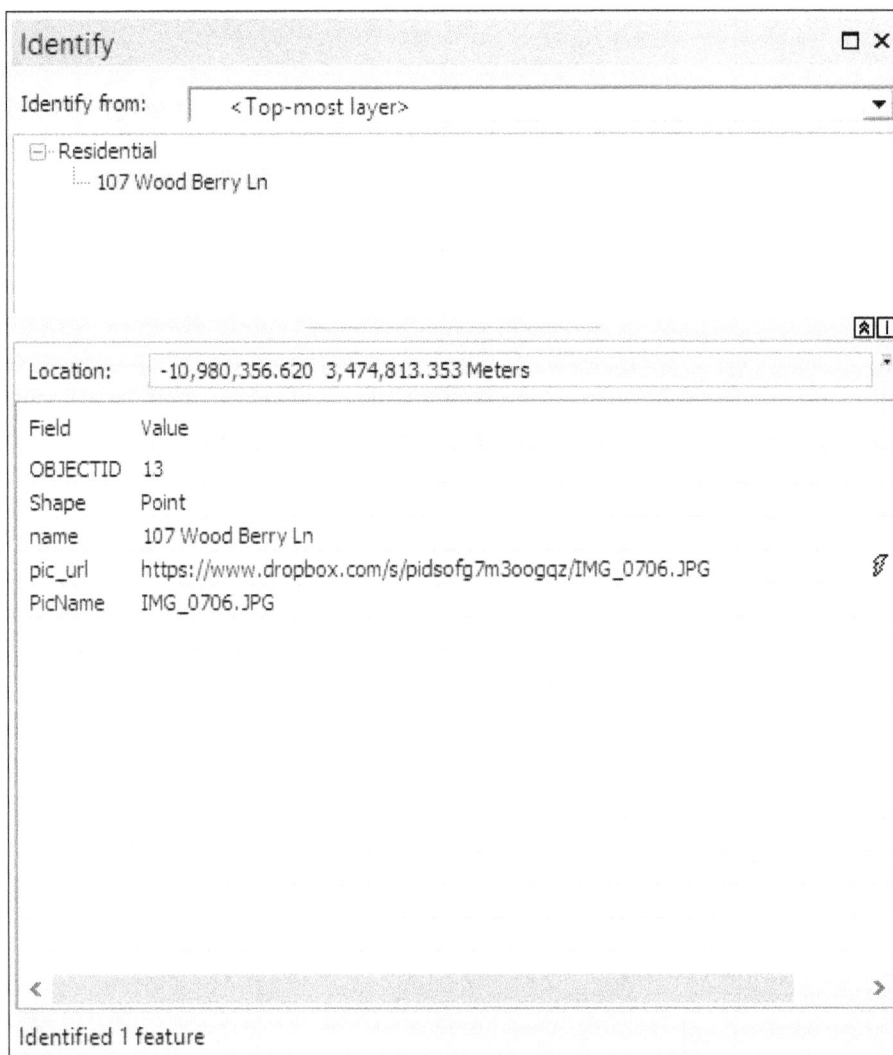

33. Check your Dropbox account inside the `RealEstatePhotos` folder, and you should see images similar to the following screenshot:

Dropbox > Apps > RealEstatePhotos

Name ▲	Modified
IMG_0694.JPG	4 hrs ago
IMG_0695.JPG	4 hrs ago
IMG_0696.JPG	5 hrs ago
IMG_0697.JPG	5 hrs ago
IMG_0698.JPG	5 hrs ago
IMG_0699.JPG	5 hrs ago
IMG_0700.JPG	5 hrs ago
IMG_0701.JPG	5 hrs ago
IMG_0702.JPG	5 hrs ago

Creating a Web Map

In the final step in this chapter, the content that was created with the
`ProcessingPhotos.pyt` Python toolbox will be imported to ArcGIS Online and
a shareable map created. This section requires an ArcGIS Online organizational
account. The steps are as follows:

1. Open Windows Explorer and go to `C:\ArcGIS_Blueprint_Python\data\`
   `Photos\StoneCreekRanch.gdb`. Create a zip file containing this
   file geodatabase.

2. Open a web browser and navigate to `http://www.arcgis.com/features/`.

3. Log in to your ArcGIS Online organizational account.

4. Select **My Content**.

5. Select **Add Item | From my Computer**, as shown in the following image:

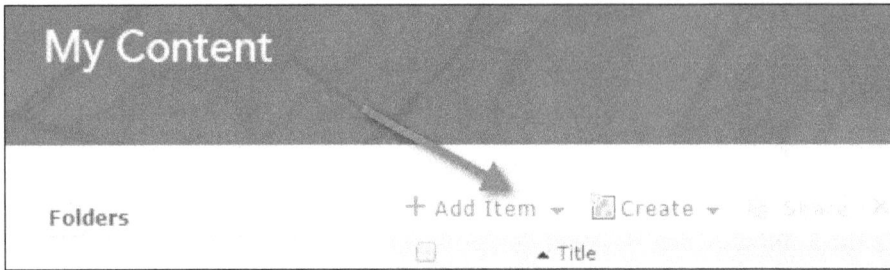

6. Select the zip file that you just created which is StoneCreekRanch.
    gdb.zip, change **Contents to File Geodatabase**, and give it a Title of
    StoneCreekRanch. Examine the following screenshot for the details:

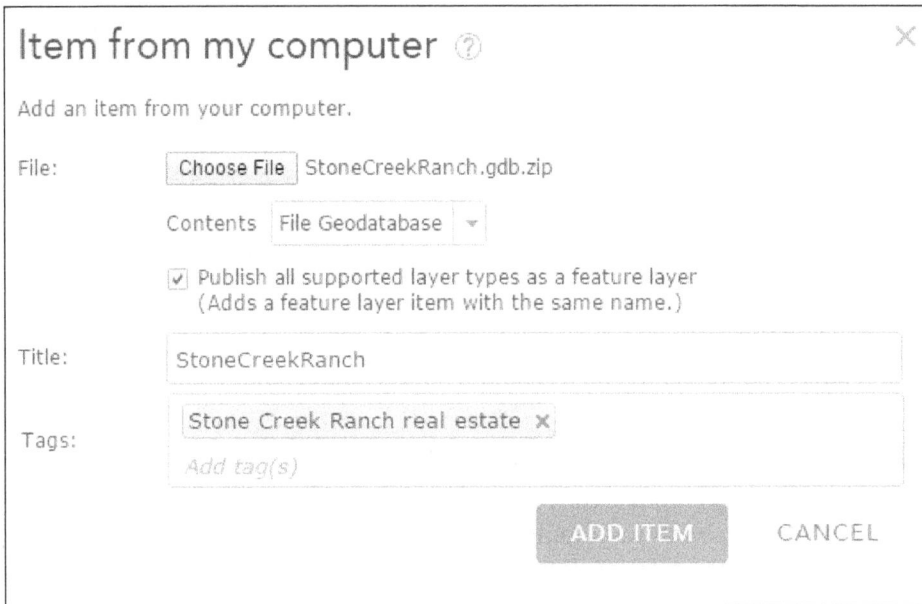

7. Click on the **ADD ITEM** button. This will create a `FeatureLayer` and `FeatureService`. Click on the **Share** button on the dialog that is displayed after this process is complete. Share it publicly:

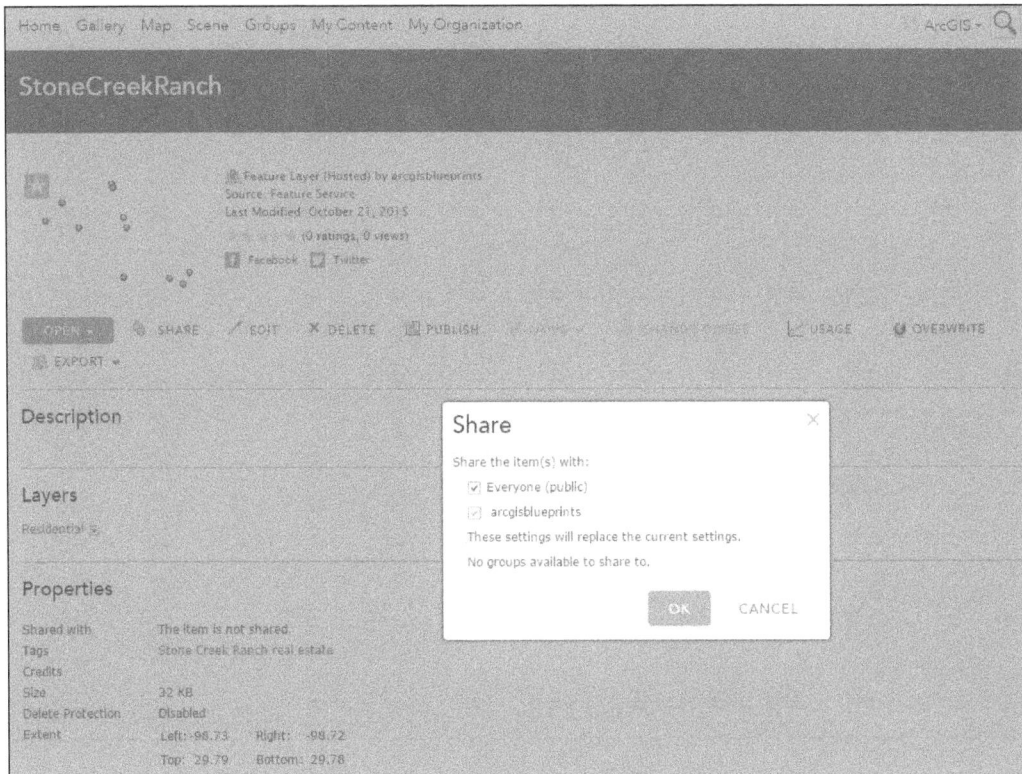

8. Select **Open** | **Add layer to new map**.

9. Change the symbology to **Single** symbol so that your map appears, as seen in the following screenshot:

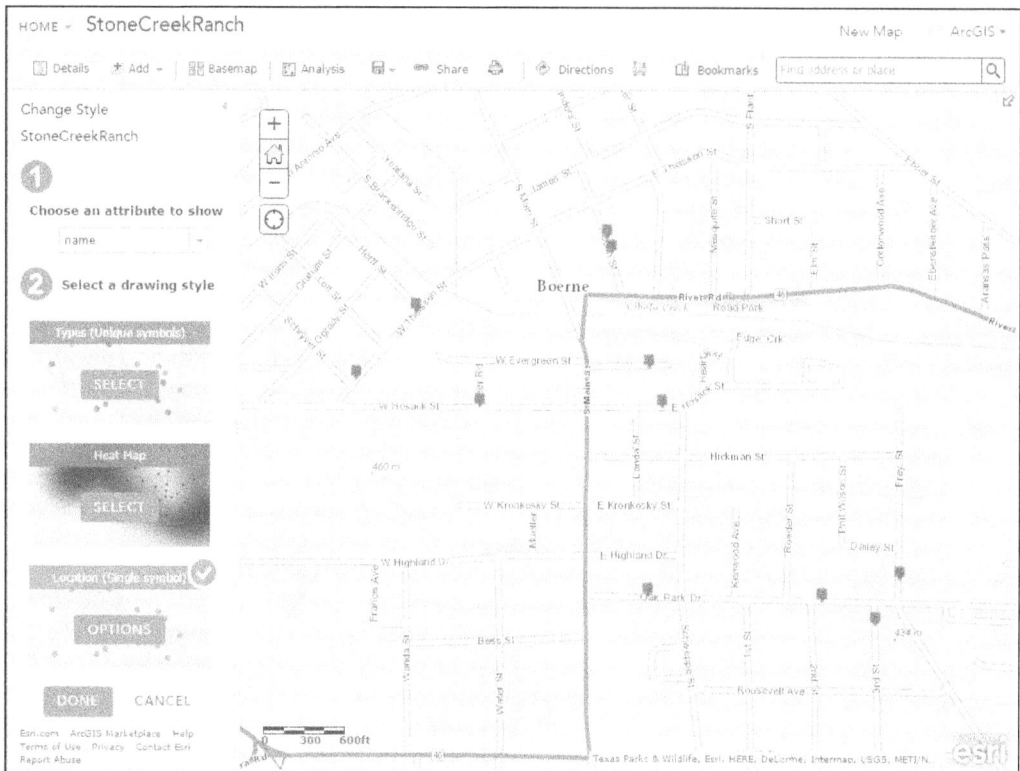

10. Click on **Done**.

11. Select the **...** (which is a **More Options**) icon to the right of the StoneCreekRanch layer and select **Configure Pop-up**.

12. Configure the popup as seen in the following screenshot and select **SAVE POP-UP**.

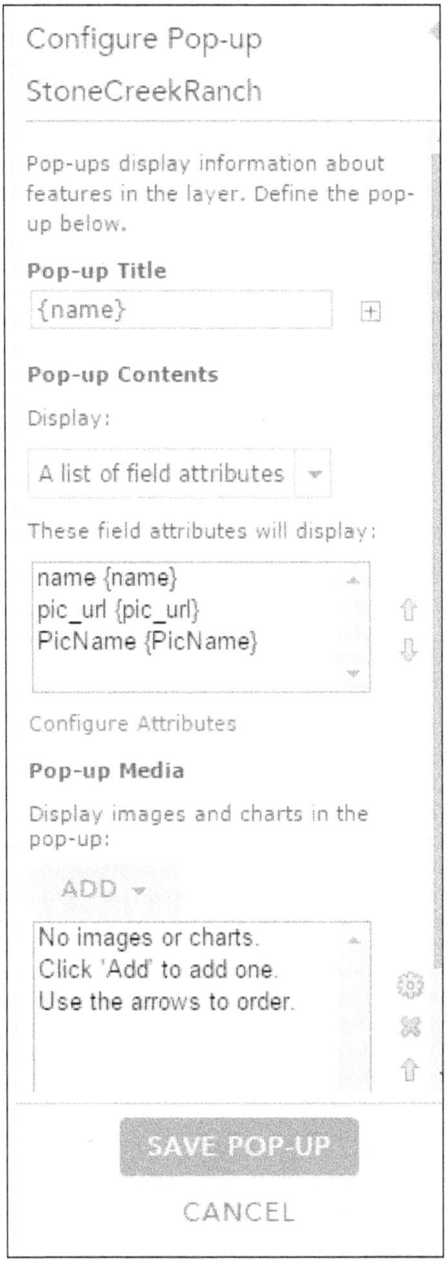
Configure Pop-up
StoneCreekRanch
Pop-ups display information about features in the layer. Define the pop-up below.
Pop-up Title
{name}
Pop-up Contents
Display:
A list of field attributes
These field attributes will display:
name {name}
pic_url {pic_url}
PicName {PicName}
Configure Attributes
Pop-up Media
Display images and charts in the pop-up:
ADD
No images or charts.
Click 'Add' to add one.
Use the arrows to order.
SAVE POP-UP
CANCEL

13. Click on of the points to see the result of the pop-up configuration. Dropbox photos can't be displayed directly in pop-up windows as of this writing.

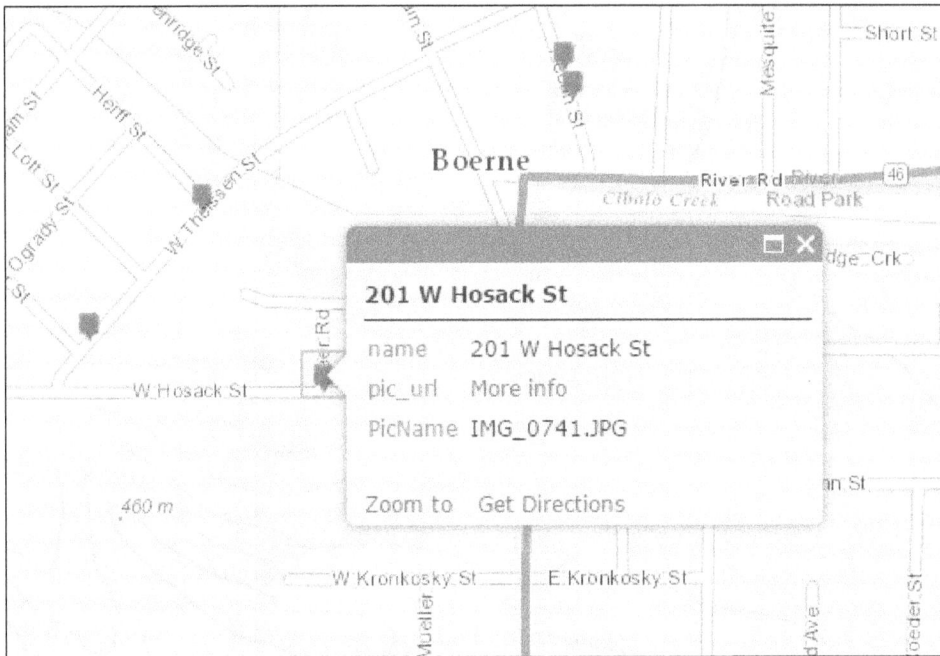

14. Clicking on the **More Info** link will display the photo in a separate window as seen in the following screenshot:

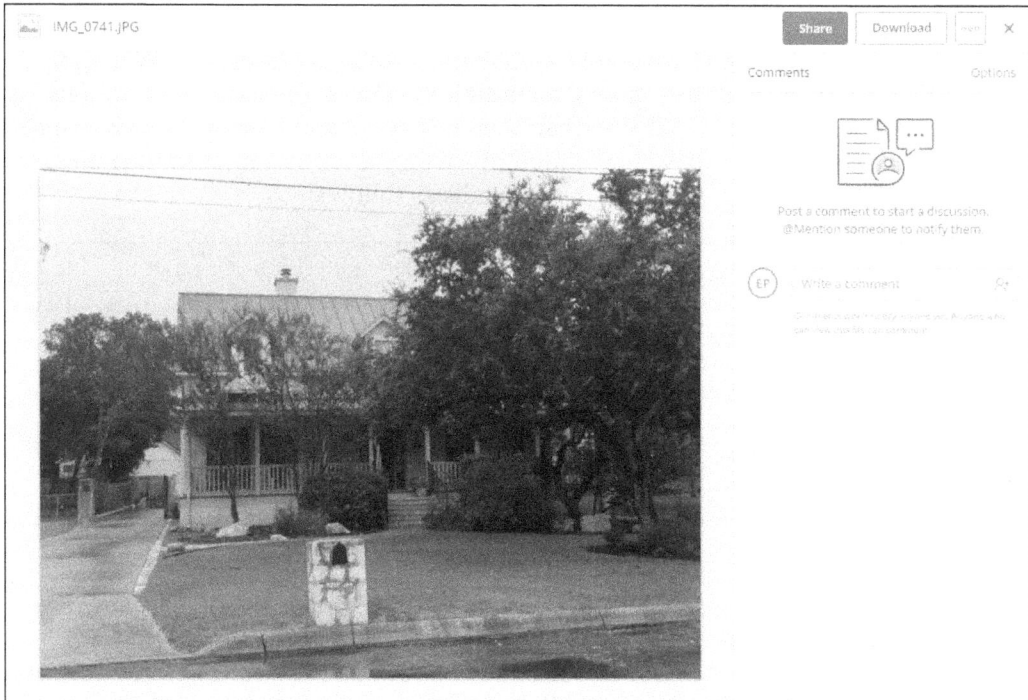

15. Click on the **Save** button on the map viewer toolbar and add the following attributes:

16. Click on **Save Map**.

17. Click on the **Share** button on the map viewer toolbar.

18. Share the map with **Everyone** (public) and then select **Embed in Website**. You can select various options, including the size of the map, map options, and symbols. The HTML code can then be copied and pasted into a website so that the map displays embedded in a web page. For a realtor, the map would supplement other property details as follows:

Summary

Because smartphones are nearly ubiquitous, they provide some unique capabilities for GIS. Photos captured with these devices include metadata that includes geographic coordinate information for each photo. This information can be read using Python and stored in a local feature class with the `ArcPy Data Access` module. Using a reverse geocoding process, it is also possible to determine the nearest address where the photo was taken. Finally, to enable this data on the web for sharing, the Python `dropbox` module can be used to upload these photos to a file-sharing platform and then integrated with the uploaded feature class data using ArcGIS Online.

Overview of Python Libraries for ArcGIS

The initial stage of any programming task requires some research into the most appropriate software libraries to use for the job. A developer must also have a good understanding of the classes that are included in the libraries along with the properties and methods that are available for these objects. Unless you have been working with a particular programming library for some time, this information will not be readily apparent and documentation is not always presented in a manner that makes it easy to understand. The goal of this chapter is to introduce you to the ArcGIS programming libraries that are available to ArcGIS Desktop Python programmers. We'll give you a high-level overview of the core `ArcPy` library along with the `ArcPy` mapping and data access modules as well as the ArcGIS REST API.

Overview of Arcpy

The `ArcPy` website package provides basic functionality that enables the creation of ArcGIS geoprocessing scripts with Python. The core functionality of this package includes many capabilities including the following:

- Execution of `ArcToolbox` geoprocessing tools as dynamic methods
- Adding, listing, removing, and validating data stores
- Describing data
- Getting and setting environment variables
- General utilities
- Graphing

- Working with fields
- Working with data stores
- Administration of geodatabases
- Geometry operations
- Getting and setting parameters
- Licensing and installation
- Listing data
- Log history
- Messaging and error handling
- Progress dialog manipulation
- Publishing
- Working with `rasters` and `NumPy` arrays
- Working with spatial references and transformations
- Tools and toolboxes
- Workspaces

In this section, we'll examine the most commonly used classes and functions in the `ArcPy` site package.

The ArcPy classes

Most of the `ArcPy` classes are somewhat generic, but can be divided into groups, including `FeatureSets` and `RecordSets`, Fields, General, Geometry, Graphing, and Parameters.

FeatureSets and Recordsets

The `FeatureSet` and `RecordSet` objects are lightweight representations of feature classes and tables, respectively. These in-memory objects contain fields as well as data. They also serve as interchange objects with a server. The constructor for both objects accepts a string that references a feature class or table. Both contain a property that returns a JSON representation of the data as well as methods to import and export the data from the object.

Fields

There are four classes related to attribute fields, including `Field`, `FieldInfo`, `FieldMap`, and `FieldMappings`. The `Field` object represents a column in a table or `feature` class. This object can be retrieved using the `ListFields()` and `Describe()` functions which we'll cover later. Read-write properties on this object provide access to the field name, alias name, domain, the editable state, if the field can contain null values, and length. Fields must contain a value, and type.

The `FieldInfo` object provides properties and methods about the fields in a layer or table view. Using this object, you can add and remove fields, control field visibility, set the split rule, set the field name, and others.

There are two objects related to field mapping: `FieldMap` and `FieldMappings`. The `FieldMappings` object serves as a container object for one or more `FieldMap` objects. Each `FieldMap` object defines a field definition and a list of input fields pulled from a table or feature class. `FieldMap` objects are added to the `FieldMappings` object.

The geometry

There is a set of generic geometry objects used to define the geometric definition of points, lines, polygons, and other geometry representations. These objects include `Point`, `Polyline`, `Polygon`, `MultiPoint`, `PointGeometry`, and the generic `Geometry` class. `Point`, `MultiPoint`, `Polyline`, and `Polygon` are self-explanatory, but the `PointGeometry` class requires some further explanation. Although being similar to the `Point` object, it does have some important differences. `Point` objects are used with cursor objects when creating or returning features from a `feature` class. The `PointGeometry` class is used in geometry operations, but not for the creation of `geometry` objects used in cursors.

Graphing

There are only two classes related to graphing in `ArcPy`: `Graph` and `GraphTemplate`. The `GraphTemplate` class uses a .tee file to construct a graph template that can then be used to create graphs from different datasets that have the same basic structure. The .tee file contains everything needed to create the graph except for the data. `ArcToolbox` contains a `MakeGraph` tool that you can use to actually create the graph using the template. Similarly, the `Graph` class also helps create graphs of different types and contains properties to define the graph title, axes, and legend information.

General

There are a number of generic classes that don't really fall into a specific group. The `Extent` class represents a rectangle that defines the geographic boundaries of an objects. This object contains spatial relationship operators that allow you to compare the `Extent` object to other geographies as well as properties that define the object.

The `Array` class is an object that can contain `Point` objects and is used to construct geometry object. The `env` object represents environment variables that can be retrieved or set. Each of the properties on this object represents an environment variable. The `SpatialReference` class has a number of properties that define what map projection options are used to define horizontal coordinates. `ValueTable` is an object that allows the creation of a `multivalue` parameter. The `Result` object simulates the **Result** window in ArcGIS Desktop and provides the ability to maintain information about the execution of tools, including `messages`, `parameters`, and `output`. The **Raster** object can be used to define map algebra expressions. There are some additional classes that fall into the general category, including `ArcSDE`, `SQLExecute`, `Index`, `NetCDFFileProperties`, and `RandomNumberGenerator`.

The ArcPy functions

There is a long list of `ArcPy` functions that provide a wide array of functionality. Perhaps the most important functions are the dynamic methods that enable you to call geoprocessing tools in `ArcToolbox` like you would any other method or function. Beyond these dynamic methods, the functions that are part of the `ArcPy` core library can be grouped by data store, describing data, environment variables, fields, general, geodatabase administration, geometry, getting and setting parameters, licensing and installation, listing data, messaging and error handling, progress dialog, publishing, raster, and tools and toolboxes.

The data store

Data for services can be registered with ArcGIS Server and can include folders or databases. Functions related to data stores include `AddDataStoreItem()`, `ListDataStoreItem()`, `RemoveDataStoreItem()`, and `ValidateDataStoreItem()`. As their names suggest, these functions allow you to add, remove, list, and validate these data stores.

Describing the data

The `Describe()` function can be used to obtain descriptive information about GIS datasets. This function accepts a single parameter that references a geographic dataset. The information returned by this function includes a variable set of properties that describe that data. The set of properties returned by this functions are correlated to the type of data being described.

Environment variables

There are a handful of functions related to managing environment variables. The `ClearEnvironment()` function resets a specific environment variable to it's default value. `GetSystemEnvironment()` gets the value of a specific environment value. `ListEnvironments()` returns a list of environment names. `LoadSettings()` and `SaveSettings()` enable you to either read environment variable settings from an XML file or write the settings to an XML file. Finally, `ResetEnvironments()` resets all the environment variables to the default settings.

Fields

The `AddFieldDelimiters()` function is an important function when creating SQL expressions for attribute queries. The delimiters used around the fields being queried are different depending on whether you're querying `shapefiles`, geodatabase files, `ArcSDE` geodatabases, or personal geodatabases. For example, with geodatabase files, the field being queried needs to be surrounded by quotes whereas braces should surround personal geodatabase fields. The `AddFieldDelimiters()` function handles the guess work of ensuring that you have the proper delimiter. Other functions that are part of this category include `ParseFieldName()` and `ValidateFieldName()`. `ParseFieldName()` parses a fully qualified field name into separate components, including database, owner name, table name, and field name. Finally, the `ValidateFieldName()` functions takes a string that represents a field name and a workspace path and returns a valid field name based on the naming restrictions of the output geodatabase.

General

There are a number of useful general functions, including `RefreshActiveView()`, `RefreshCatalog()`, and `RefreshTOC()`, which force a refresh of these objects in **ArcMap**. In some cases, it's important to refresh these objects when the data has changed in some way. Other commonly used general functions include `ListPrinterNames()`, which returns a list of printers to the computer where the script is running, and `CreateRandomValueGenerator()`.

The `Exists()` function can be used to perform a test, to see if a dataset exists before continuing with a geoprocessing operation. `ValidateTableName()`, accepts a table name as a parameter as well as a workspace path and returns a valid table name for the workspace. Similarly, `ParseTableName()` parses a table name into its components, including database owner and table.

Geodatabase administration

There are a small number of geodatabase administration functions: `AcceptConnections()`, `DisconnectUser()`, and `ListUsers()`. The `AcceptConnections()` function allows a script to enable or disable the ability to connect to a geodatabase. `DisconnectUser()` can be used to disconnect a user, and `ListUsers()` generates a list of the users connected to a geodatabase.

Geometry

There are a small number of `ArcPy` functions related to converting geometry data. The `AsShape()` function converts either an Esri `JSON` or `GeoJSON` object to an `ArcPy` `Geometry` object. The `FromWKB()` and `FromWKT()` functions create `Geometry` objects from well-known binary and well-known text formats.

Getting and setting parameters

There are a number of get and set functions related to parameters. Perhaps, the most well-known function in this category is `GetParameterAsText()`, which is used to retrieve a specified parameter as a text string through the use of an index position from the list of parameters. `GetParameterValue()` returns the default value of a desired parameter. `GetParameterCount()` returns a count of the number of parameters values for a specified tool. `GetParameter()` returns a `Parameter` object for the specified parameter. `GetParameterInfo()` returns a list of `parameter` objects for a given tool. There are a couple set functions including `SetParameter()` and `SetParameterAsText()`. The `SetParameter()` function sets a specified parameter by index using an object and is used when passing objects from a script to a script tool. `SetParameterAsText()` sets a specified parameter property by index using a string value.

Licensing and installation

The functions in this category allow you to work with products and extensions. `ProductInfo()` returns the current product license. `SetProduct()` can be used to set the ArcGIS Desktop license. `CheckProduct()` checks to see whether a license is available. There are three functions related to extensions: `CheckExtension()`, `CheckInExtension()`, and `CheckOutExtension()`. You can get a list of installation types including server, desktop, and engine with the `ListInstallations()` function.

Listing data

There are many list functions that return a Python list containing data of some sort. These functions are most often used as the first step in a multistep process where the first step is simply to generate a list of data that will then be used in a geoprocessing operation. These include `ListDatasets()`, `ListFeatureClasses()`, `ListFields()`, `ListFiles()`, `ListIndexes()`, `ListRasters()`, `ListTables()`, `ListVersions()`, `ListWorkspaces()`, `ListSpatialReferences()`, and `ListTransformations()`.

Messaging and error handling

All tools produce messages as they are executing. Most messages are informational in nature, but warnings and errors can occur as well. Messages are divided into severity levels that indicate whether a message is informational only, a warning, or an error. In addition to the messages that are generated by a tool, you can also add your own messages to the stack. There are a number of `get` functions such as: `GetMaxSeverity()`, `GetMessage()`, `GetMessageCount()`, `GetMessages()`, `GetReturnCode()`, `GetSeverity()`, and `GetSeverityLevel()`. Most of these functions either retrieve messages or the severity level associated with the message.

There are also three `add` functions such as: `AddMessage()`, `AddWarning()`, and `AddError()`. These three functions correspond to the different severity levels and enable you to add your messages to the stack also being generated by the tool itself.

The progress dialog

The progress dialog can be controlled through a set of functions, including `SetProgressor()`, `SetProgressorLabel()`, `SetProgressorPosition()`, and `ResetProgressor()`. The `SetProgressor()` function creates a `progressor` object. Using this object, you can then pass information to the progress dialog box. In addition, you can control the appearance of the progress dialog by choosing either the default `progressor` or the step `progressor`. The label and position of the status bar can be set through the `SetProgressorLabel()` and `SetProgressorPosition()` functions. Finally, there is a `ResetProgressor()` function that resets the `progressor` back to its initial state.

Publishing

There are three `ArcPy` functions related to creating **Service Definition Draft (SDDraft)** files for various types of ArcGIS Server services. The Service Definition Draft file is a file type used as an interchange file between ArcGIS Desktop and ArcGIS Server. These include `CreateGeocodeSDDraft()`, `CreateGPSDraft()`, and `CreateImageSDDraft()`. These functions correspond to the types of services being created.

Raster

There are two functions related to converting `rasters` and `NumPy` arrays. The `NumPyArrayToRaster()` functions converts a `NumPy` array to a raster, whereas `RasterToNumPyArray()` convert a raster to a `NumPyArray`.

Tools and toolboxes

Toolboxes can be added and removed using various functions, such as `AddToolbox()`, `ImportToolbox()`, and `RemoveToolbox()`. Any custom or third-party toolboxes must be imported before you can use them in your scripts. Server tools can also be imported. The `AddToolbox()` and `ImportToolbox()` functions are equivalent, so you can use either for this purpose. The `ImportToolbox()` functions accepts an input that references the custom toolbox to be imported along with an optional module name. If you need to remove a toolbox, you can use the `RemoveToolbox()` function. A related function that is part of this category is the `IsSynchronous()` function which is used to determine if a tool is running synchronously or asynchronously. All non-server tools will be synchronous. This simply means that results are automatically returned. Any asynchronous functions will be related to ArcGIS Server tools where the data may not be returned immediately. In the case of an asynchronous, tool you must set your script up to wait for the results to be returned before continuing. This can be accomplished with the Python `time.sleep()` method.

Overview of the ArcPy mapping module

The `ArcPy` mapping module, part of the `ArcPy` site package, provides some really exciting features for map automation, including the ability to manage map document and layer files as well as the data within these files. Support is also provided to automate map export and printing as well as the creation of map books and publication of map documents to ArcGIS Server map services.

The capabilities of the module include the following:

- Managing map document and layer files
- Managing the data within map document and layer files
- Changing layer symbology and properties
- Inserting layers into a data frame or group layer
- Moving layers in a data frame or group layer
- Zooming to selected features
- Working with time-enabled layers in a data frame
- Creating reports
- Changing the map extent
- Finding and fixing broken data links
- Printing maps
- Exporting maps to PDF files
- Exporting maps to image files
- Updating the layout view
- Building a map book with Data Driven Pages
- Publishing a map document to an ArcGIS Server service

In this section, I'll present an overview of the available functionality provided by the classes and functions in the `ArcPy` mapping module.

ArcPy mapping classes

The classes available in the `ArcPy` mapping module can loosely be grouped into several categories, including map documents and associated datasets, Data Driven Pages, managing time-related layers, element classes related to a layout view, PDF document creation and editing, and symbology.

Mapping documents and associating dataset classes

There are four classes in the `ArcPy` mapping module related to map documents and their associated data: `MapDocument`, `DataFrame`, `Layer`, and `TableView`. These four classes are probably the most essential and often used objects in the module.

The MapDocument class

The `MapDocument` class is probably the most essential object in the `ArcPy` mapping module and is required to some degree in most of the scripts you write using this module. A reference to this object is required for most scripts, so it's usually one of the first lines of code in a geoprocessing script. The constructor for this object accepts either a string that contains the keyword CURRENT or a path to a map document file.

There are a number of properties on this object that expose a variety of functionality, including getting the active data frame or active view, to the author of the document, the last date when the document was exported, printed, or saved, a description, path to the file, title, and a few others. You can also determine if the map document is Data Driven Pages enabled, and, if so, a `DataDrivenPages` object can be returned with the `dataDrivenPages` property.

In addition, there are a number of methods on this object that allow you to perform various operations, such as fixing broken data links, saving the map document, and working with thumbnails. Two methods, `findAndReplaceWorkspacePaths()` and `replaceWorkspaces()`, can be used to fix broken data links. The `findAndReplaceWorkspacePaths()` method replaces an old workspace path with a new workspace path for all layers and tables in the map document. The `replaceWorkspaces()` method is used to change the workspace type of all layers and tables in the map document. The `save()` and `saveACopy()` methods are used to save the map document, and `makeThumbnail()` and `deleteThumbnail()` are used when working with thumbnail images of a map document.

DataFrame

A `DataFrame` in a map document serves as a working area and container for datasets, including layers and standalone tables. Most of the properties on the `DataFrame` object are `read/write` and include the ability to work with the geographic extent, change the size and position of the element in layout view, get or set the name, scale, spatial reference, description, and a few others. If you have time-enabled layers in the data frame, the read-only property `time` provides access to the `DataFrameTime` object that we'll discuss next.

There are two methods on the `DataFrame` class: `panToExtent()` and `zoomToSelectedFeatures()`. The `panToExtent()` object accepts an `Extent` object as a parameter and pans the map to the geographic extent provided. The `zoomToSelectedFeatures()` method zooms the map to the extent of the selected set of features.

The Layer class

The `Layer` class provides a reference to layers in a map document or `layer` file, and provides many properties and methods to work with the object. I'll examine the most commonly used in this section.

There are a number of ways that you can create an instance of the `Layer` object including a `Layer(lyr_file_path)` constructor that creates instance of this object by passing a path to the `.lyr` file as a parameter to the constructor.

Some of the properties on this object provide access to common properties found in the `Layer` **Properties** dialog in **ArcMap** that is displayed by right-clicking on a layer and selecting **Properties**. However, many of these properties are not exposed to scripting through the `Layer` object. The properties that are exposed allow you to get and set the definition query, description, label classes, minimum and maximum scale, name, turning labels on and off, transparency, and the visibility of the layer.

There are a number of `is` properties that allow you to test the layer type. These include `isFeatureLayer`, `isGroupLayer`, `isNetworkAnalystLayer`, `isRasterLayer`, and `isServiceLayer`.

A set of properties dealing with the source dataset of the layer is also present, such as `datasetName`, `dataSource`, and `workspacePath`. `datasetName` returns the name of the layer's dataset the way it appears in the workspace, not the table of contents. The `dataSource` property returns the complete path for the layer's data source including the workspace path and dataset name combined. `workspacePath` returns a path to the workspace for the layer or `ArcSDE` connection file.

The `Layer` class also includes a number of methods that allow you to work with the extent of the layer or selected features from a layer, fix a layer that becomes broken due to a new workspace path or type, and save the layer file or save it to a new copy.

The TableView object

The `TableView` object allows you to manage standalone tables in a map document file. The constructor for this object accepts a parameter that includes a full path to the workspace where the table exists and should also include the name of the table.

There are only a handful of properties on this object, most of which relate to dataset path and naming. The `workspacePath` property returns a path to the table's workspace or connection file. The `datasetName` returns the name of the table in the workspace, and `dataSource` returns the table's source path. One helpful property is `definitionQuery`. This property provides the ability to limit the displayed records to only records that match a specific query.

There are also a couple of methods on this object that allow you to fix a broken data source, including `findAndReplaceWorkspacePath()` and `replaceDataSource()`. These are the same methods that are available on the `Layer` object and also perform the same functionality.

Data Driven Pages classes

The Data Driven Pages functionality is handled through the `DataDrivenPages` class. The Data Driven Pages functionality in ArcGIS enables you to create a series of maps for a geographic area for the purpose of creating a map book. **ArcMap** includes a Data Driven Pages toolbar that you can use to create this series of maps without having to write any code. You could also elect to automate the entire process through a Python script without using the toolbar. However, the creation of a map book is best accomplished through a combination of the Data Driven Pages toolbar and scripting with `ArcPy` mapping using the `DataDrivenPages` class. The toolbar could be used to author the Data Driven Pages functionality in the map document, and the scripting would handle any custom requirements, such as changing titles for each map in the series as well as exporting the maps to pdf files. The methods and properties on the `DataDrivenPages` class enable you to work with the individual pages in a map document that already has Data Driven Pages enabled.

Classes related to managing time layers

There are two classes in the `ArcPy` mapping module related to time-enabled layers: `DataFrameTime` and `LayerTime`.

The DataFrameTime class

Time-enabled layers in a data frame can be controlled through the `DataFrameTime` object. This object can be used in scenarios where map documents have already been published with time-aware layers, including the use of the **Time Slider Options** dialog to set various properties. It can also be used in situations where map documents don't already have time-enabled layers, but the intent is to add them through a script.

Properties of the `DataFrameTime` object allow you to get and set the start time, current time, end time, time window, and time window units. It also includes a read-only property to obtain the time step interval. The only method on the `DataFrameTime` object is `resetTimeExtent()`, which resets the time extent of the data frame.

The LayerTime class

This class provides the ability to manage time-enabled layers. It provides information about how time is stored and configured. The properties on the class are read-only and allow you to retrieve information about the start and end times for the layer, the fields being used to store the start and end times, the time format, time zone, whether the time information is observing daylight saving time, and the time step interval. There are no methods associated with the `LayerTime` class.

Element classes associated with the layout view

Element classes in `ArcPy` mapping represent everything that you add to the layout view in **ArcMap**. Using these element classes, you can make changes to the layout view through your scripts including changing the size and position of elements and altering the data associated with an element. In addition to the classes discussed here, the previously discussed `DataFrame` element can also be included with these.

The LegendElement class

The `LegendElement` class provides properties for the positioning of the legend on the page layout and modifying of the legend title, and also provides access to the legend items and the parent data frame. A `LegendElement` class can be associated with only a single data frame. The methods available on this class enable you to update or remove legend items, adjust the column count, and obtain a list of the legend items.

The GraphicElement class

The `GraphicElement` class is a generic object for various graphics that can be added to the page layout, including tables, graphs, Neatlines, markers, lines, and area shapes. This object provides a limited set of properties that allow you to reposition and resize the elements on the layout as well as set the name. In addition, there are two methods on this class: `clone()` and `delete()`. The `clone()` method creates a copy of the element, whereas `delete()` is used to remove the element from the layout.

MapsurroundElement

The `MapsurroundElement` can refer to north arrows, scale bars, and scale text and like `LegendElement`, is associated with a single data frame. Properties on this object enable repositioning and resizing on the page.

PictureElement

`PictureElement` represents a raster or image on the page layout. The most useful property on this object allows you to get and set the data source that can be extremely helpful when you need to change a picture such as a logo in multiple map documents. For example, you could write a script that iterates through all your map document files and replaces the current logo with a new logo. You can also reposition the object.

TextElement

`TextElement` represents text on a page layout, including inserted text, callouts, rectangle text and titles, but does not include legend titles or text that is part of a table or chart. Properties enable modifying the text string, which can be extremely useful in situations where you need to make the same text string change in multiple places in the page layout or over multiple map documents, and of course repositioning of the object is also available.

PDF document creation and editing

Although there is only one class related to creating pdf documents in the `ArcPy` mapping module, we'll explore a second way that you can create pdf documents when we discuss the `ExportToPDF()` function in the next section.

PDFDocument

You can manipulate existing PDF documents or create new PDF documents using the PDFDocument class. You can merge pages, set document open behavior, add file attachments, and create or change document security settings. The PDFDocumentOpen() function is used to open an existing PDF file for manipulation. The PDFDocumentCreate() function creates a new PDF document. These functions are often used in the creation of map books.

You'll need to use PDFDocumentCreate() to create a new PDF document by providing a path and filename for the document. The PDF is not actually created on disk until you insert or append pages and then call PDFDocument.saveAndClose(). The appendPages() and insertPages() functions are used to insert and append pages.

PDFDocumentOpen() accepts a parameter that specifies the path to a PDF file and returns an instance of the PDFDocument class. Once you can make modifications to PDF file properties, you can add or insert files and can attach documents. Make sure that you call PDFDocument.saveAndClose() after all operations to save the changes to disk.

A number of properties can be set on a PDF document through the PDFDocument object, including getting a page count, attaching files, updating the title, author, subject, keywords, open behavior, and the layout. You can also update the document security by calling PDFDocument.updateDocSecurity() to set a password, encryption, and security restrictions.

Symbology

There are a number of classes in the ArcPy mapping module that provides a limited ability to make changes to the symbology of an application including GraduatedColorsSymbology, GraduatedSymbolsSymbology, RasterClassifiedSymbology, and UniqueValuesSymbology classes.

GraduatedColorsSymbology

This class provides a limited ability to change the appearance of a layer's graduated color symbology. Layer symbology can be applied to layers in a map document or layer file. Properties on this object enable you to get and set the class break values, labels, number of classes, field used to create the symbology, description, and normalization. The only method on this object is reclassify(), which resets the layer's symbology. For access to a more complete set of symbology properties and settings, you would need to make the changes in **ArcMap**, save the changes to a layer file, and then use the UpdateLayer() function in ArcPy mapping.

GraduatedSymbolsSymbology

The `GraduatedSymbolsSymbology` class is similar to `GraduatedColorSymbology` but deals with graduated symbols instead of graduated colors. Like `GraduatedColorSymbology`, this object also provides access to a limited set of properties that you can use to change how graduated symbols are symbolized. The properties and methods are the same as described on the `GraduatedColorsSymbology` class.

RasterClassifiedSymbology

`RasterClassifiedSymbology` allows limited access to properties that can be used to change the symbology of a raster layer. This object is similar to the `GraduatedColorsSymbology` and `GraduatedSymbolsSymbology` objects we discussed earlier, in which it provides access to only a limited set of properties such as the class break values, labels, descriptions, number of classes, the value field, and others.

UniqueValuesSymbology

This class provides access to properties that can be used to control a layer's unique value symbology. This class is similar to the other symbology objects we have already discussed, in which it exposes a limited number of properties for controlling things such as the field used for the values, labels, descriptions, and others.

Arcpy mapping functions

The `ArcPy` mapping functions can be divided into sections that control the export and printing of maps and managing map documents and layers.

Exporting and printing maps

There are a handful of functions related to exporting maps to various image file formats. These include `ExportToAI()`, `ExportToBMP()`, `ExportToEMF()`, `ExportToEPS()`, `ExportToGIF()`, `ExportToJPEG()`, `ExportToPNG()`, `ExportToSVG()`, and `ExportToTIFF()`. Each of the functions accepts somewhat different parameters, but all will by default export the layout view in **ArcMap** to an image file. Instead of exporting the layout view, you can also elect to export a specific data frame by passing a reference as a parameter to the function.

Yet another export function is the `ExportToPDF()` function that can be used to export either the layout view or a data frame to a PDF file. As we'll discuss later in this section, there are also two additional functions that can be used to work with pdf files.

There are two functions related to printing the layout view or a specific data frame. The `ListPrinterNames()` function gathers a list of the available printers to the computer where the script is running. Using the list returned, you can then pass a specific printer to the `PrintMap()` function to print either the layout view or a specific data frame to a printer. If you don't pass a printer name to the `PrintMap()` function, it will attempt to find a printer saved with the map document or the default system printer if a printer hasn't been saved with the map document.

There is one additional function in this category that is used to export reports. This is the `ExportReport()` function, which exports a formatted, tabular report using data in the map document file. It uses a report template file that has been previously created.

Managing map documents and layers

There is a wide range of functions to manage map documents and layers. Various function types, including managing layers and tables, working with pdf files, working with ArcGIS Server services, generating lists, and working with ArcGIS Server are available.

Creating lists

A handful of list functions can be used to generate lists of bookmarks, data frames, broken data sources, layers, layout elements, map services, style items, and table views. These functions each return a Python list of data that can also be filtered in various ways. For example, the `ListLayers()` function accepts two optional parameters including a wildcard and data frame that can be used to restrict the list of layers that is returned. By default, all layers in the map document or layer file are returned, but it is often necessary to limit the returned objects. All the list functions contain similar optional parameters that can be used to limit the returned list.

Managing layers and tables

Layers and tables can be added, removed, and updated. You can add layers to a map document or group layer using either `AddLayer()` or `AddLayerToGroup()`. In addition, the `InsertLayer()` function can be used to add a layer to a map document or group layer with more precision. It uses a reference layer to precisely define the location of the layer to be added. The `MoveLayer()` function, which is used to move a layer to a new location within a specific data frame , also uses a reference layer. Standalone tables can be added or removed from a map document using `AddTableView()` or `RemoveTableView()`. The symbology and properties of a layer can be updated through the `UpdateLayer()` method. There is also an `UpdateLayerTime()` function that can be used to update the properties of a time-enabled layer.

Working with pdf Files

In the previous section, the ExportToPDF() function was introduced. There are two additional functions related to working with pdf files including PDFDocumentCreate() and PDFDocumentOpen(). Both are commonly used in the creation of map books. PDFDocumentCreate(), as its name suggests, is used to create new pdf files, whereas PDFDocumentOpen() can be used to open an existing pdf file. Both return an instance of the PDFDocument class.

Working with ArcGIS Server services

Map documents can be published to ArcGIS Server as services. The ArcPy mapping module provides several functions related to this conversion and publication process. Before a map document can be published as a service, it must go through a conversion process. The first step in the process is to create a Service Definition Draft file (SDDraft). This can be accomplished with the CreateMapSDDraft() function. This function also returns a Python dictionary containing errors, warnings, and information messages. Any errors must be resolved before publication. This can also be accomplished with the AnalyzeForSD() function. After any errors have been resolved, there are two geoprocessing tools that you can use to publish the file as a service.

Other functions related to working with ArcGIS Server include ConvertWebMapToMapDocument(), which can be used to convert a web map in JSON format to a map document. In addition, the CreateGISServerConnectionFile() function creates a connection file for accessing ArcGIS Server.

There are also some deprecated functions that you shouldn't use but that are still technically part of the module including AnalyzeForMSD(), ConvertToMSD(), DeleteMapService(), ListMapServices(), and PublishMSDToServer(). These functions have either been replaced by new functions or tools, or are now provided through the ArcGIS REST API.

Overview of the Arcpy data access module

The ArcPy data access module, known as arcpy.da, provides capabilities for working with tables and feature classes. Through the use of various cursor objects, you can select, insert, update, and delete records from tables and feature classes. The data is held as an in-memory copy of the data. This module also supports edit sessions, NumPy array conversions, and support for versions, domains, and subtypes.

ArcPy data access classes

The primary classes in this module deal with the various types of cursors that can be created. Cursor objects are the in-memory copy of data pulled from a table or feature class. The data access module includes three types of cursor objects including SearchCursor, InsertCursor, and UpdateCursor. Each has a corresponding constructor function that is used to create the object. There are some additional classes that support edit sessions, domains, versions, and replicas.

The SearchCursor class is used to create read-only access to tables and feature classes. The constructor function for this class provides parameters for defining the feature class or table associated with the object, a list of fields to return, and an optional WHERE clause to limit the records returned in this object.

The InsertCursor class is used for situations where new records need to be added to a table or feature class. The constructor function for this class provides parameters for defining the feature class or table associated with the object along with a list of field names to be returned. The only method on this class is insertRow(). This method inserts a new row into a table or feature class.

The UpdateCursor class is used to edit or delete records in a table or feature class. The constructor function for this class provides parameters to define the feature class or table associated with the object along with a list of field names to be returned and an optional where clause that can be used to limit the records returned. There are two primary methods on this class: deleteRow() and updateRow().

All cursors support the concept of geometry tokens that allow you to return a portion of the geometry for a feature class rather than the default of returning all geometry. In situations where you have highly detailed polygon or polyline datasets, this can increase the performance of cursors by limiting the amount of data returned when the object is created.

Another important object in the data access module is the Editor class. Through this class, you can enable edit sessions against tables and feature classes. The functionality provided through this class is the same as that provided through the Edit toolbar in ArcMap.

Several additional classes including Domain, Version, and Replica provide a limited set of read-only properties for these objects. For example, the Domain class has read-only properties that will return the coded values or range for a domain, name, split policy, merge policy, domain type, and description. There are read-only, so you can't make any changes to the Domain, Version, or Replica objects.

Arcpy data access functions

The data access functions can be grouped into categories that provide lists of data, `NumPy` array conversion capabilities, and a utility function to generate data names in a `Catalog` tree.

List functions

There are a handful of list functions that return a list of data. These include `ListDomains()`, `ListFieldConflictFilters()`, `ListReplicas()`, `ListSubtypes()`, and `ListVersions()`. Most are self-explanatory, but the `ListFieldConflictFilters()` requires some explanation. This function lists the fields in a versioned feature class or table that have field conflict filters applied.

NumPy Array conversion functions

Tables and feature classes can be converted to `NumPy` arrays through the `TableToNumPyArray()` and `FeatureClassToNumPyArray()` functions. Existing `NumPy` arrays can also be converted to tables and feature classes through the `NumPyArrayToTable()` and `NumPyArrayToFeatureClass()` functions. There is also an `ExtendTable()` function to join a `NumPy` array to another table based on a common attribute field.

The `Walk()` function generates data names in a `Catalog` tree and can navigate top-down or bottom-up. Each folder or workspace in the tree contains a Python tuple object consisting of the directory path, directory names, and filenames. This function is similar to the Python `os.walk()` function, but with the added capability of being able to investigate the contents of a geodatabase structure. The `os.walk()` function doesn't have this ability.

An overview of the ArcGIS REST API

The ArcGIS REST API provides access to ArcGIS Server and ArcGIS Online services to any language that can make requests and handles the returned responses. Python is one such language along with many others. To make use of this API, you must understand what requests can be made, how to structure those requests, and how to process the responses.

The operations provided through the API include the following:

- Consume ArcGIS Server and ArcGIS Online services
- Publish and manage services
- Create and share ArcGIS Online or portal services
- ArcGIS Server and ArcGIS Online administration

The REST API can be categorized into sections including using Esri-provided services, using your own services and services published by others, managing services, and administering services and portals.

Basics of using the ArcGIS REST API

All resources and operations exposed by the REST API are accessible through a hierarchy of endpoints or Uniform Resource Locators (URLs) for each GIS service published with the ArcGIS Server. When using the ArcGIS services portion of the REST API, you typically start from a well-known endpoint, which represents the server catalog.

You need to understand some basic concepts of the ArcGIS REST API before putting it to use through Python. Specifically, you need to know how to construct a URL and how to interpret the response that is returned. All resources and operations in the ArcGIS REST API are exposed through a hierarchy of endpoints. For now, let's examine the specific steps you need to understand to submit requests to the API through Python. The services directory can be used to generate a URL that can be used in your requests.

The first step is to determine the well-known endpoint. This represents a server catalog that is a set of operations that ArcGIS Server can perform along with specific services. The default endpoint for ArcGIS Server takes the form: `http://<server>/arcgis/rest/serviceshttp://<server>/arcgis/rest/services`.

The next step is to go to the /rest/services endpoint to see the content of the ArcGIS Server instance. For example, open a browser and navigate to `http://sampleserver1.arcgisonline.com/arcgis/rest/services` and you will be presented with a list of folders displayed as links, as seen in the screenshot:

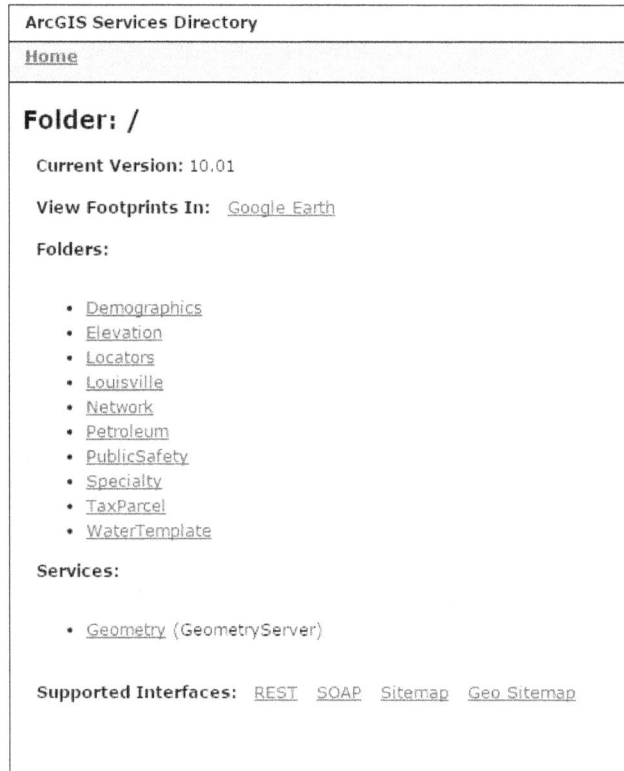

An ArcGIS Server instance does not have to have folders, but it is a good way of grouping services. You can click on a specific folder to see the services that are contained within.

Each service will have a name such as ESRI_Census_USA along with a type such as MapServer. The service type is listed in parentheses to the right hand side of the service name. In your browser with the sampleserver1 instance up, click through the various folders and services and note how the URL changes.

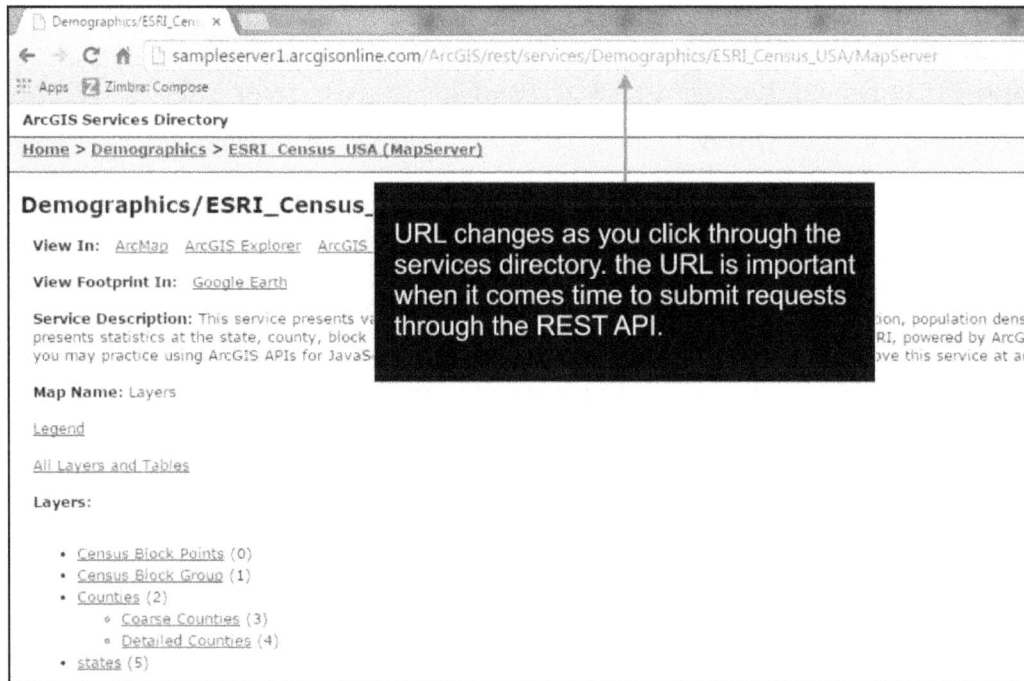

As you click on through the various links for a services directory, note how the URL in the address bar changes. This URL is very important because it provides you with the content that will be submitted through a Python request.

Now it's time to understand the documentation for the ArcGIS REST API. However, don't spend a lot of time on the documentation right now because we'll be going through many of the capabilities of the API as we move through the book.

Now properly construct the URL for the request. This is a very important step. The syntax for the request includes the path to the resource along with an operation name followed by a list of parameters. The operation name is what operation will be performed against the resource. For example, you might want to export a map to an image file. The question mark begins the list of parameters. Each parameter is then provided as a set of key or value pairs separated by an ampersand. All of this information is combined into a single URL string. A syntax example is provided as follows:

```
http://<resource-url>/<operation>?<parameter1=value1>&<parameter2=value2>.
```

As we'll see later in the book, you can use the Python requests module to simplify this. The requests module allows you to define the list of parameters as a Python dictionary and then it handles the creation of the URL query string including URL encoding.

The response that is returned can be in various formats including `.html`, `.json`, `.amf`, an image, and many others. To define how the response should be structured, you'll need to use the `f` parameter. JSON is a very popular output format and can easily be handled in your Python code.

The Services Directory contains dialog boxes that you can use to generate parameter values. You can find links to these dialog boxes at the bottom of the services page. Click on one of the links to see the dialog box. This is illustrated in the following screenshot:

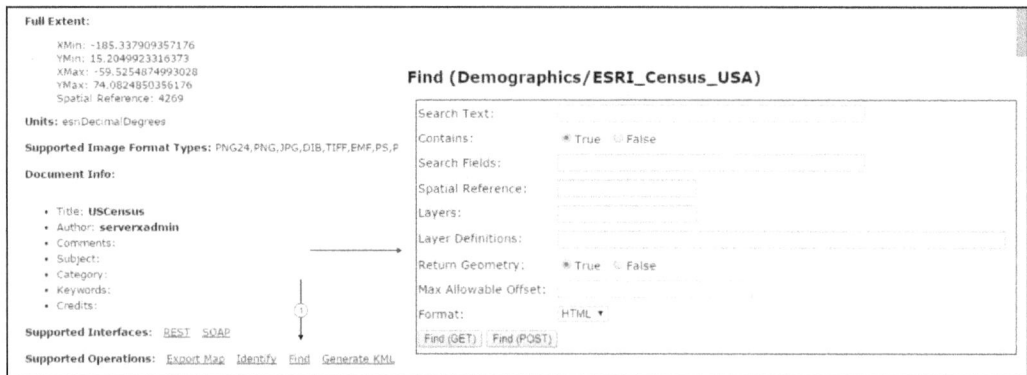

This is the preferred and most common way of making a URL request to ArcGIS Server. In this case the entire request is encoded in the URL. However, it does have a limitation of 1024 characters. Therefore, if you have a request that will exceed this number of characters you'll need to use the `Post` method.

Now that you have at least a basic idea of how to construct a REST API query, let's discuss the capabilities provided by various sections of the API.

Esri services

The REST API provides access to Esri provided services including those provided by ArcGIS Online. Service types include ready-made maps and `basemaps`, geocoding, directions and routing, demographic and lifestyle attributes, and spatial analysis. Some of these services do require credits. When using services that require the use of pre-purchased credits you'll need to pass a token as part of the request for information from a service.

Mapping services include `basemaps` of different varieties and sources including street maps, topographic maps, and hybrid maps. Esri also provides a World Geocoding service that can be used for address matching and reverse geocoding. A number of network analysis services, including routing, closest facility, service area, and others, can be used to accomplish network tasks. The `GeoEnrichment` services provides access to demographic and lifestyle attributes. Other services include spatial analysis and elevation analysis.

Your own services

Services that you have published as part of your own ArcGIS Server instance can be accessed through the REST API, as can services that others have provided and made available. There are many capabilities exposed by the REST API, so we'll just cover them at a high level for now. Generally, we can divide the capabilities into service-related functionality and functions that are more utilitarian in nature.

Service-related functionality

The REST API enables you to work with features, maps, geocode, geodata, geometry, geoprocessing, globe, image, network, schematic, and stream services. A wide array of operations is possible with each of these services. We'll discuss some of the capabilities provided.

For map and feature services, you can add attachments, export maps and tiles, retrieve features, find features, generate KML, render symbology, define HTML popups, identify features, retrieve a legend, perform queries, and more. You will find an example of a URL string used to perform a query against a layer as follows:

```
http://sampleserver1.arcgisonline.com/ArcGIS/rest/services/
Specialty/ESRI_StateCityHighway_USA/MapServer/1/query?where=STATE_
NAME='Florida'&f=json.
```

Feature services have the added capability of being able to perform edits, including adding and removing features, updating features, and deleting features. Here, you will find an example of a URL string used to delete a feature in a feature service:

```
http://services.myserver.com/ERmEceOGq5cHrItq/ArcGIS/rest/services/
SanFrancisco/311Incidents/FeatureServer/0/deleteFeatures.
```

```
http://services.myserver.com/ERmEceOGq5cHrItq/ArcGIS/rest/services/
SanFrancisco/311Incidents/FeatureServer/0/deleteFeatures.
```

Geocoding services provide the ability to geocode and reverse geocode addresses. Geocoding functionality provides the ability to map a single address or batch geocode a set of addresses. Reverse geocoding accepts a point and returns a set of address candidates. There is also a suggest operation that will provide a list of suggested addresses based on typed input from the user.

A geometry service is included with every ArcGIS Server instance and provides operations for many geometric operations including buffering, calculation of areas and lengths, generalization, intersection, projection, union, and many others. These operations work with individual geometry objects typically defined in a JSON format. An example of using the buffer operation is provided here:

```
http://sampleserver6.arcgisonline.com/ArcGIS/rest/services/Utilities/
Geometry/GeometryServer/buffer?geometries=-117,34&inSR=4326&outSR=432
6&bufferSR=3857&distances=1000.
```

Geoprocessing services represent geoprocessing tasks that have been created in ArcGIS Server. Operations provided through the API related to geoprocessing services include the ability to execute a task, cancel a job, retrieve the result of the task, and others.

Globe services published with ArcGIS Server provide information about the service including the service description as well as the layers published with the service. This includes individual layers as well as tiles.

Raster data can be accessed through an image service. This can include a single raster or multiple raster served as a single image through mosaicking. An image service supports accessing the mosaicked image, its catalog, and also the individual rasters in the catalog. Operations provided include export image, query, identify, download, measure, computer histograms, add, update, delete, upload, get samples, computer class statistics, and compute tie points.

Network service operations include solving closest facility tasks, routes, and a service area problem. You can also access network service information including the service description and the network layers associated with the service.

Schematic services support working with diagrams. Using this service, you can create, edit, delete, and save diagrams. Additional operations include loading, locking, querying, exporting, and updating diagrams, among others.

Stream services enable real-time applications where the datasets are frequently changing. This does require the ArcGIS `GeoEvent` extension for ArcGIS Server that must be licensed and installed. The stream service resource provides basic information about the service, including event attribute fields, geometry, type, and `WebSocket` resources. Operations include broadcast and subscribe. The broadcast operation serves as an endpoint for a stream service, and the subscribe operation serves as a connection point to a stream service.

Utility functions

The REST API includes a small number of utility functions to manage ArcGIS Server. The **Catalog** resource is the root note of an ArcGIS Server instance and can be used to retrieve the folders and services published. `ServerInfo` is a resource that provides information about the server including version information, whether the server is using token-based authentication and the token services URL. The `generateToken` resource generates an access token to access services that are token secured. The info resource provides information, metadata, and a thumbnail about services. Other operations include the export web map task, a refresh service, and a set of upload operations to upload data.

Managing your organization

Using the REST API, you can manage your organization's ArcGIS Online account as well as the Portal for ArcGIS. Operations enable you to work with users, groups, and content. User operations include basic user information gathering, adding and removing users, sending a user invitation, searching for users, updating user information, getting and setting user tags, and enabling and disabling login access. Group operations including creating and deleting a group, joining a group, reassigning a group, updating a group, adding and removing users from a group, leaving a group, obtaining group information, and more. There are many content-related operations including creating services and folders, adding and deleting items, sharing and unsharing items, analysis of files before publication, the generation of output files, and much more.

Administering your server

You can programmatically administer your ArcGIS Server instance or Portal using the REST API. Using operations provided by the API, you can work with the site, clusters, services, security, system, data, uploads, logs, KML, info, and reports.

Site operations allow you to create, join, export, import, and delete a site. In addition, you can generate tokens, register, unregister, rename machines, work with SSL, start and stop machines, and edit machines.

Cluster operations include starting and stopping a cluster, editing the protocol for a cluster, deleting a cluster, retrieving the services in a cluster, and adding and removing machines from a cluster.

There are many service operations including starting, stopping, editing, and deleting services, retrieving service statistics and service types, and adding and cleaning permissions. The items associated with a service also have operations including editing item information, uploading item information, deleting item information, and working with the service manifest. You can also federate and un-federate a service.

Security operations associated with the API including working with users, roles, security configuration, tokens, and working with the primary site administrator. You can add and remove users, update users, get a list of users, assign and remove roles, and get privileges. Role operations include adding, removing, and updating roles, searching for roles, getting roles for specific users, getting a list of users within a role, adding and removing users in a role, and assigning privileges to a role.

System operations allow you to update server properties, register, edit, clean, and unregister directories, edit configuration stores, work with web adaptors, retrieve job information, clear the cache, and edit the services directory.

Data operations include registering and unregistering a data item, finding data items, validating data items, starting, stopping, removing, and validating a data store, and updating the datastore configuration.

Upload operations including uploading an item, registering an item, and working with individual items. For individual items you can upload a part, commit an item, delete an item, or retrieve item parts.

Log operations including editing log settings, querying logs, counting error reports, and cleaning logs.

There is a single operation related to KML files. This is the Create KMZ operation which will create a KMZ file on the server from an input KML file.

Finally, there is a set of operations related to usage reports. These include editing usage report settings, creating a usage report, editing or deleting a usage report, and querying report data.

Administering Portal

The REST API includes operations that can be performed programmatically and that can't be performed using the Portal for ArcGIS website. Operations for Portal for ArcGIS include system and security operations.

System operations for Portal for ArcGIS include creating a site, working with licenses, working with web adaptors, directory operations, database operations, and system properties. License operations include updating the license manager, releasing a license, and working with entitlements including getting entitlements, importing entitlements, and removing entitlements. Web adaptor operations include unregistering a web adaptor and updating a web adaptor's configuration.

Security operations include working with users and groups, updating token configuration, setting up OAuth, configuration operations, and working with SSL certifications. User operations include creating users, searching for users, and refreshing user membership. Group operations include searching groups, refreshing group membership, getting users within a group, and getting a list of groups for a particular user. You can also update the token configuration. OAuth operations provided by the REST API include changing the application id, getting the application information, and updating the app information. Configuration operations include updating the security configuration, updating the identity store, and testing the identity store. Using SSL operations, you can update the web server certificate, generate a certificate, import an existing certificate, and export or delete a certificate.

Administering ArcGIS Online hosted services

The administration of ArcGIS Online hosted services using the REST API falls into two administrative categories: map services and feature services. Map services can be administered through operations including editing a service, checking the status of a service, refreshing a service, updating tiles, and getting tile creation information. For feature services, you can check the status of the service, refresh a service, add, update, or delete the definition, and work with the individual feature layers in the service.

Conclusion

ArcGIS includes a number of programming libraries that can be used to automate your geoprocessing scripts or develop functional applications. In addition, you can also integrate other non-GIS libraries into your projects to support ancillary tasks. While the primary focus of most ArcGIS Desktop development efforts with Python in the past has been centered on the `ArcPy` site package and its supporting mapping and data access modules, an increasing amount of functionality is now being delivered through the ArcGIS REST API, which can be called from Python. In this book, we'll use `ArcPy`, its supporting modules, the ArcGIS REST API, and some supporting libraries to build domain-specific applications using Python.

Index

GraduatedSymbolsSymbology 338
RasterClassifiedSymbology 338
UniqueValuesSymbology 338

T

time-enabled data frame
creating 36-42
time-enabled data layer
creating 36-42
time layers
DataFrameTime class 335
LayerTime class 335
tweepy
Tweet geographic coordinates, extracting
with 255-264
Tweet geographic coordinates
extracting, with tweepy 255-263
Twitter account
URL 255

U

user interface
creating, with wxPython 155-175

V

visualization product
creating 143-150
VisualizeMigration tool
coding 42-46

W

Web Map
creating 313-320
Windows Task Scheduler application
design, defining 254
wxPython
about 153, 154
URL 157
user interface, creating with 155-175

[PACKT] **open source** ✺
PUBLISHING   community experience distilled

Thank you for buying
ArcGIS Blueprints

About Packt Publishing

Packt, pronounced 'packed', published its first book, *Mastering phpMyAdmin for Effective MySQL Management*, in April 2004, and subsequently continued to specialize in publishing highly focused books on specific technologies and solutions.

Our books and publications share the experiences of your fellow IT professionals in adapting and customizing today's systems, applications, and frameworks. Our solution-based books give you the knowledge and power to customize the software and technologies you're using to get the job done. Packt books are more specific and less general than the IT books you have seen in the past. Our unique business model allows us to bring you more focused information, giving you more of what you need to know, and less of what you don't.

Packt is a modern yet unique publishing company that focuses on producing quality, cutting-edge books for communities of developers, administrators, and newbies alike. For more information, please visit our website at www.packtpub.com.

About Packt Open Source

In 2010, Packt launched two new brands, Packt Open Source and Packt Enterprise, in order to continue its focus on specialization. This book is part of the Packt Open Source brand, home to books published on software built around open source licenses, and offering information to anybody from advanced developers to budding web designers. The Open Source brand also runs Packt's Open Source Royalty Scheme, by which Packt gives a royalty to each open source project about whose software a book is sold.

Writing for Packt

We welcome all inquiries from people who are interested in authoring. Book proposals should be sent to author@packtpub.com. If your book idea is still at an early stage and you would like to discuss it first before writing a formal book proposal, then please contact us; one of our commissioning editors will get in touch with you.

We're not just looking for published authors; if you have strong technical skills but no writing experience, our experienced editors can help you develop a writing career, or simply get some additional reward for your expertise.

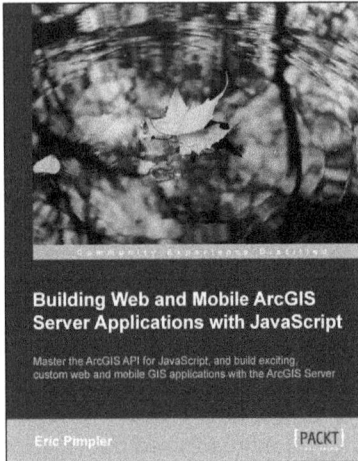

Building Web and Mobile ArcGIS Server Applications with JavaScript

ISBN: 978-1-84969-796-5   Paperback: 274 pages

Master the ArcGIS API for JavaScript, and build exciting, custom web and mobile GIS applications with the ArcGIS Server

1. Develop ArcGIS Server applications with JavaScript, both for traditional web browsers as well as the mobile platform.

2. Acquire in-demand GIS skills sought by many employers.

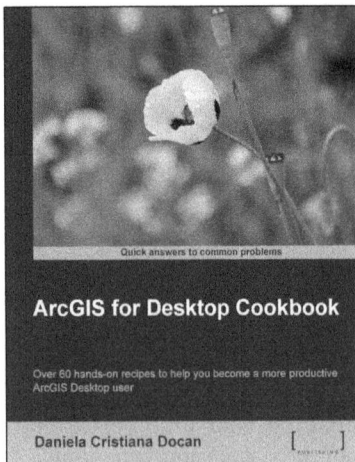

ArcGIS for Desktop Cookbook

ISBN: 978-1-78355-950-3   Paperback: 372 pages

Over 60 hands-on recipes to help you become a more productive ArcGIS Desktop user

1. Learn how to use ArcGIS Desktop to create, edit, manage, display, analyze, and share geographic data.

2. Use common geo-processing tools to select and extract features.

3. A guide with example-based recipes to help you get a better and clearer understanding of ArcGIS Desktop.

Please check **www.PacktPub.com** for information on our titles

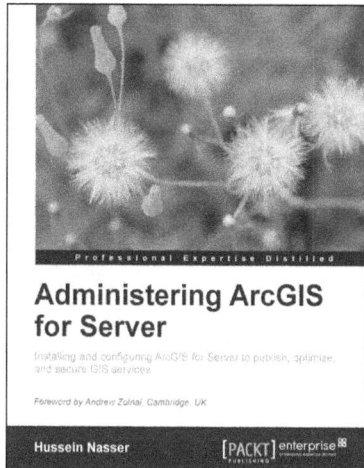

Administering ArcGIS for Server

ISBN: 978-1-78217-736-4   Paperback: 246 pages

Installing and configuring ArcGIS for Server to publish, optimize, and secure GIS services

1. Configure ArcGIS for Server to achieve maximum performance and response time.

2. Understand the product mechanics to build up good troubleshooting skills.

3. Filled with practical exercises, examples, and code snippets to help facilitate your learning.

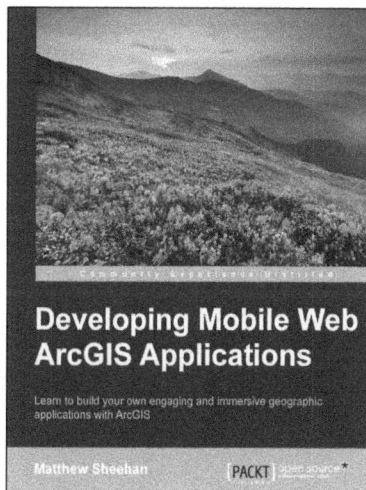

Developing Mobile Web ArcGIS Applications

ISBN: 978-1-78439-579-7   Paperback: 156 pages

Learn to build your own engaging and immersive geographic applications with ArcGIS

1. Create multi-utility apps for mobiles using ArcGIS Server quickly and easily.

2. Start with the basics and move through to creating advanced mobile ArcGIS apps.

3. Plenty of development tips accompanying links to functional maps to help you as you learn.

Please check **www.PacktPub.com** for information on our titles

www.ingramcontent.com/pod-product-compliance
Lightning Source LLC
Chambersburg PA
CBHW080712220326
41598CB00033B/5392